Emergent Spatio-temporal Dimensions of the City

Fabian Neuhaus

Emergent Spatio-temporal Dimensions of the City

Habitus and Urban Rhythms

Fabian Neuhaus
Centre for Advanced Spatial Analysis
University College London
London, Cardiganshire, UK

ISBN 978-3-319-09848-7   ISBN 978-3-319-09849-4 (eBook)
DOI 10.1007/978-3-319-09849-4

Library of Congress Control Number: 2014957727

Springer Cham Heidelberg New York Dordrecht London
© Springer International Publishing Switzerland 2015
This work is subject to copyright. All rights are reserved by the Publisher, whether the whole or part of the material is concerned, specifically the rights of translation, reprinting, reuse of illustrations, recitation, broadcasting, reproduction on microfilms or in any other physical way, and transmission or information storage and retrieval, electronic adaptation, computer software, or by similar or dissimilar methodology now known or hereafter developed.
The use of general descriptive names, registered names, trademarks, service marks, etc. in this publication does not imply, even in the absence of a specific statement, that such names are exempt from the relevant protective laws and regulations and therefore free for general use.
The publisher, the authors and the editors are safe to assume that the advice and information in this book are believed to be true and accurate at the date of publication. Neither the publisher nor the authors or the editors give a warranty, express or implied, with respect to the material contained herein or for any errors or omissions that may have been made.

Printed on acid-free paper

Springer International Publishing AG Switzerland is part of Springer Science+Business Media (www.springer.com)

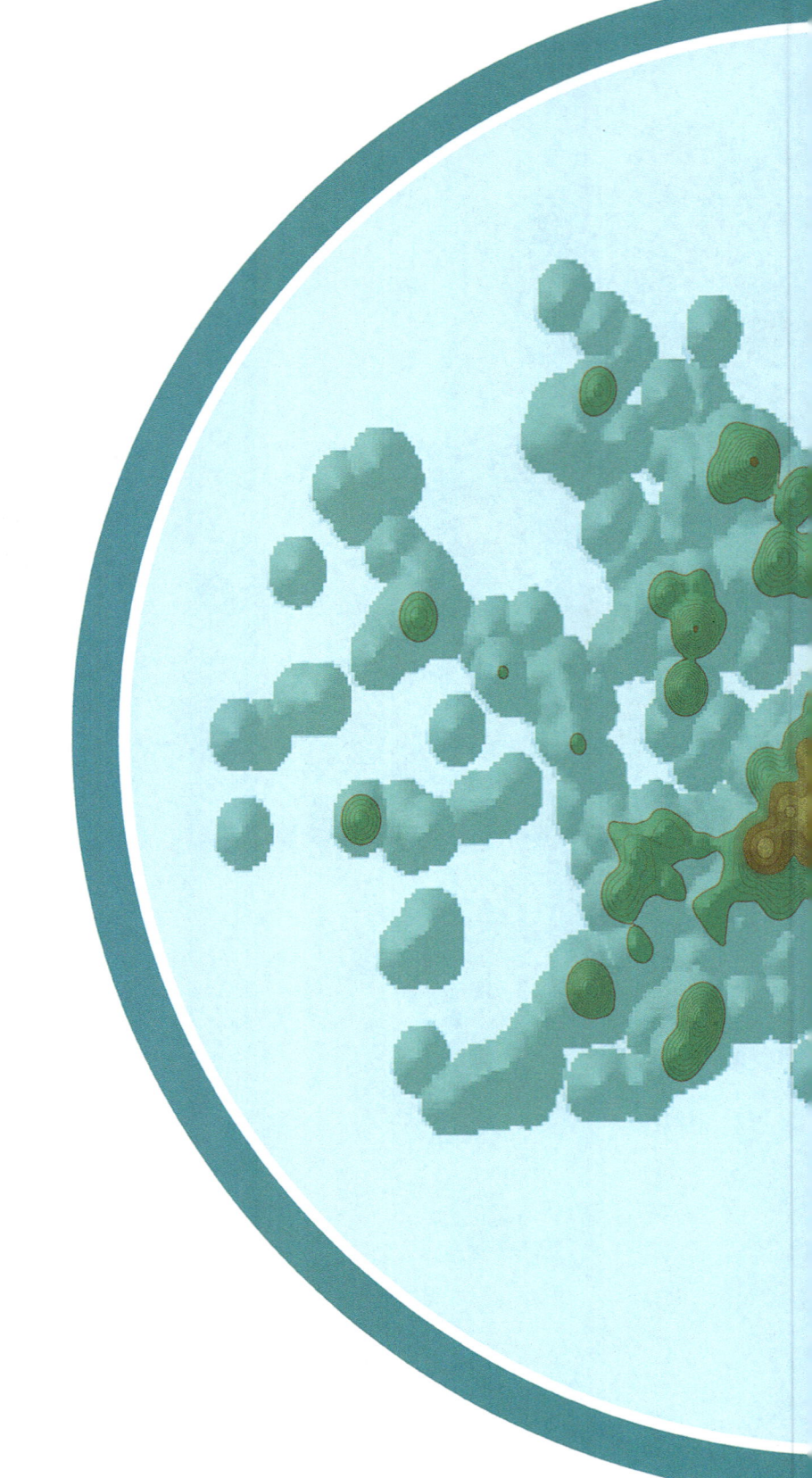

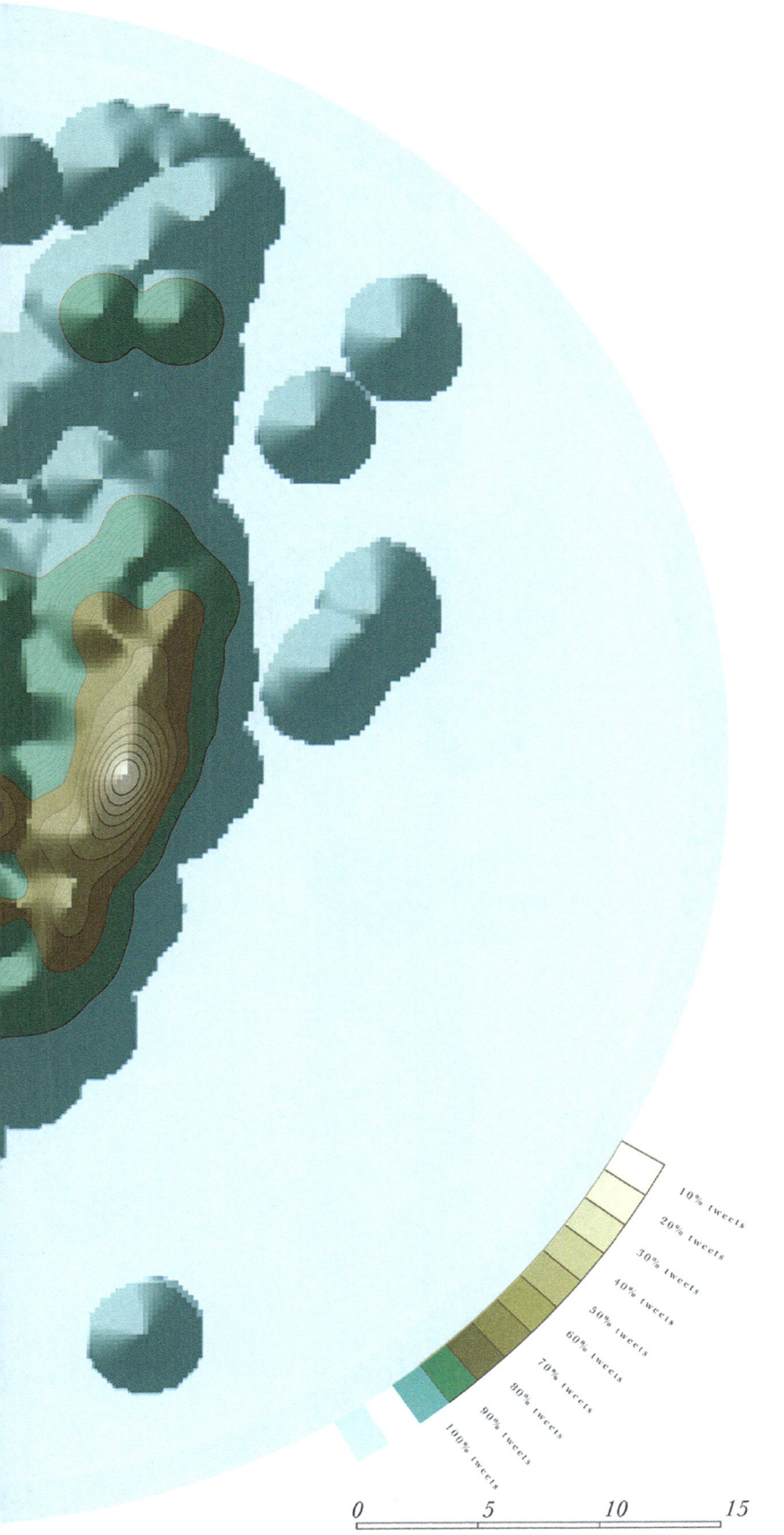

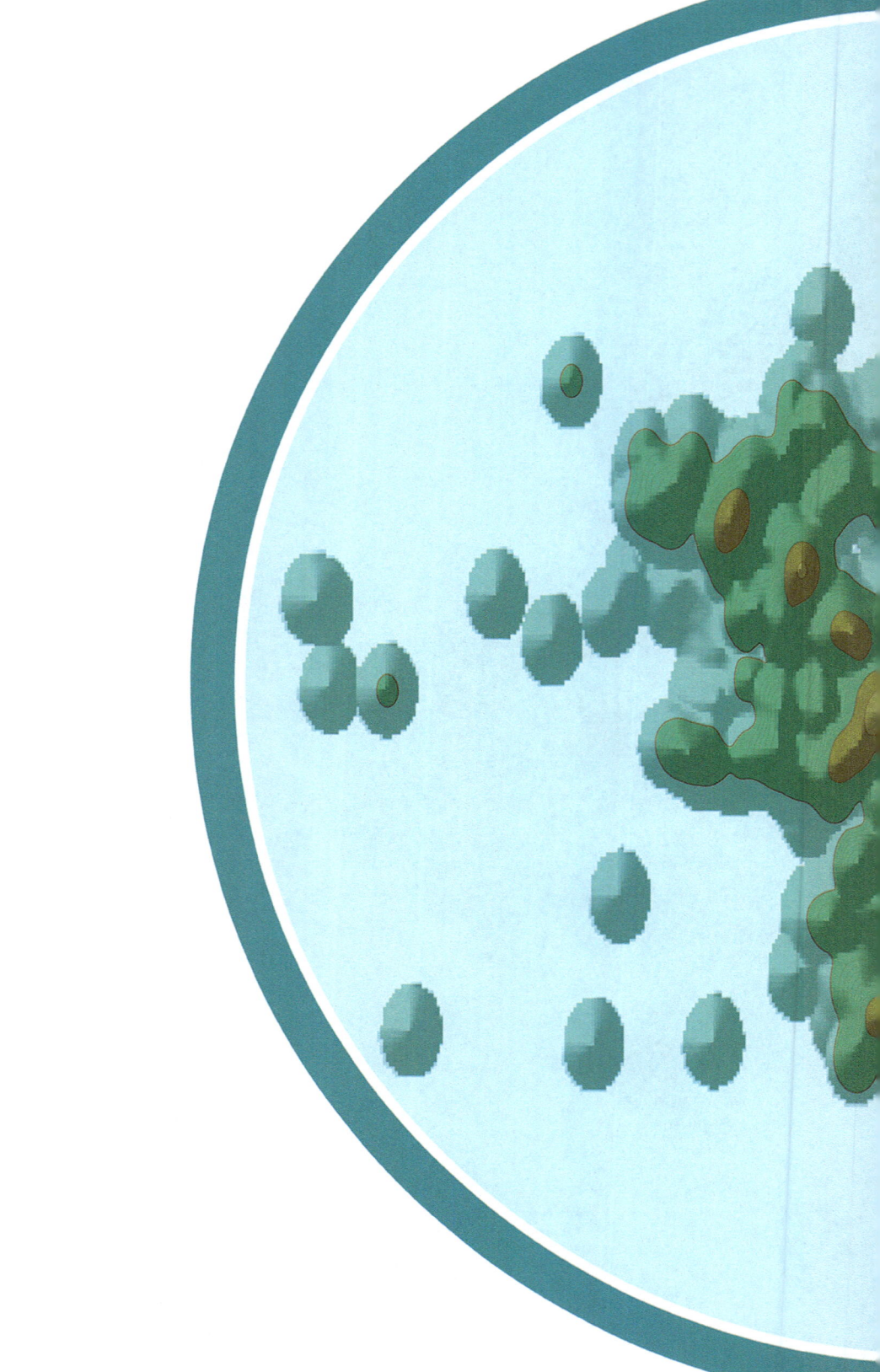

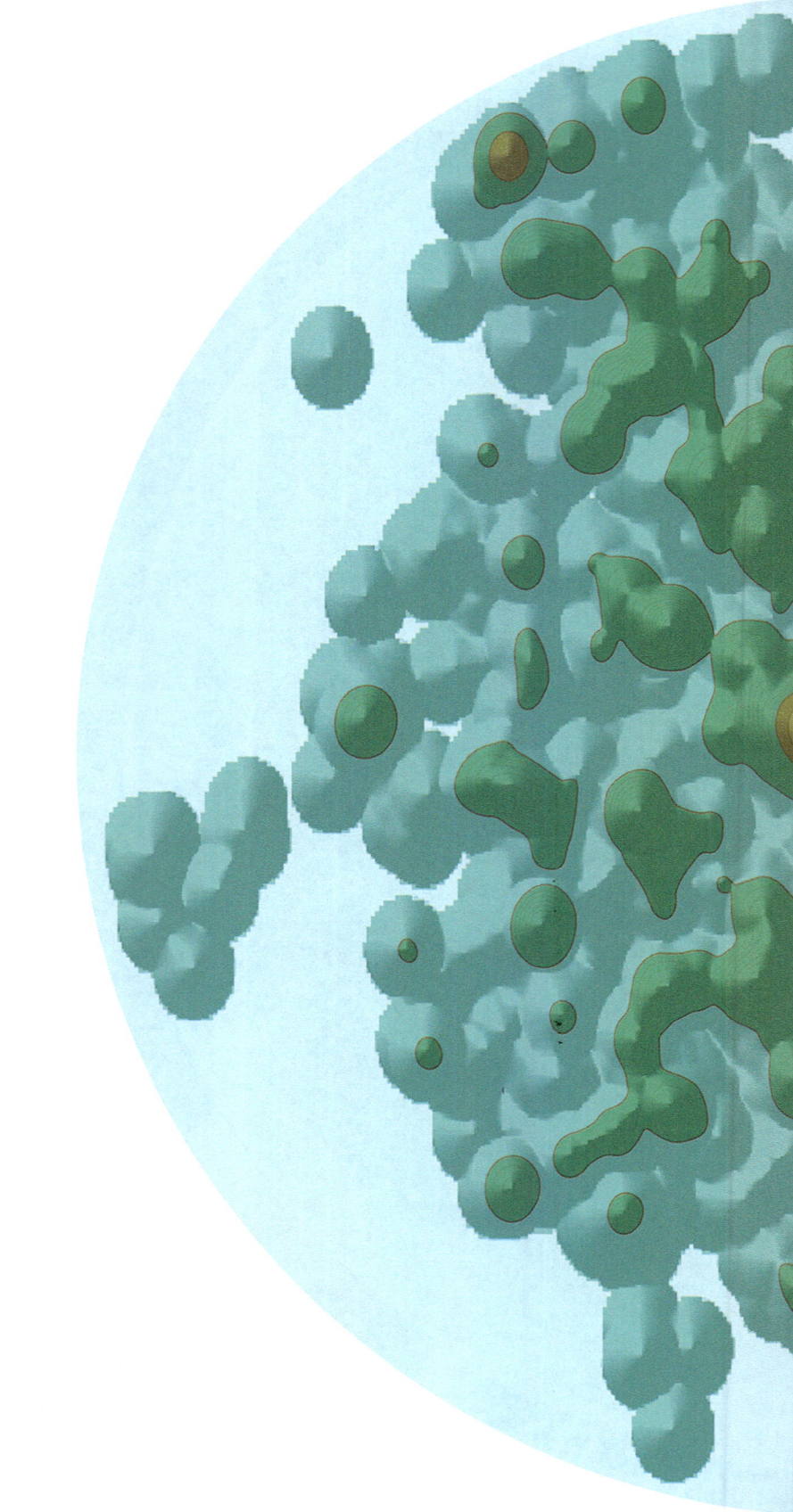

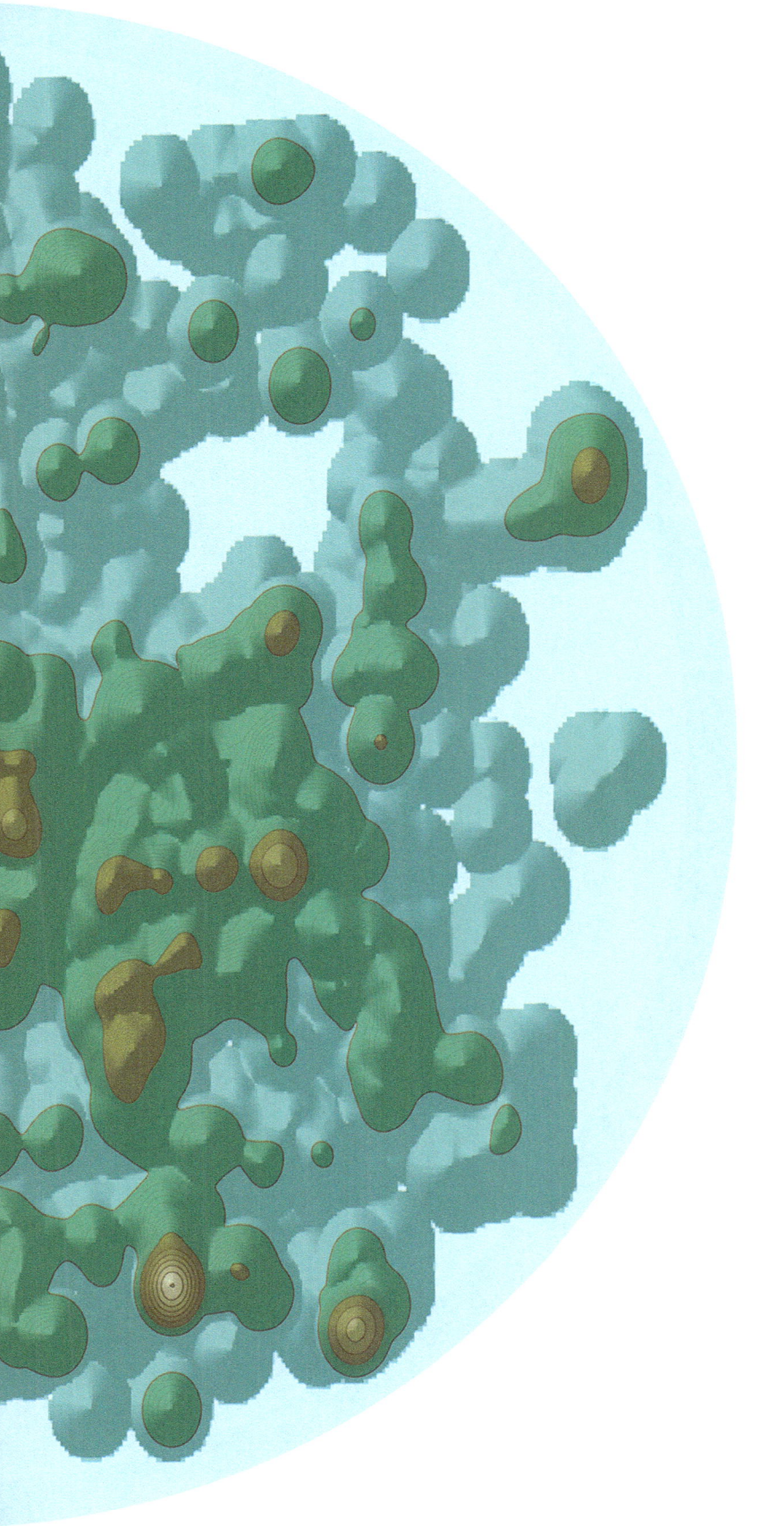

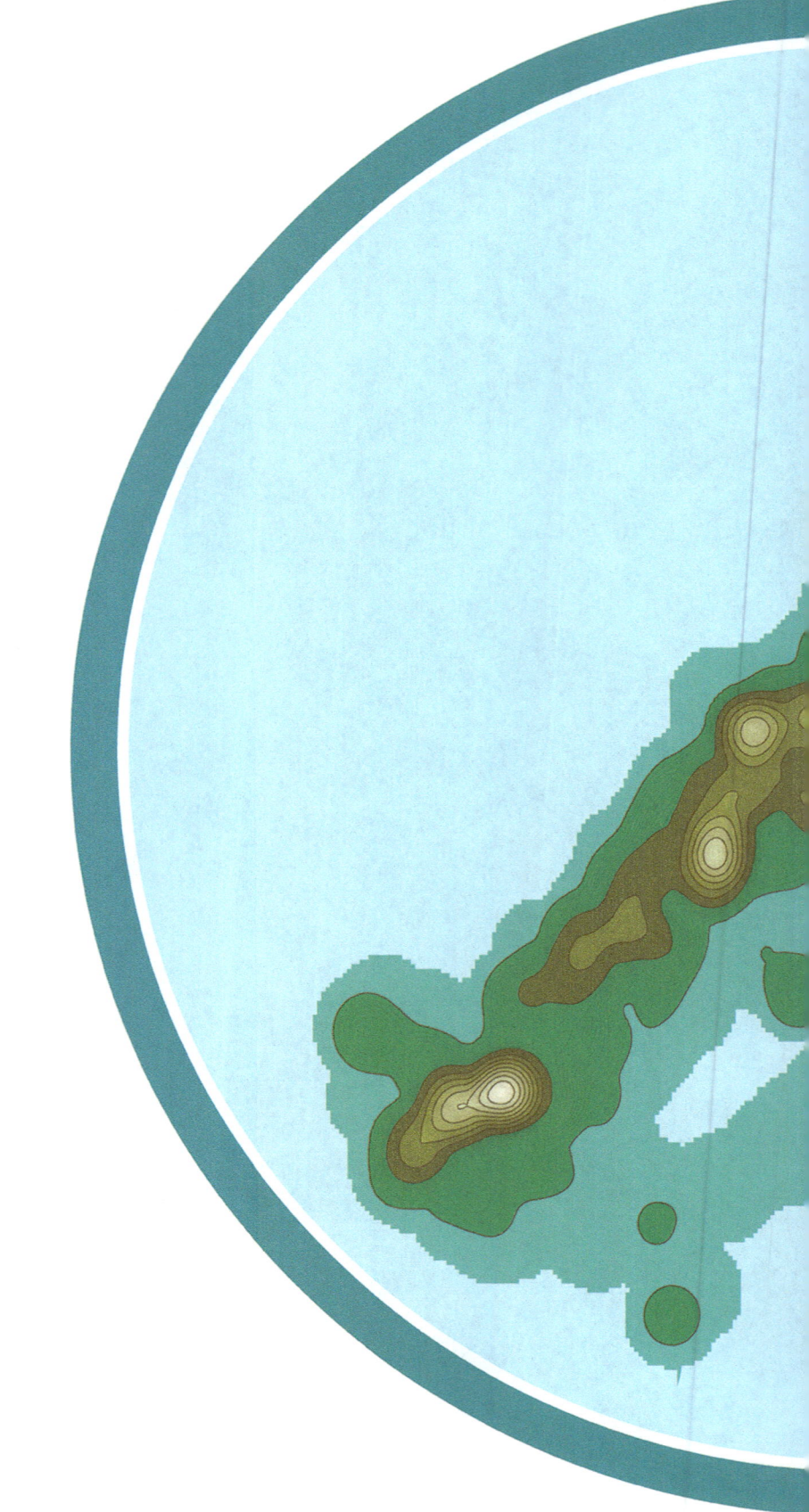

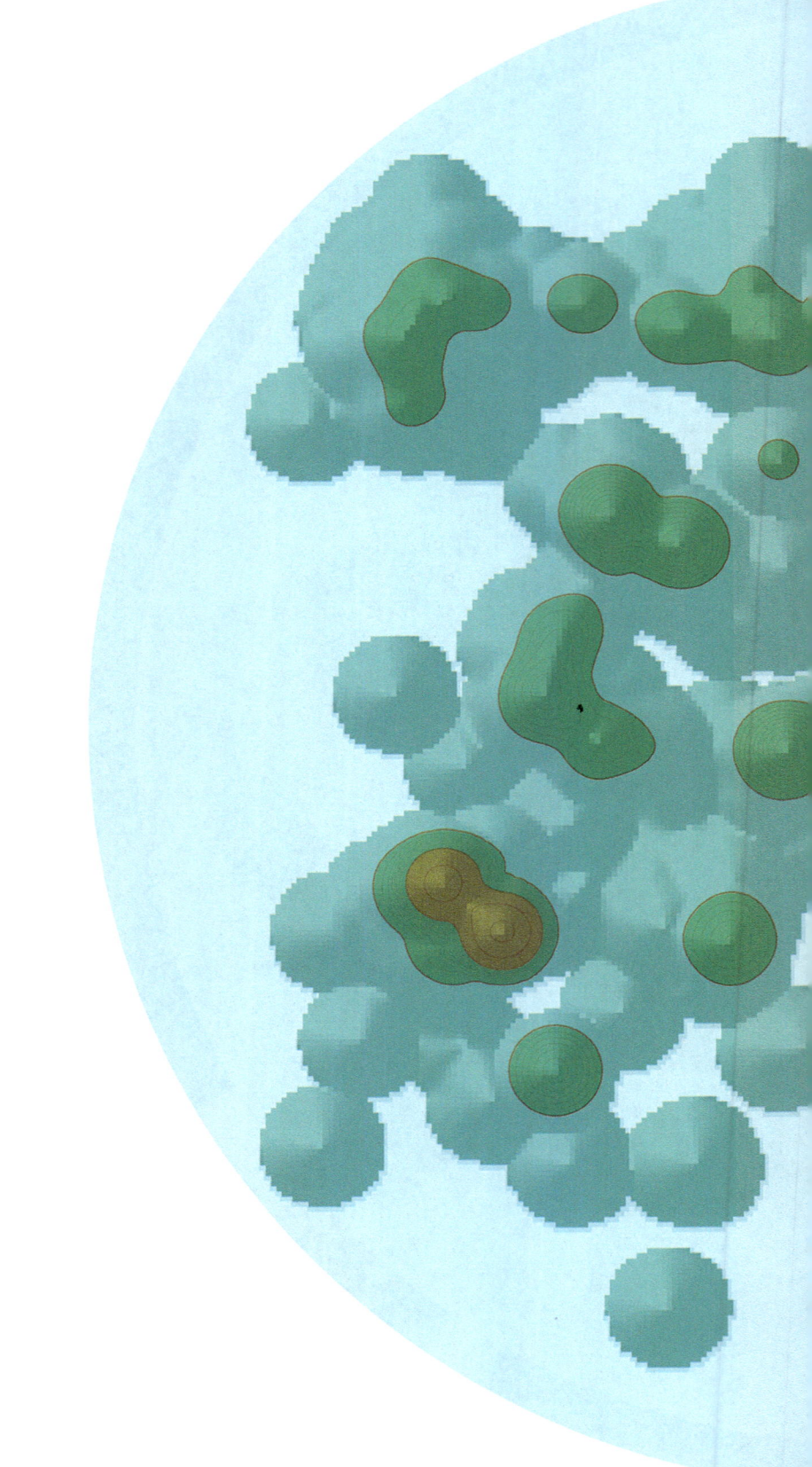

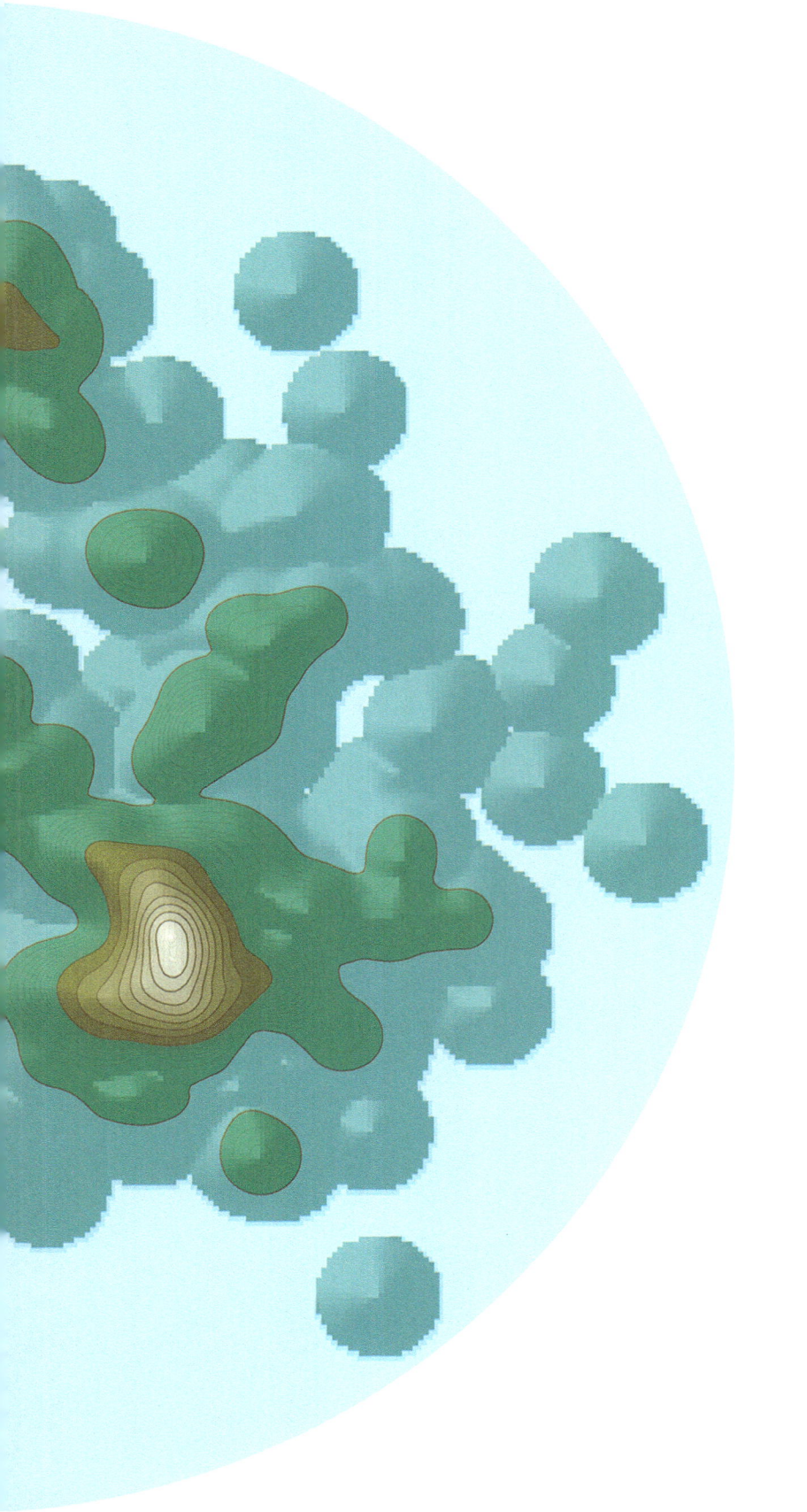

Dedicated, in regards to the habitus *as the present past and the present future, to my late grand father, my late father, Malik and Hannah.*

Foreword

The City is not What it Seems

This book is of fundamental importance because it lays down the basis for a new way of perceiving and understanding cities and hence, potentially, for designing better cities in the future.

Fabian Neuhaus is a most unusual thinker in that he stands at the turning point between our dominant contemporary urban paradigms and the next one that is just emerging. He is essentially a mutant, still prone to quote, in his writings, from classic authors such as Lynch, Alexander, Debord, Bourdieu and even Durkheim, although the cutting-edge culture he actually belongs to – and is in the process of shaping – is a completely new intellectual landscape, requiring new methods of observation, new experimental tools and methodologies and a fundamentally new philosophical approach.

As he rightly points out in his text, our reading of the city is still essentially spatial rather than temporal. Even those observers who think of the city as a complex system of ever changing activities rather than as a fixed physical form tend, in the end, to represent it, by default, as a spatial layout. It takes a considerable mental effort to start to think and see differently: in effect, we have to get it into our heads that the city we see is not the real city. Or, to put it succinctly, the city is not what it seems. Our customary ways of seeing the city since Modernism are now so deeply rooted in our unconscious that they operate autonomously, automatically. We have to make the effort to deliberately de-activate these automatisms and to start looking and thinking counter-intuitively.

By revealing the city as a space-time continuum of virtual communications, as exemplified by the transmission of twitter messages, Fabian is exploring a territory of urban investigation that, until now, only a few poets and science fiction authors such as William Gibson were able to conceive of. It was in the latter's preminiscent novel Neuromancer that the term "cyberspace" was coined for the first time, suggesting that the city of the digital age is now made of two equally important superimposed layers: a "lower" layer of material reality, comprised of the physical bodies of human beings as well as buildings, open spaces and other man-made artefacts such as infrastructure networks, and an "upper" layer of immaterial digital communications, comprised of trillions of bits of information being continuously sent, received and responded to, both by human beings and by artificial intelligence agents. The spatial distinction between these two layers, one above and one below, although highly evocative, is of course purely metaphorical: telecommunication signals freely penetrate human flesh as well as the walls of buildings; they do not "float" above us any more than the soul, assuming there is such a thing, which can be said to hover, like a halo, above our heads. One does not need to "jack up", to borrow the term used by Gibson in his novel, into the realm of digital telecommunications. Materiality and immateriality coexist. They are intimately intertwined. They are fundamentally copresent and together constitute the urban milieu.

Paradoxically, within this copresence, the signals we get from one "layer" may be significantly different from the ones we receive from the other, and it is in this paradox that lies the crux of Fabian's new understanding of the city.

Taking a specific case helps to understand the issue. If one looks at a traditional map of a city, one acquires a particular understanding of what it is, or at least what it appears to be. One sees the location of houses, schools, office buildings, hospitals, public buildings of various types, railway stations, airports, stadia, parks, etc. In other words, one sees where people are expected to be sleeping, working, eating, playing, etc. A land-use map will make this distribution of activities clear. We are educated to believe that this form of representation of the distribution of activities is legitimate and that, although it may not give a perfectly true account of human activity at a particular moment in time, it nevertheless represents a reliable statistical truth. A geological map will show the subsoil upon which these activities are implanted, and conversely, an aerial photo or, more accurately, a satellite view can show, in addition, the changing atmospheric conditions above ground. The extraordinary novelty of the last decade or so is that such maps are now available synchronously in the so-called real time rather in delayed time. Traditional maps, which took a long time to produce and disseminate, used to show us where things were supposed to be rather than where they actually were. Now the lapse between the time at which the territory is being mapped and the time at which the map can be seen has been considerably reduced and might, theoretically, become zero. Google's "Street View" shows us any street on this planet perhaps not exactly as it is "right now", but at least as it was a few days or a few months ago rather than a few decades ago. Moreover, we are shown not just inert objects and buildings, the containers where activities are meant to take place and where people are meant to be, but the people themselves. This shift to quasi-simultaneity constitutes not just a quantitative evolution in the nature of mapping but a fundamental qualitative leap. We are getting closer to an understanding of what people are really doing, to what is precisely happening in our cities. And this might lead to a few surprises!

Whilst all forms of mapping are evolving and becoming increasingly simultaneous, the most significant leaps are taking place within the realm of telecommunications and, most particularly, within the so-called social media, the most current manifestations of "cyberspace". Fabian's research focus has led him to analyse the real-time data gathered from the social media platform "Twitter". This reveals that traditional mapping techniques not only do not tell us everything but can actually be highly misleading: in contradiction to the highly aggregated and static information contained in land-use maps, people may be conducting certain activities at places and times that are very different from what we expect. This realisation of the true life of the city, the truth beneath the surface, can only be reached if one has access to large amounts of real-time geolocated data at a suitable level of disaggregation, and Fabian has amply demonstrated that we are getting close to that point.

It is important to realise that entering the world of "big data" does not mean having access to more of the same information, at a higher degree of resolution, but, in all probability, potentially being exposed to a different kind of information. The underlying message may be very different, as Antonioni revealed in his film "Blow up" where what appears at first to be a tranquil garden setting turns out, when zoomed in, to be the scene of a crime. Similarly, in "The conversation", Brian de Palma explores the same forensic process of zooming in, this time dealing with sound, acoustic data rather than visual material. As in cryptography, the apparent message is not the real message.

The data sets that Fabian now has access to and for the analysis of which he is developing radically new deciphering methods are on the brink of new discoveries and lead to completely new questions concerning the nature of the city and its future. If, for instance, it becomes apparent that the researchers in Silicon Valley's high-tech computer industries do most of their creative thinking whilst sitting and chatting with their friends in coffee shops rather than when sitting at their desks, then the design and construction of sophisticated office facilities for them to work in becomes irrelevant and totally superfluous. If it becomes apparent that people's cycles of thinking activity and working patterns follow a much more complex sets of rhythms than the customary regular pendulum swings from place of residence to workplace in the morning and back home in the evening, it follows that the provision of communication infrastructures, both in terms of conventional modes of transportation and new modes of telecommunication, needs to be looked at much more carefully than current planning mechanisms are capable of doing: precise information concerning the true dynamics of the changing work culture and its optimal environments could lead to much more appropriate and efficient solutions than the ones that are currently the norm and have led to the very high levels of infrastructure overprovision and redundancy that are characteristic of our wasteful age.

It is conventional wisdom to claim that our cities are undergoing a continuous process of change, but the truth is that our cities are in reality extremely resistant to change. Both in terms of habitus and type, i.e. in terms of activities and form, cities exhibit high levels of inertia: they tend, above all, to perpetuate the status quo. It can be argued that the activities that take place in cities, their layout and the basic typology of their architectural building blocks have not fundamentally changed in the last 5,000 years, since urban agglomerations first came into existence. The contemporary city is not so different from, say, Pompeii: the urban fabric is still made up of a generic mass of residential dwelling units out of which a set of key public spaces and buildings emerge. In both cases, the urban structure is made up of a network of streets enabling the movements of pedestrians and vehicles, and buildings are lined up along these street armatures. Buildings are now higher perhaps, but the basic urban typology has not changed. If a citizen of Pompeii were to be magically transplanted to contemporary Rome, New York or Shanghai, would not find it so difficult to recognise and "read" the contemporary city and the activities taking place within it, as something totally familiar.

So, in effect, whilst our technologies have evolved very fast, as well as our social, economic, political and cultural superstructures, the largest and most significant of all human artefacts, the one that shapes our lives, remains essentially the same. Now that more than half of the world population is urbanised, it is ironic that the concept of the city and its design remain so anachronistic.

Only a finely tuned understanding of the activities taking place within the city can lead to finely tuned responses. The planning and design tools we have used until now in urbanism are far too broad brushed. They were largely based on unsubstantiated evidence and very broad generalisations. In cybernetics, the "law of requisite variety" states that in order to be able to "control" a system, the control mechanism must inherently have at least the same "degree of variety" (i.e. information content) as the system to be controlled. It is only now, through the type of advanced research that Fabian is doing, that our design tools and methodologies and, more importantly still, our ways of thinking and our philosophical concepts are about to acquire such a degree of sophistication.

Hong Kong
June 1, 2014

Colin Fournier

Preface

This book is based on the work undertaken for a PhD at the Centre for Advanced Spatial Analysis and was and accepted as thesis by University College London in 2013. I would like to thank my supervisors at CASA, Professor Mike Batty and Dr Andrew Hudson-Smith, for their support.

The topic of cycles, rhythm and repetition in activity pattern in the city is something I have been developing already in the earlier publications *Cycles in Urban Environments* (Neuhaus 2010) and *Studies in Temporal Urbanism* (Neuhaus 2011). The topic is further developed introducing the term *habitus* to summarise a whole range of observations, theories and practices. This allows to formulate a more concise alternative perspective on urban spatial practices.

I would like to acknowledge support made available for this text. Many people made this research possible and to all of them I am very grateful. I would like to thank my past and present colleagues for input, feedback and comments during discussions and seminars. In particular, the help and support provided and the questions asked by Dr Duncan Smith, Dr Andrew Crooks and Dr Anders Johannson. They all helped to shape and focus the content of this book. I am especially grateful to Dr Timothy Webmoor for his discussions, input, and specific suggestions regarding theoretical aspects, aspects of *temporality* in particular.

Prior to this research, a number of people have supported me in developing my academic interests, which feed into the project that lead to this publication, and to whom I am very grateful for pointing the way. I would like to thank specifically Dr Niklaus Kohler, who encouraged me to follow my interests to pursue research in my field of interest. It was whilst working with his team at the University of Karlsruhe on scenario tools for urban decision-making systems that I discovered the wider topic of process, time and space. I would also like to thank Professor Luca Selva who mentored large parts of my academic career so far.

I was able to present selected parts of this research at conferences in GSA 2010, Denver, Visualisation in the Age of Computerisation, Oxford, CUPUM 2011, Lake Louise, CA, GeoCom 2011, London, RGS 2009 and 2010 and 2011, London, CRESC 2010 and 2011, Manchester, 7th Virtual Cities and Territories 2011, Lisbon, SACRPH 2011, Baltimore, Spaces and Flows 2012, Detroit to list a few and preparing for the presentation, as well as feedback and comments from the conference participants, was helpful in shaping the arguments and overall presentation of the subject.

Finally, I would also like to thank my family, especially Sandra and Malik, for their continuous support and understanding. Without their encouragement throughout, as well as the occasionally imposed distraction, this work would never have been completed. It was in exchange with them that I discovered the topic of repetition, routine and cycles, which developed into my main research interest and later on into this publication.

The fieldwork was supported through the provision of GPS units by the UCL Centre for Transport Studies and Garmin. The basis for the New City Landscape project was developed in collaboration with Dr Andrew Hudson-Smith and Steven Gray and is in part supported by a JISC grant at CASA.

This work was supported financially by Bürgergemeinde Biel, ffn, UCL Advances, CASA, UCL Graduate School, the Sydney Perry Foundation and the Stapley Trust.

Contents

1	**Urban Rhythms**	1
	1.1 Rhythm and Cycle	1
	1.2 Reference	2
	1.3 Habitus	4
	1.4 Context	5
	1.5 Ethics	6
	1.6 Outline	7
	1.7 About Working on Time and Space	10
	Bibliography	11
2	**Urban Machine and Time-Space**	13
	2.1 Urban Machine	15
	2.2 Time-Space	25
	Bibliography	35
3	**Body Space and Spatial Narrative**	37
	3.1 Body Space	38
	3.2 Spatial Narrative	45
	Bibliography	52
4	**Urban Diary**	55
	4.1 Introduction	55
	4.2 Lines of the Everyday: Tracking	59
	4.3 Between Object and Subject: Technology	61
	4.4 The Image of the City as Routine: Mental Maps	65
	4.5 Sampling the Urban Diary	72
	4.6 Individual Spatial Experience: The Interviews	73
	4.7 The Individual Spatial Shape	77
	Bibliography	80
5	**New City Landscape**	83
	5.1 Introduction	84
	5.2 Digital Social Networks: A Recent Phenomena	88

5.3	Metadata and Data Mining	93
5.4	From Twitter to New City Landscapes	106
5.5	Technical: Twitter API, Criteria and Storage	108
5.6	Data: Numbers, Sample, Implications and a Comparison	110
5.7	Kernel Density, Contours and Colouring	115
5.8	Networks	119
5.9	Ethics: Privacy, Ownership, Agreement and Protection	121
5.10	The Collective Data Pool	124
	Bibliography	126

6 Structuring Time ... 131
- 6.1 Clock Time ... 132
- 6.2 Urban Diary Times ... 133
- 6.3 Schedules and Planning ... 150
- 6.4 NCL Times ... 169
- 6.5 Rhythmic Time ... 181
- Bibliography ... 185

7 Structuring Space ... 189
- 7.1 Cartesian Space ... 190
- 7.2 Urban Diary Spaces ... 192
- 7.3 New City Landscape Spaces ... 207
- 7.4 Timed Space ... 211
- 7.5 Social Space ... 216
- 7.6 Rhythmic Space ... 232
- Bibliography ... 235

8 Temporality: The Rhythmic City ... 237
- 8.1 Introduction ... 237
- 8.2 The Experience of Time and Space ... 238
- 8.3 The Concept of Clock Time and Cartesian Space ... 247
- 8.4 Reflection on the Time and Space Context ... 251
- 8.5 Routine in Time and Space as Habitus ... 253
- 8.6 Towards an Integrated Temporality ... 256
- 8.7 The *Habitus* ... 259
- Bibliography ... 263

9 Appendix ... 265
- 9.1 Glossary ... 265
- 9.2 External Resources ... 269
- Bibliography ... 270

Index ... 295

List of Figures

Fig. 2.1	La Citta Nuova	17
Fig. 2.2	Metropolis	18
Fig. 2.3	Der Lauf der Dinge	19
Fig. 2.4	The Valley Section	20
Fig. 2.5	Ciudad Lineal	21
Fig. 2.6	Diagram city structure	22
Fig. 2.7	La Ville Radieuse	23
Fig. 2.8	Human machine	24
Fig. 2.9	Hopi space and time	29
Fig. 2.10	Time-space cube	30
Fig. 2.11	GeoTime vis of Twitter data	32
Fig. 2.12	Human activity patterns	33
Fig. 2.13	Held and Hein	34
Fig. 3.1	The horse in motion	43
Fig. 3.2	Barcelona Pavilion	45
Fig. 3.3	The view from the road	46
Fig. 3.4	What shape are you, six	48
Fig. 4.1	Forerunner 405	57
Fig. 4.2	UD London centre	61
Fig. 4.3	Global Positioning System	62
Fig. 4.4	GPS tracker models	64
Fig. 4.5	UD London all	65
Fig. 4.6	UD London one	66
Fig. 4.7	Representing identities	68
Fig. 4.8	Los Angeles mental map	69
Fig. 4.9	Children's mental maps	70
Fig. 4.10	UD participant mental map	71
Fig. 4.11	A UD participant's schedule day	75

Fig. 4.12	A UD participant's schedule week	75
Fig. 4.13	Mental map vs. GPS track	76
Fig. 4.14	UD London	78
Fig. 4.15	What shape are you—four	79
Fig. 5.1	Network Graph	85
Fig. 5.2	Social Connections vs. Traffic Volume	87
Fig. 5.3	Persona Web Application	93
Fig. 5.4	Google Latitude	96
Fig. 5.5	Network Graph Location	98
Fig. 5.6	Geotagged Image Locations	100
Fig. 5.7	Vancouver Geotagged	101
Fig. 5.8	Moscow Geotagged	102
Fig. 5.9	Social connections World Wide	103
Fig. 5.10	Twitterlandschaft	104
Fig. 5.11	Good Morning	105
Fig. 5.12	Just Landed	105
Fig. 5.13	Tweet-O-Meter	109
Fig. 5.14	NCL Table	111
Fig. 5.15	NCL 24 Sample Cities	113
Fig. 5.16	NCL Collection Area	114
Fig. 5.17	NCL Cities Data in Numbers	115
Fig. 5.18	NCL Cities Comparison	116
Fig. 5.19	NCL Contour Lines	117
Fig. 5.20	Barcelona New City Landscape	118
Fig. 5.21	NCL Network Graph	121
Fig. 5.22	Twitter Interface	123
Fig. 6.1	UD Diary Graph Month	136
Fig. 6.2	UD Diary Graph Week	137
Fig. 6.3	UD Diary Graph Day	138
Fig. 6.4	UDp-35 Distance from Home WW	141
Fig. 6.5	UDp-35 Distance from Home WE	142
Fig. 6.6	UDp-30 Distance from Home WW	143
Fig. 6.7	UDp-30 Distance from Home WE	144
Fig. 6.8	UDp-06 Distance from Home WW	145
Fig. 6.9	UDp-06 Distance from Home WE	145
Fig. 6.10	UDp-all Distance from Home WW	147
Fig. 6.11	UDp-all Distance from Home WE	147
Fig. 6.12	Recorded Schedules three	152
Fig. 6.13	Schedule Sketch	154
Fig. 6.14	Activity Percentage	155
Fig. 6.15	Activities Count Blocks	156
Fig. 6.16	Activities Count Detail	157
Fig. 6.17	Times Count Weekday	159

List of Figures   xxxix

Fig. 6.18	UDp-07 Schedule Sketch	161
Fig. 6.19	Activity Week	161
Fig. 6.20	UDp-17 Schedule Sketch	163
Fig. 6.21	Garden City	165
Fig. 6.22	The London Communities Map 1942	166
Fig. 6.23	Causes of Mortality	171
Fig. 6.24	Low-Dimensional Multivariate Data	172
Fig. 6.25	Statistical Analysis of Circular Data	172
Fig. 6.26	The Sydney Kidney	174
Fig. 6.27	NCL City Time per Day	175
Fig. 6.28	NCL Average Times	176
Fig. 6.29	NCL Report Times	177
Fig. 6.30	NCL Time Over the Week	179
Fig. 6.31	NCL Twitter Topics	180
Fig. 6.32	NCL Tsunami Arrival	182
Fig. 7.1	The virtual global grid	191
Fig. 7.2	The GPS track	193
Fig. 7.3	Anchor locations Golledge	194
Fig. 7.4	UD London in colour	195
Fig. 7.5	Central London three	197
Fig. 7.6	Basel three	198
Fig. 7.7	What shape are you—all	200
Fig. 7.8	What shape are you—three	201
Fig. 7.9	What shape—glow	202
Fig. 7.10	City path tracks in comparison LBP	204
Fig. 7.11	Plan for Plymouth with paths	205
Fig. 7.12	Basel growth and paths	206
Fig. 7.13	NCL morphology types	208
Fig. 7.14	NCL morphology centre	209
Fig. 7.15	NCL morphology Island	209
Fig. 7.16	NCL morphology feature	210
Fig. 7.17	NCL morphology airports	210
Fig. 7.18	GPS track colour coded	212
Fig. 7.19	Space-time aquarium	213
Fig. 7.20	GeoTime path	214
Fig. 7.21	24 animation stills	215
Fig. 7.22	NCL map of Munich	216
Fig. 7.23	The Naked City	219
Fig. 7.24	City Islands	220
Fig. 7.25	London mental map—three	222
Fig. 7.26	Basel mental maps—three	223
Fig. 7.27	UDp-04 path sketch	225
Fig. 7.28	UDp-04 city sketch	227
Fig. 7.29	UDp-35 and UDp-17 path and city sketches	228

Fig. 8.1	UD-Bcircles01	240
Fig. 8.2	UDp-36commute	242
Fig. 8.3	UDp-35commute	243
Fig. 8.4	UDp-02commute	245
Fig. 8.5	UDp-04commute	245
Fig. 8.6	UDcompRoutine01	252

Chapter 1
Urban Rhythms

Abstract From the observation of personal activities, recurring patterns can vaguely be identified. It is more a feeling and a habit, rather than a conscious decision, to approach a task in a similar manner every time. Embedded in a string of decisions, buried under constant adjustments to external demands, there seems to be an operational compass guiding individual actions attached to a much larger sociocultural field of interlinked conditions.

Is it then possible to describe such individual routines in terms of conditions reflecting time and space? And to what extent does this have significance in an urban setting? These questions have led to the research discussed in the following text and will be examined in detail on various levels of scale and connectedness. To do so, various different perspectives are adopted to test and explore the concept of repetition and its implications spatially and socially in an urban context.

From the observation of personal activities, recurring patterns can vaguely be identified. It is more a feeling and a habit, rather than a conscious decision, to approach a task in a similar manner every time. Embedded in a string of decisions, buried under constant adjustments to external demands, there seems to be an operational compass guiding individual actions attached to a much larger sociocultural field of interlinked conditions.

Is it then possible to describe such individual routines in terms of conditions reflecting time and space? And to what extent does this have significance in an urban setting? These questions have led to the research discussed in the following text and will be examined in detail on various levels of scale and connectedness. To do so, various different perspectives are adopted to test and explore the concept of repetition and its implications spatially and socially in an urban context.

1.1 Rhythm and Cycle

The beat of the urban environment is most directly apparent in connection with transport systems and built infrastructure: the regular interval of trains arriving on the platform, the frequency of bus stops or take off and landing of airplane at busy airport. It often leads to the interpretation of the city as a physical machine, which ultimately then is understood as a provision of service. Cities tick in this way. Does this conceptualisation of the city or urban form really represent what we are hoping our cities to be? Or is it mainly a leftover of modernist ideas which happens to suit the expectation and the realisation of individualism that we come to demand of our cities today?

1.2 Reference

Three main theoretical works serve as a basis for the examination of the topic of repetition and serve as theoretical guidance. Amongst important writings, Henri Lefebvre published *Rhythmanalysis* (Lefebvre 2004), a text, originally published in French in 1992, in which the topic of repetition is explored in detail from a theoretical and philosophical perspective. These investigations into rhythms in the context of the everyday touch topics of language, music and of course space. What will be adopted and explicitly implemented through fieldwork investigations in this text is Lefebvre's emphasis on a connected approach to rhythms. His explorations start from the body experience of rhythms, the heartbeat and the breath or steps as basic counter measurements and reference systems. The emphasis is put on an integration of rhythms across all scales and fields as an interlinked system guiding all areas of life. The aspects of natural cyclical pattern on the urban environment and the design of cities were earlier explored in more detail in the publication *Cycles in Urban Environments: Investigating Temporal Rhythms* (Neuhaus 2010).

An additional important aspect that features implicitly in *Rhythmanalysis* is human agency and the production of space, discussed in more detail in *The Production of Space* by Lefebvre and Nicholson-Smith (1991). This aspect of practice is picked up here and serves as the initial method for interpreting the observations we make on daily routines and rhythms in urban life.

Such a standpoint implies an intense investigation on the level of the individual in a qualitative sense. Lefebvre examines the options to investigate the topic in his introduction. He decides against the bottom-up approach, from the individual to the collective, and starts from an abstract theoretical definition of rhythm. However, here, a combined approach is proposed where the theoretical level is supported by both individual and collective case studies.

The linkage between individual experiences and the urban context is a key point of investigation and draws its theoretical inspiration from the writing of Lynch (1960), particularly from his *The Image of the City*. Using mental maps of the city drawn by participants, Lynch develops a collective description of the urban context of a particular place based on individual experience and memory. Ultimately he summarises the collected observations in specific meaningful elements of the urban environment. These elements *Node, Link, Landmark, Path* and *District* can, as Lynch proposes, be employed to summarise various other places.

1.2 Reference

Using citizens' memories of travel as research data has inspired the fieldwork undertaken for the case studies here. The mental or cognitive map provides a tool to capture an individual view of a variable context. In many ways, the resulting method can provide access to an individual perspective, very much in the sense that Lefebvre emphasises the importance of practice and the individual agency. At the same time, the setting is purely urban and spatial.

Whereas, in Lynch's work, time featured as a naturally given element, here the emphasis will be put on the aspects of time which are as important to the individual memory as space. Aspects of movement and *getting around town*, learning about surroundings, are here understood as interwoven with routines and time patterns, and therefore, a focus will be put on how these aspects influence the individual experience.

This leads to the third initial text reference taken from the field of human geography, or more specifically time geography. The concept of *path trajectories* as a conceptualisation of movement in time and space developed first by Torsten Hägerstrand serves as a theoretical but also visual, starting point. Hägerstrand (1970) explains his concepts of *constraints* limiting possible behavioural options in a time-space framework in the paper *What about People in Regional Science*. From his work on lifetime patterns, for example, in *Survival and Arena* (Hägerstrand 1978), the concept of the *path* is developed as being the linear trajectory that individual movement produces in time and space, with the body at its very apex in the here and now. The *path* is trailing behind in the past and will be observable, and it becomes possible to analyse it.

Time geography has developed a range of tools to investigate the time-space problem in connection to movement. The most important model is the time-space cube or also called the time-space aquarium. This representation renders the space and time information in an abstract 3D representation where x and y horizontally represent the spatial dimension and z, vertically, represents the time dimension. This visualisation, developed by Hägerstrand at the Lund School, has been described and visualised by Carlstein et al. (1978).

The application thereof, especially using new technologies both to record and capture the data as well as to analyse and visualise the results, has gained new relevance now some 25 years on. However, the theory, whilst starting with the individual, denies their agency by putting the emphasis on the overwhelming *constraints* with no differentiated concept of power. This to some extent conflicts with the other two key texts, and the debate and negotiation between the different concepts will form an important part of the later chapters.

In combination, the three concepts complement one another in many ways, and this will be utilised for the investigation into the observed routines and patterns that stem from individual habits. This spatial practice combined with the cyclical time component results in a *habitus*, a time-space pattern influenced by cultural, social and natural aspects of the environment.

Whilst these three key concepts serve as a starting point, a whole range of secondary literature is drawn upon to support or argue the case of *habitus* in an urban context.

1.3 Habitus

The observed repetition and the cyclical nature of most natural processes have led to the use of a term that covers the whole range of different aspects involved when talking about everyday routine. Whilst the recurrence of a similar moment, the same task or action is emphasised, the context, the interlinkage, the consequence and the origin of such a pattern are also important.

Observing one repetitive activity, say, for example, the lunch break at 1 pm, it quickly becomes apparent that it is not an isolated case, but something the before and the after are closely connected to. To be able to go for lunch, one has to be engaged in a different activity at first in order to change activity and *go* for lunch. Furthermore, one has to be hungry to actually *eat* lunch. Then one does not just eat: one *has* a proper lunch break, maybe meeting up with a business partner, friends or family. Lunch is *social*. And maybe if one happens to visit a different place, lunch is slightly different; it could be at 12 noon, as often in mainland Europe, or much later at 3 pm, for example, around the Mediterranean. Lunch is also a *cultural* event.

This is true for most routines. Even if at first they appear to be very personal and fitting with an individual lifestyle, they are most likely not unique. A whole group of people, if not a whole nation, is doing the very same thing, often at the same time. Rush hour in the morning or in the evening is a great example for this. Such a large-scale synchronisation is beyond the individual, and the influences of *scale, nature, time, society and culture* are key to how these cycles apply.

To summarise these aspects, the term *habitus* is borrowed from anthropology, specifically as described and defined by Bourdieu (1990). With this the aim is to describe these repetitive patterns of activity and anchor them in a wider context. The emphasis is put on what is happening beyond the individual and how a single activity is part of a whole sequence. It is about embedding the single journey to work in a tsunami of migration across the city.

Habitus as used in sociology and anthropology is also reintroduced[1] by Marcel Mauss (2006). Mauss defined habitus as those aspects of culture that are anchored in the body or daily practices of individuals, groups, societies and nations. It includes the totality of learned habits, bodily skills, styles, tastes and other non-discursive knowledge that might be said to *go without saying* for a specific group. It can be said to operate beneath the level of rational ideology. This concept was used later on by Bourdieu (1990, p. 54) where he summarises *habitus* as:

> The *habitus*, a product of history, produces individual and collective practices - more history - in accordance with the schemes generated by history. It ensures the active presence of past experiences, which, deposited in each organism in the form of schemes of perception, thought and action, tend to guarantee the "correctness" of practice and their constancy over time, more reliable than all formal rules and explicit norms.

---

[1] The concept reaches back to Aristotle who used it as *hexis* translated as *state* or *habitus*, implying the duration both in space and time (Urmson 1973; Malikail 2003).

This definition of *habitus* is used as the basis for the discussion of routine and rhythm of everyday activities under the aspect of time, culture, society and space. Specifically the aspects of time in the data collected in the various field studies, and in this context the emphasis on the particular interpretation of the term *habitus*, are of importance. The data collected in the fieldwork are passive recordings of everyday activities. The actions and motivations are not specific to the research, but are picked up by following the participant's activities, recording what has happened, as opposed to recording given tasks. In other words, the study objective is not the participant's decision-making process, but his or her individual actions.

Bourdieu put strong emphasis on the definition of *habitus* in the context of time, more specifically history. History is used by Bourdieu to summarise past experiences. They are not static, but constantly redefined, and as history shapes the *habitus*, it is the *habitus*. Bourdieu argues that practice is neither defined in the moment nor a product of history, e.g. *habitus*, that led to it, but it is the *interrelation* of these two, both defined by the *habitus*.[2] He states:

> The *habitus*—embodied history, internalized as a second nature and so forgotten as history - is active presence of the whole past of which it is the product. As such, it is what gives practices their relative autonomy with respect to external determination of the immediate present (Bourdieu 1990, p. 56).

1.4 Context

This research focuses on cycles and rhythms in the urban environment. Day and night, the rush hour, weekends, train timetables, paydays or yearly celebration days are examples of repetitive patterns occurring in the city. Such patterns could theoretically be the result of spatial and social configurations, and they are seen here as based around the organisation of the urban environment. This view on the city is nothing new. For example, in *The Social Logic of Space*, Hillier and Hanson (1984) collected a large set of examples demonstrating the connection between social configuration and the morphology of the built form in a static setting.

As the hypothesis of this research, cycles as repetitive activities are believed to be a third dynamic element within the urban system of objects (the first element) and

---

[2]Bourdieu refers to Emile Durkheim in the case of the history reference. He quotes him as: "In each one of us, in different degrees, is contained the person we were yesterday, and indeed, in the nature of things it is even true that our past *personae* predominate in us, since the presence is necessarily insignificant when compared with the long period of the past because of which we have emerged in the form we have today. It is just that we don't directly feel the influence of these past selves precisely because they are so deeply rooted within us. They constitute the unconscious part of ourselves. Consequently we have a strong tendency not to recognise their existence and to ignore their legitimate demands. By contrast, with the most recent acquisitions of civilisation we are vividly aware of them just because they are recent and consequently have not had time to be assimilated into our collective unconscious" (Durkheim, 1977. p. 11, as cited in Bourdieu 1990, p. 56).

their interrelationships (the second element). In any urban setting, these repetitive patterns are the main source of identity, provide orientation and are a main creator of memory. Barry Curtis even regards memory as:

> one of the key ingredients in the creation of place (Barry Curtis quoted in Borden 2001, p. 63)

and he reflects on this by adding:

> Memory is rarely without contradictions, and it must be compromised in order to function. (Barry Curtis quoted in Borden 2001, p. 63)

1.5 Ethics

The research presented here deals with ethics on a number of levels and has taken a proactive approach to engage with the implications of ethics and privacy concerns in connection to the fieldwork and data collection. The research complies with the University College London ethical guidelines, and it has been approved by the Universities Ethics Committee.

The importance of ethical considerations for this project is inherently connected to the nature of the research interest and the focus on time and space. For the data collection, the field work utilises GPS both as tagged by mobile devices and tracking using specialised GPS devices delivering a high resolution of both spatial and temporal information. The research interest is to look closely at individuals' routines and habits and connecting them to a wider context in an urban sense. Studying these very personal elements involves looking at private habits, of which sometimes even the individual is only vaguely aware. It can be said that the routines and habits are closely connected to the individual's character and reflect personal beliefs, status, culture and so forth as a reflection of societies imprint on the individual.

In aggregate, this information and the fact that the data reflects repetitive activities is essentially private information per se. Furthermore, the data is based on both time and space identifications, putting it in the wider context of a cultural convention of practice. This is data at an extremely detailed level, but at the same time, it is an extreme abstraction. The combination of the two makes the subject vulnerable to a very high degree. The detail could potentially be used or misused for specific targeting of the individual and the abstraction leaves plenty of room for misinterpretation and potential misjudgements about the individual. This vulnerability needs to be considered and protected by the research in order to, on the one hand, prevent any harm to participants and, on the other hand, to ensure credibility and good practice for the research as such.

The research work has shown that the routines are extremely strong overall in the context of everyday practice. This has been confirmed by other research work focusing on similar topics: see, for example Barabási et al. (2010), where the researchers conclude a predictability of location to be in the range of 80–92 %.

The ethical implications are similar in both areas—the GPS tracking of the Urban Diary project and the Twitter data collection of the New City Landscape project. However the guidelines and the standards are very different. Whilst the Urban Diary project can rely on the detailed standards and practice of general social science research involving human subjects, the New City Landscape research part is operating in an emerging field of online digital social networking data and the collection and the analysis thereof, for which very few cases and no standards or guides are available as yet.

1.6 Outline

The text is divided into eight chapters. The chapters are grouped in pairs to form four modules. Module one represents an extended literature review, module two covers the fieldworks and the applied methodologies, in module three the results of the analysis are presented, and in module four the findings are discussed. Each chapter has a short introduction covering the main theoretical and practical aspects before the chapter is presented, discussed and concluded.

Chapters 2 and 3 form the first part, the literature review. Here the material of concepts and existing texts relevant to the text are presented and discussed. Whilst the material on cycle and rhythm literature is not very large, many related fields contribute texts that touch on relevant aspects. It was the intention specifically to include research and theory developed in other fields in order to base the concept of rhythms and repetitive activity patterns in a broader context. As a result of this, the relevant body of literature is large and only a selection can be discussed. In addition, literature is added in the relevant chapters to deepen the discussion on the individual topics. In this sense the literature review is not conclusive, but is intended to open the discussion and frame the concepts discussed here.

The four topics of *Urban Machine*, *Time-Space*, *Body Space* and *Spatial Narrative* provide a structure to hold the entire range of aspects together and act as a grid for reference.

Section 2.1 forms a wider review of relevant literature relating to urban systems. It investigates the impact of modernist ideas on city planning and the perception of function as a defining characteristic. Examples are drawn from a range of sources to create a picture of how today's city came to be understood as a machine. The machine here largely stands for an abstract model of repetition in the sense of clockwork. The machine is examined as part of planning, under the aspect of the function or usage, but also in terms of experience and models of power.

In the same chapter, Sect. 2.2 covers conceptions of time and space as a result of the social configuration of the city, as well as critically reviewing the founding concepts of time geography. The first part discusses the different units of the calendar type time organisation as modelled on natural repetitive phenomena. This establishes those rhythms in a larger context, but ultimately in relation to the city as the place. Together with time, the concept of *place* and *space* also have to be integrated in a holistic view. The second part discusses approaches in time geography to deal with the time phenomena in general. A special emphasis will be placed on the discussion of the time geography time-space model and the related visualisation.

Chapter 3 forms the second part of the literature review, again covering two sections, with the first one being Sect. 3.1. Here the rhythmic structure as a function of the human body as well as investigating the body-city relationship is discussed. The body is discussed as the ultimate point of reference to actually *measure* any repetitive pattern. Very much in the sense of Lefebvre's *Rhythmanalysis*, this establishes the human capacity to perceive cycles and relate to them. The second part of this section discusses the connection to cycles external to the body and focuses especially on the relationship of the body and the city. It examines the relationship overall as a model, but also specifically related to individual or a group of cycles, discussing the possibility of defining space in relation to the body or body extension.

Section 3.2 moves on from the concept of body space towards outlining a concept of structure as a sequence of events constituting an overall story: a story unfolding as a course of activity in the context of the everyday. The narrative is discussed as an element of structure to describe and capture the nature of ongoing processes. It is utilised to provide a framework that can actually integrate the numerous different aspects examined as part of everyday experience.

In module two, the fieldwork is presented in Chaps. 4 and 5. Both cover the practical data collection part, each for a separate project. In Chap. 4, this is a study to track individual participants, recording the spatial extension of their everyday lives. It includes two locations—London, UK, and Basel, CH—each with 20 participants. Also covered in this section are method and technology employed in the fieldwork. This included GPS tracking but also mental or cognitive maps and interviews.

In separate sections, the study setting, the tracking, the technology, the cognitive maps and the interviews are presented and discussed. In the introduction to GPS technology, the possibilities and the problems associated with it are highlighted. Also examined are best practice and different options to validate the data.

With cognitive maps, the aim is to collect data on participants' personal memory of their individual urban environment. The technique of using mental maps in a number of example studies will be discussed. In the interview section, a way of using semi-structured interviews with the participants to collect additional data on individual experiences and perception is presented. Different topics were covered in the interviews. This includes personal routines over different time periods, travel preferences, orientation in the urban environment, identity of places and the individual, memory of travel as well as location (in the form of mental maps) and finally general experiences with the GPS device. The data will be discussed in relation to the technical section on GPS tracking. Finally the discussion focuses in detail on the study's different samples involved in the different fieldwork studies.

Continuing module two, Chap. 5 presents the second data collection, which consists of social networking data mined from the Twitter microblogging platform.

1.6 Outline

This section provides a description of the methods and tools with a detailed discussion of the ethical considerations. With NCL, the focus is on the geolocated Twitter data collected for 20 major cities from around the world. The data is used to investigate city activity patterns overall, based on location and time.

Section 5.2 reviews the rise and fall of digital social networks from Myspace and Bebo to Virtual Worlds and Facebook, alongside the location-based services Foursquare, Gowalla and Twitter. Following this overview, the methods of data mining and crowd sourcing are discussed with, in the latter part, a focus on location-based data sets. This then leads to a discussion of the concrete technical aspects of the collection of Twitter data, discussing different strategies and possibilities.

This is followed by a short discussion of the complications with the Twitter data sample in general and specifically the location-based data. The next sections are concerned with aspects of data analysis in time-space as well as networks. At the end, the aspects of privacy and ethics are discussed. This extends the general discussion of the ethics relevant to this publication, due to the urgency and grey areas in this field if applied to online social networking data.

The third module covers the presentation of the findings from both studies *Urban Diary* and *New City Landscape*. The findings are grouped into two main aspects of *time* and *space* in a separate chapter each.

Chapter 6 *Time* focuses on the aspects of the temporal in bringing together both data sets under the aspect of *clock time* but also *experienced time*. Thus both focus on the individual and the collective as well as discuss cultural aspects of time and time conception. Here the processing and visualisation of the data with a focus on the impact on morphology with respect to sets of flows are included. The different data sets, interviews, mental maps and tracking will be discussed in relation to the individual level. The NCL work focuses on the collective level. As a second stage, the individual and collective parts are related and a synthesis is generated. The main focus will be on the relationship between activity and temporal patterns.

The first subsection focuses on the individual data collected using the GPS technology. A main part is the visualisation, the rhythm in the data and the continuity of the activity narratives. The data is presented both with regard to overall measurement of activity and as a measure of distance from home. With Sect. 6.3, the experiential and planning aspect of time as well as the lead-in to the cultural dimension is introduced.

This leads to the collective time data set mainly based on the *NCL* data collected for the 20 major cities. Here the focus is on overall activity pattern, describing the sample of individuals as a group representing a certain location. For this, we introduce the use of *Time Rose diagrams* to visualise the continuity as well as the rhythmic pattern. Moving on to *Social Time*, the *NCL* data is presented under the aspect of topics and the changes of topics over time relating to both individual and collective activity as time patterns.

Overall this section extends the time aspects of the literature review and critical discussion of the results in the context of the presented theories. A specific focus here is on the time geography concepts and anthropology discussing the value and applicability of *habitus*, *constraints* and *path*, to name a few.

In Chap. 7, the results from the perspective of space, again including both Urban Diary and New City Landscape data sets, are presented. The starting point is *Cartesian Space*. Discussed are the aspects of space on this basis, questioning its relevance for the different data sources. In this context, the focus is put on the implications for the interpretation of the data regarding the comparison if this spatial system is used as the reference. The sections in this chapter are used to test the aspects of space in various settings from examples of *globally* mapped out tracks and traces, as well as cities, to the animated visualisations of time-based mapping. These can be discussed under the heading of space for both visualisation and also for the interpretation of space. Furthermore in Sect. 7.5, space is looked at as it is described and registered by the individual. Here the data is sourced from the mental maps, using the experience of space as communicated by the participants. This leads to the discussion of urban spaces as *self-territory*, with an individual perspective composed of personal cities, and how this links the spaces from both individual and collective perspective. With *Urban Islands*, the focus shifts to the idea of how the important locations act as *anchor points* and relate to one another in analogy to the *Naked City*, as proposed by Debord (1957), and the creation of different types of spaces according to travel patterns.

Finally in Chap. 8, the *habitus* is positioned in relation to the concepts of *time* and *space*, highlighting the production capacity of the *habitus* regarding these two concepts. It will also be positioned with regard to the existing *constraints* model.

Besides the *habitus*, discussed earlier in this introduction as *the present past*, the final chapter also focuses on the term *temporality* as the description of *in between time* manufacturing room for *agency* and allows for the *production* of *time* and *space*.

The previous findings are discussed in regards to a description of *temporality* based on time-space concepts, specifically as patterns of routine and rhythm. This will be based on the concept of the *habitus*, which has been filled with additional meaning from the findings of the fieldwork data. Examples drawn from the fieldwork will be discussed under the heading of both concepts *time* and *space* simultaneously, combining the aspects previously discussed separately.

This brief overview lets us dive right in and start to discuss the current debate of existing ideas in the present and the concept of time-space in an urban context.

1.7 About Working on Time and Space

Over the course of this research work, the approach has evolved. Whilst the starting point and motivation were set out clearly at the beginning, as stated earlier in this chapter, the intensive engagement with the subject has continuously developed a changing position. This kind of shift is one of the leads guiding through the text in the following. It is not a sudden change, but a gradual one that only in the concluding Chap. 8 emerges in a fully formulated new position.

The main shift circles around the understanding of time and space and the "out of curiosity" motivated questioning thereof formulating the research interest for this project. Already with the initial setup of the methods and research aims, a curious duality is implied. A certain critique on the static concepts of the two topics time and space stands opposing the decision to employ ultimately in this very static concept rooted technologies of GPS and GIS.

Whilst this seems to imply a contradiction, on looking back, it did make perfect sense at the outset. Furthermore the two tools are the ultimate methods to engage with the chosen topics in the field of, namely, urban planning, architecture and geography. There are no other readily available tools for this kind of work currently available. Critique has been long formulated, but no efficient alternative has evolved so far. It can be speculated this is the case because of the inherent contradiction in the terms time and space itself, the frequent use of them as either opposites or complementary aspects.

Nevertheless the developed methods for this research have been, possibly in anticipation of such inconsistencies, devised from GPS- and GIS-driven investigations paired with additional elements such as cognitive map and interviews. This includes a variety of lines of investigation and points of views on the chosen topics. It is such a cross-informed method which in the end has enabled the research to evolve and transform and lead to informed speculations on the role and applicability of time-space concepts including its tools, GPS and GIS and finally in the conclusion the proposition of a third theoretical term the *habitus*.

Bibliography

Barabási AL, Qu Z, Song C, Blumm N (2010) Limits of predictability in human mobility. Science 327(5968):1018–1021
Borden I (ed) (2001) The unknown city: contesting architecture and social space: a strangely familiar project. MIT, Cambridge
Bourdieu P (1990) The logic of practice. Polity Press, Cambridge
Carlstein T, Parkes D, Thrift NJ (eds) (1978) Making sense of time. Edward Arnold, London
Debord G (1957) The naked city. (Reprinted in Sadler S (1998) The situationist city. MIT, Cambridge, MA, p 60)
Hägerstrand T (1970) What about people in regional science? Pap Reg Sci 24(1):7–24
Hägerstrand T (1978) Survival and arena. In: Carlstein T, Parkes D, Thrift NJ (eds) Timing space and spacing time. Volume 2: human activity and time geography. Edward Arnold, London
Hillier B, Hanson J (1984) The social logic of space. Cambridge University Press, Cambridge
Lefebvre H (2004) Rhythmanalysis: space, time and everyday life. Continuum, London
Lefebvre H, Nicholson-Smith D (1991) The production of space. Basil Blackwell, Oxford
Lynch K (1960) The image of the city. MIT, Cambridge
Malikail J (2003) Moral character: hexis, habitus and 'habit'. Internet J Philos 7:1–22
Mauss M (2006) Techniques of the body (1935). In: Schlanger N (ed) Techniques, technology and civilisation (Marcel Mauss, 1872–1950; edited and introduced by Nathan Schlanger). Durkheim Press/Berghahn Books, New York, pp 77–95
Neuhaus F (2010) Cycles in urban environments: investigating temporal rhythms. LAP Lambert Academic Publishing, Saarbrücken
Urmson JO (1973) Aristotle's doctrine of the mean. Am Philos Q 10(3):223–230

Chapter 2
Urban Machine and Time-Space

Abstract In this first part of the literature review the section "Urban Machine", forms a wider review of relevant literature relating to urban systems. It investigates the impact of modernist ideas on city planning and the perception of function as defining character. Examples are drawn from a range of sources to create a picture of how today's city came to be understood as a machine. The machine here largely stands for an abstract model of repetition in the sense of clockwork. The machine is examined as part of planning, under the aspect of the function or usage, but also in terms of experience and models of power.

Further on in section "Time Space", covers conceptions of time and space as a result of the social configuration of the city, as well as critically reviewing the founding concepts of time-geography. The first part discusses the different units of the calendar type time organisation as modelled on natural repetitive phenomena. This establishes those rhythms in a larger context, but ultimately in relation to the city as the place. Together with time, the concept of place and space also have to be integrated in a holistic view. The second part discusses approaches in time-geography to deal with the time phenomena in general. A special emphasis will be placed on the discussion of the time-geography time-space model and the related visualisation.

In this first part of the literature review, Sect. 2.1 forms a wider review of relevant literature relating to urban systems. It investigates the impact of modernist ideas on city planning and the perception of function as defining character. Examples are drawn from a range of sources to create a picture of how today's city came to be understood as a machine. The machine here largely stands for an abstract model of repetition in the sense of clockwork. The machine is examined as part of planning, under the aspect of the function or usage, but also in terms of experience and models of power.

Further on Sect. 2.2 covers conceptions of time and space as a result of the social configuration of the city, as well as critically reviewing the founding concepts of time geography. The first part discusses the different units of the calendar type time organisation as modelled on natural repetitive phenomena. This establishes those rhythms in a larger context, but ultimately in relation to the city as the place. Together with time, the concept of *place* and *space* also has to be integrated in a holistic view. The second part discusses approaches in time geography to deal with the time phenomena in general. A special emphasis will be placed on the discussion of the time geography time-space model and the related visualisation.

When looking at patterns of organisation in nature, one of the structuring elements is the repetitive pattern of day and night. The rotation of the earth creates a defining rhythm of light and darkness which impacts on most life on earth. A number of cycles are then derived from these structuring elements. For example, the normal human sleep pattern, as Strogatz (2004) showed in *Sync*, directly connects to the rhythm of day and night. In fact, Strogatz shows that basically all body functions are kept in sync by the day and night pattern. Even though one cycle influences others, this is not to say that the underlying concept should be identified as an organisation of hierarchies. Rather, in the sense that Alexander (1966) writes in *A City is not a Three*, the spatial organisation of the city is nonhierarchical in the sense of a top-down tree diagram. The two cycles are not to be read as one inside the other. They are simply in sync. Any sorting criteria can be applied to group the patterns temporally, but in the long run such separations will not withstand the ever-changing nature of the subject.

Constant change of the environment can be a source of anxiety and unrest. To mitigate such unrest, the cycle is used as an element of structure. Reliable repetition provides stability and security. The routine provides a sense of familiarity as well as orientation. For identification with a location such as the home, such processes are essential. Globalisation has however mixed up the individual sense of place. As Doreen Massey put it in an interview at the Tate Britain in London,

> The definition of place is up for challenge because of the globalisation and the resulting rivalry between global and local (Massey 2009).

Things are no longer what they used to be, and places are connected to remote places on the other side of the globe without physical or experiential connection.

Another perspective on cycles is the impact of repetition on everyday life. As a general meaning, the term cycle is often understood as a constraint, limiting *free roaming* and the *free will* of individuals. This perspective is iconically visualised in the work of Carlstein et al. (1978a). Carlstein presents the repetition pattern as a constraint on the human activity during the day. The models are based on the Hägerstrand (1970) model of constraint. Hägerstrand and Carlstein have worked on these time geography problems closely together at Lund. Examples Carlstein lists are shop opening hours and school times. Time windows open and close and therefore require other activities to be guided along them. However, in this paper, this view is modified towards an understanding of *constraints* as limitations imposed by the repetition rather than the organisation of events. This therefore turns the perspective inside out and takes the standpoint that constraints are the regularity and *free* activities are the exception.

Similar to the individual routine interrupted by globalisation, the definition and organisation of the city as a place is challenged by an explosion of growth never seen before. We are talking *mega cities* (Burdett and Sudjic 2008) of more than 30 million residents. Most of them have grown through informal settlement. The existing concepts of place, urban and city are clearly unable to deal with these new demands and conditions. Since the fall of the medieval city walls, the clear concept of the city as a unit has gradually dissolved, as reflected in current debates around the identity of the suburbs and urban sprawl. In 2008 estimates showed that half the world's population lived in urban areas (UN News Centre 2009). The projection that over 75 % of the world's population will be living in urbanised areas by 2050 (Burdett and Sudjic 2008) brings a new dimension to this.

In this context the envisioned potential of this approach lies in the activation of the inherited structure of the cycle. In this and the next chapters, this potential is outlined with a description of references and sources. In Chap. 2, summarise investigations around the planning and perception of the contemporary city as a result of the modernist movements and the creation and maintenance of the configuration of space and time, respectively.

In Urban Machine, the focus lies on modernist city planning as a direct result of industrialisation. The construction of space is examined as a form allowing implementation of power with the help of clearly defined routines, resulting in strong cycles to drive the urban movement. Section 2.2 examines time as clock time and the aspect of cycles and patterns in calendars and clocks, as well as time geography to define a time-space concept.

Chapter 3 contains Sects. 3.1 and 3.2, investigating, respectively, the body as constituted from repetitive functions, the body-city relationship and the creation of sequence. Body Space looks at natural body rhythms that govern the functions and, through this, human behaviour. In the second part of the chapter this process is looked at in relation to the city and finally the body-city relationship is discussed. The fourth section Urban Narrative outlines a conceptualisation of the proposed approach and summarises a structuring approach to organisation.

The concept of cycles to some extent relies on itself. As Lefebvre explains in *Rhythmanalysis*:

> We know that rhythms are slow or lively only in relation to other rhythms (often our own: those of our walking, our breathing, our heart) (Lefebvre 2004, p. 10).

2.1 Urban Machine

Cities are a multidimensional construct of social activities, processes and configurations taking shape or shaping through relationships, whether imagined, projected, actual, civic or public (Grosz 1998). Cities are also places where flows and power networks intensify and, on the one hand, manifest in physical form but, on the other hand, through their very nature, dominate change. The city ticks somehow. The multitude of activities is far beyond a single person's comprehension, whether as an actor or as a receptor. In our general perception, the city is about organisation (facilities) to enable us to go about our business. We want to have clean water out of the tap at any point in the day, the transport between two destinations to be reliable in time and the city to deal with our waste. All this is preferably provided in a way that we as users neither notice nor are confronted with (Borden 2000, p. 104). We pretend that the city is a black box to serve our needs whilst trying to create as much distance between us and *it* as possible. We are rarely aware that we are actually part of the function of the city, as long as the machine works the way we expect it to. It works, and it works along our routines, whilst it has its own cycles of flow and production. The machine adapts and the city, in many ways, could quite literally be described as a machine. Metaphors such as *it works* or *it breaks down* are strong and widely used, referring to services, functions and events.

The late nineteenth century, with the industrial revolution in full swing, was the time of the machines. Machines were everything: they were adored by a great number of people, including scientists, architects and artists. Machines embodied a functional, efficient, powerful and organised vision. This is summarised in the term Machine Age, sometimes used by historians to summarise the late nineteenth–early twentieth century. Le Corbusier, for example, was outspoken in his admiration of machines such as the automobile and the ocean liner. The fascination was very strong, and references to these machines can be found in many of his projects. He even wrote:

> A house is a machine for living in (Le Corbusier and Goodman 2008).

Another expression *Form follows function* coined by Sullivan (1896) in *The tall office building artistically considered* can be seen as relating to machines. It describes reduction to the essential elements to allow the desired functionality. It is about optimising resources and output. This theory influenced design, architecture, planning and engineering for the rest of the twentieth century. This is from architecture easily transferred to cities, where the *functioning* turns into a dogma.

Others were imagining the city literally. For example, Antonio Sant'Elia, the Italian artist, articulated machine-like dreams. In his utopian city visualisations, machines play the lead role and serve to guide—in a primarily visual way—the imagined city and the composition as a whole. See, for example, Fig. 2.1. Most of Sant'Elia's images were produced between 1910 and 1917, some 100 years ago, yet they bear a remarkable visual resemblance to the contemporary city. In his manifesto for Futurist Architecture[1], he describes the new style:

> We must invent and rebuild the *Futurist City* like an immense and tumultuous shipyard, agile, mobile and dynamic in every detail; and the Futurist house must be like a gigantic machine.

Several movies pick up on this topic. For example, *Modern Times* (1936) by Charlie Chaplin draws on the same metaphors, but at the same time, it is a documentation of the life within, or even a critique of the metaphor of, the machine. The idea of the city as a machine has replaced the image of a medieval city that is dark, narrow and alive but out of control. The industrial city as a machine had

---

[1] The Manifesto can be accessed online at http://www.unknown.nu/futurism/architecture.html

2.1 Urban Machine

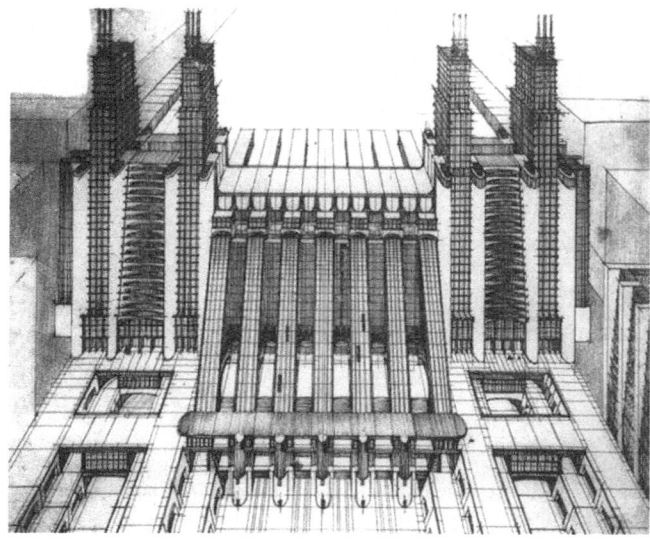

Fig. 2.1 Perspective drawing from La Citta Nuova by Antonio Sant'Elia, 1914. It shows the project for a central station in Milan, combining a train station and an airport connected via elevators

an internal function, and each piece was understood to be fulfilling a role, where there is a very strong sense of control. To some extent, this is still how the city is imagined, as a huge interlinked machine that someone is in charge of.

One thing leads to another—so says the common proverb. This is part of our daily experience. A lot of the actions we take will have some form of impact on how we do something afterwards. Our decisions are not only driven by what we want to do but by the consequences it might have. This can be summarised in the term *planning*, or *projecting* as we will examine it later.

Still, there are moments when things do not go according to plan. These too will influence everything thereafter. These aspects apply to the whole range of scales. On the level of the city infrastructure, an incident can have the same consequences as a chain reaction. An accident on a road in central London will disrupt the journeys of thousands of commuters. Greater events, such as 10 cm snow[2] can bring the city to a standstill. However, things work somehow in most days and this is all residents are concerned with. The city can be imagined as gigantic machinery with hundreds of thousands of little elements—switches and circuits that work in sync. The most quoted visualisation in this context is probably Metropolis, the city machine, as

---

[2]London was covered under snow resulting from a snowstorm in early February 2009. It was featured in two blog posts on urbanTick http://urbantick.blogspot.com/2009/02/london-beats-differently-cycles.html and http://urbantick.blogspot.com/2009/02/how-disruption-reminds-us-of-routine-we.html

Fig. 2.2 A screenshot taken from the film Metropolis by Fritz Lang, 1927. Showing the *new tower of Babel* in between a high-rise vision of the growing metropolis. The view features several layers of car traffic as well as air traffic between the buildings

shown in Fig. 2.2. Hardly anyone cares about what actually happens behind the scene of the real city and how it all works together; maybe no one even knows. For a large city it is hard to imagine that there is one person that really knows why and how everything interlocks. In the genre of the movie, this topic has been explored continuously, with a more recent example being Ridley Scott's Blade Runner (Scott 1982).

Furthermore, this aspect of interlocking events has been the subject of great works in the arts. The artists Fischli and Weiss (1988) created the famous movie *Der Lauf der Dinge* (engl. The Way Things Go) in 1987. See screenshot in Fig. 2.3. Similar to a chain reaction, a motion is unleashed that travels through a setting, constantly changing its form, shape and character. One element bumps into the next one, which in turn rolls on to knock over the following element, which causes a weight to fall and so on. The movie manages to build up a tension carried by curiosity for the next transformation. A very recent interpretation of the theme from 2009 is the clip *Nearness*[3] produced by Berg and Timo. In this version a fundamental shift has taken place. From the very physical and body/object centred original, the latest interpretation has replaced the physical aspect with the use of technology. The moving objects don't touch one another any longer; with the use of RFID (radio-frequency identification), radio waves, mobile phone signals and light, the next step is executed. Even though some of the fundamental aspects of the original *Der Lauf der Dinge* are missing, it does very much resemble the daily life of interlocked actions. It is not so much curiosity, but familiarity that builds up the tension in this new example. We can relate to this familiarity in the daily experience

---

[3]The video clip can be accessed on vimeo at: http://vimeo.com/6588461

2.1 Urban Machine

Fig. 2.3 A screenshot taken from the film *Der Lauf der Dinge (How things go)* by Fischli Weiss, 1987. It was set out as an experiment where each element triggered the following element to become active and so forth

of the city, from the use of the mobile phone to the touch of the smart cards to the use of GPS.

A similar shift has taken place in the city, from physical to virtual. A symbol for this is the automobile. With its introduction, a new kind of individual mobility became imposed on the city. It resulted in some of the most dramatic changes in urban form, but similarly changed the conception of the urban. This saw the rise of the individual utilising the city to serve and function. Alastair Bonnett describes it in his essay on public transport as:

> First in America, but later in other societies, the presence of automobiles became the ultimate symbol both of urban freedom and urban modernity (Bonnett 2000, p. 27).

Much was, and still is, projected into this individual travelling machine. The idea of *anywhere* and *immediately* became the norm. Subsequently the same projections have been imposed onto the city, and the demand for the round the clock availability of urban infrastructure feels *natural* in the western world. To achieve this, attention was paid to individual activities that subsequently were categorised into groups of functions—the functional city is the result of this. Plans were to be based on the categories *housing, work, recreation, transport and divers* as recognised by the CIAM (Congres Internationaux d'Architecture Modern) in 1933 (Shane 2005, p. 110). In terms of urban design, the functional approach has a very long tradition and is not a new concept, as Shane (2005) argues.

> The goal is to achieve maximum efficiency with minimum investment of energy and resources ... seize on the new opportunities offered by the Industrial Revolution (mechanization of the life-world) to provide happiness to the greatest number, following Bentham's utilitarian tenets (which also inspired his Panopticon) (Shane 2005, p. 107).

The formalisation and rationalisation of urban spaces have always been part of planning approaches. From early Chinese cities, to Roman layouts, to garden cities, to new towns, the city was often compressed into a single perspective. This approach is tightly interwoven with structures of power, representation and truth. These aspects are also inherent in the modernist movement, although this involved a shift from a personal representation of power such as a King to a more institutionalised reign of the plan (drawing or proposition) as the central holder of truth and power. Alongside this, the architect/planner as the creator of the plan slipped into a unique position as the single person of Enlightenment authority. Within this context, the term functional city could have a slightly different meaning. It is a more scientific meaning that imposes a great deal of rationality and logic, a concept of the master directing all the different sequences—similar to a conductor.

Only in the very late twentieth century did new descriptions of the city emerge, linked to organic structures. For example, in *The City Shaped* (Kostof 1991, p. 43), there is a clear distinction made between unplanned and planned cities. The two terms are also found in this publication as *unplanned evolution* and *instinctive growth* (Kostof 1991, p. 43). The historian F. Castagnoli made the distinction as follows:

> The irregular city is the result of development left entirely to individuals who actually live on the land. If a governing body divides the land and disposes of it before it is handed over to the users, a uniformly patterned city will emerge (Castagnoli as quoted in Kostof 1991, p. 43).

The planned, designed city versus the grown, spontaneous city, as a categorisation, might already be a child of Sullivan's dogma, which biases the categorisation at the start. However, there is a growing interest in the organic structure of the city. Some well-known urban planners in the past have come from a biology background, such as Geddes (Marshall and Batty 2010); see, for example, Fig. 2.4. Nevertheless, even though the spatial separation of function in urban planning is today regarded

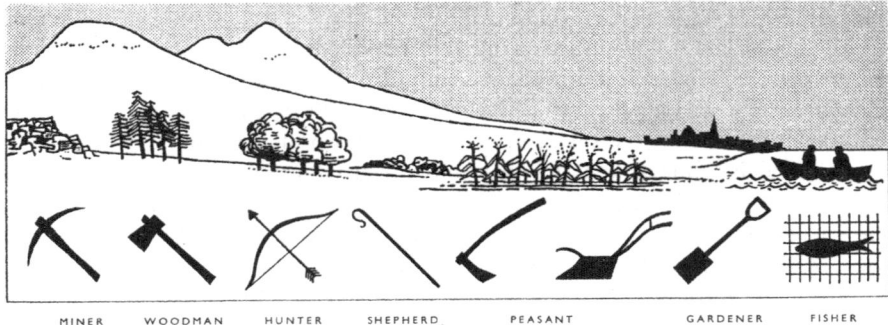

Fig. 2.4 The Valley Section with basic occupation by Patrick Geddes. Inspired by the River Tay in Scotland and later the Ganges in India, Geddes studied life from the mountains to the water. He characterised the specialisation and professionalism according to distance along the valley from an urban centre of trade

2.1 Urban Machine

as problematic in terms of flows, congestion and energy consumption, for example, the concept continues fundamentally to influence planning.

To go back to the modernist concept of city planning and the idea of separation of functions, there is an additional aspect to the city machine. Rhythms do result from the interaction between separated functions, but in addition, as introduced earlier, many of the modernists' concepts were based on machines quite literally. This is articulated in *Cities, Design and Evolution*:

> Modernist city planning, then, is the same as classical city planning except for being distinguished by its use of modern technology - railways, motorways, steel and reinforced concrete (Marshall 2009, p. 33).

This illustrated to some extent a city that was actually *built* around machines. It does not quite define the role of the city, but we can assume that it was thought to be more than just the packaging, but an extension of the machine.

Arturo Soria y Mata proposed a linear city in 1892. It could be the first modernist city plan. The project *Ciudad Lineal* is based on a train line, resulting in a linear city, which could potentially be extended into the countryside or across continents (Marshall 2009, p. 34). See Fig. 2.5 as illustration.[4] The city machine was born and this illustrates beautifully the way the city was thought of. The pulse of the transport network still plays a big role today in the constitution of the city's pulse. The Urban Machine generates a certain rhythm. The pace of departure of the public transport, the frequency of stops, but also the location of stations spatially drives this rhythm. Any live tracking transport site gives a good idea of the pulse of the transport network. As a result of functional separation, travel became the driving activity of the city dynamic. Whilst the separation through functional parts treated to kill it, this beat kept the city alive.

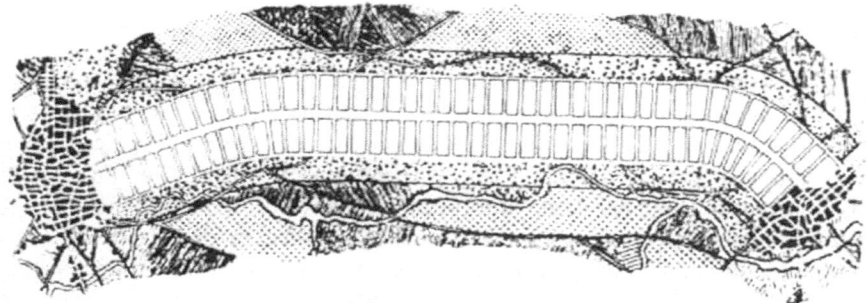

Fig. 2.5 Ciudad Lineal, 1892 by Arturo Soria for Madrid. A design for a linear city capable of extension at either end. The structure here is clearly defined by the serving middle spine of transport, in this case, a proposed tramway under which regime the complete morphology is guided

---

[4]Image is taken from Wikipedia, available at http://en.wikipedia.org/wiki/File:Ciudad_lineal_de_Arturo_Soria.jpg

Fig. 2.6 Diagram by Kevin Lynch, taken from *Good City Form*, showing a conceptual setup for a city structure as a flow diagram of sorts

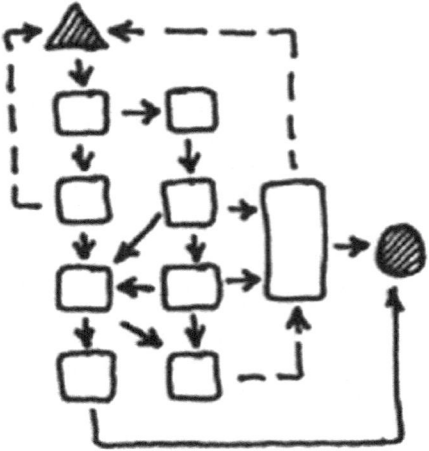

In *Good City Form*, Lynch (1981) speaks of a slightly different *City as a Machine* model. See Fig. 2.6. First he describes the machine:

> "A Machine also has parts, but those parts move and move each other." He goes on with "The whole grows by addition. It has no wider meaning; it is simply the sum of its parts" (Lynch 1981, p. 81).

However he then continues to put some distance between his machine city model and the general perception of machines powered by electricity or steam and made of shining metal. Lynch wants to use his model in the context of a *functional* city for:

> ...wherever settlements were temporary, or had to be built in haste, or were being built for clear, limited, practical aims, as we see in so many colonial foundations (Lynch 1981, p. 82).

In this sense he locates this model directly within the modernist idea of the functional city. However, it seems that he stresses the aspect of time in the terms *temporal* and *haste*. This highlights the importance of time, even in the machine model. Lynch gives examples of the colonisation of America to illustrate the machine model. Here, the aspect of form and structure becomes a relevant topic. As seen before, the aspect of power or *truth* is again central to the approach. In the laws of the Indies of 1573, the Spanish emperor gave clear instructions on how the new cities in conquered land had to be built. They were all based on a grid system with clearly defined locations for the important buildings for a *functional* city. Again, the aspect of power to implement the structure is put forward, but also the simplicity of expanding the structure is stressed. The implementation of power and hierarchy is not mentioned by Lynch. The functioning of the structure is the primary concern and this is thought to be achieved through simplification and purity of form. From this setting it is not far to the rectangular grid cities resulting from land allocation and land speculation in modern America. Lynch goes on through the history of urban

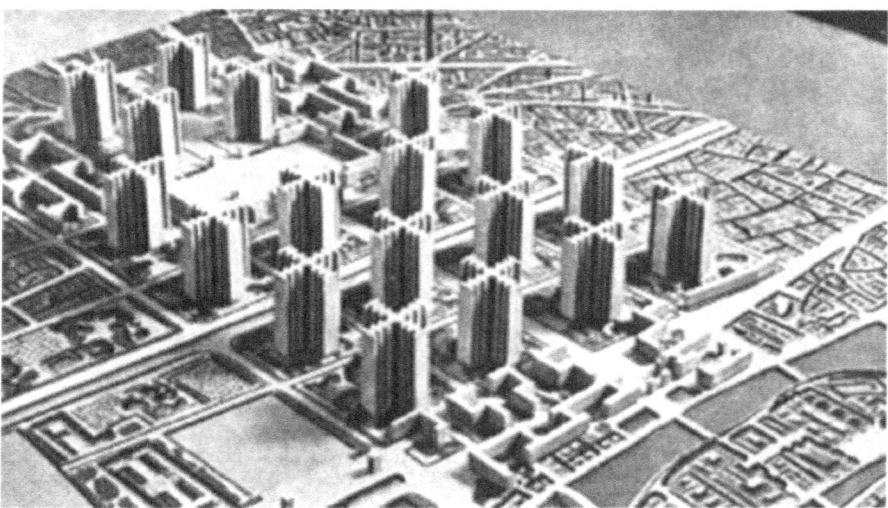

Fig. 2.7 *La Ville Radieuse* by Le Corbusier, 1933, as a project for Paris. With this design Le Corbusier introduced a completely new morphology pattern with a reinterpretation of relationships between open/green space, road and building but also usage and activity (FLC/ ADAGP, Paris and DACS, London 2014)

design and names the Radiant City by Le Corbusier (Fig. 2.7), Soria y Mata's linear utopia, but he also identifies the machine in Peter Cook's Archigram projects and in the work of Soleri or Friedman. He concludes:

> In less sweeping terms, the machine model lies at the root of most of our current ways of dealing with cities ... (Lynch 1981, p. 86).

During the 1980s and especially the 1990s, the computer became an important tool of the Urban Machine. Layered, and later also rendered, computer models seemed the ultimate extension of the Urban Machine far beyond the physical. With CAD (Computer-Aided Design) software in architecture and GIS (Geographic Information System) in geography, the machine began to create itself. Planners began to adopt these tools and functionality became preimplemented in planning processes (Shane 2005). The success of video games such as *Sim City* made this approach especially popular.

During the machine period, the human body was subject to the mechanical imagination. It was a time where sport and competing became important, and the training of the human, mostly male body, as a machine was convenient.

The artwork of Fritz Kahn (Borck 2007) falls into this period and illustrates the ideas beautifully; see Fig. 2.8. Metaphors have probably always been used to explain human body events: phrases like *butterflies in our stomach, eardrums and eyeballs*, the heart is *broken* or our *mind's eye*. These mental visualisations can illustrate feelings to help make them more understandable for others, since they are very personal and experienced individually. The industrial age was concerned

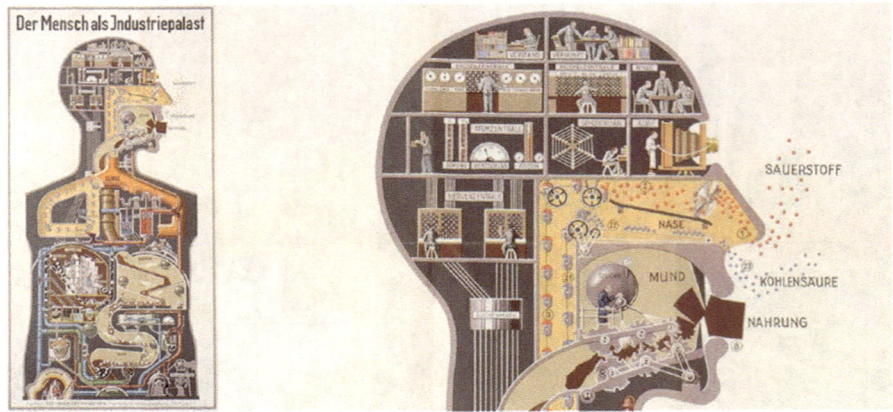

Fig. 2.8 Fritz Kahn transformed the human body in his artwork and reinterpreted the functions as machines

with efficiency, and production was reaching unprecedented levels. In this context it is easier to understand how people have tried to push the human body. To the mechanical understanding, the unpredictable aspects of the human body became a threat that medical science tried to overcome. But some other aspects of understanding of the body were important at the time. Industrial evolution also introduced the human body to new forms of movement. The train and the car meant that dramatically different speeds could be experienced and time and distance in relation to the body had to be newly defined. The big change was the fact that flying was now possible. The human body was able, with the help of the machine, to fly in the air, just like birds.

However, this linking between human function and motion is not limited to the art world. A good example of intense relationship building is ergonomics, especially early ergonomics, which initially focused on task optimisation in relation to factory work. The optimisation task was for movement of the human body to fit with the rhythm of the machine. It was mainly developed by Frederick Winslow Taylor as described in Baxmann (2011). Taylor used photography and video to break a motion into elements and reassemble it to fit with the rhythmic machine movement structure. The early ergonomists' ideas however would reach a lot further than their practice, aiming to integrate the individual into the beat of a great social machine (Baxmann 2011, p. 116). The conveyor belt in the factory, defining an assembly line, shaped a new world defined by technological rhythm. In particular, it sharply defined the workers' movements, to the extent of synchronising many individuals into a working collective. In many texts this leads to theoretical social utopias aiming to produce a new type of labourer, all synchronised. The rationalisation of spatial movement to a large extent also spilled over beyond the factory walls out into the social context of individual labourers. Together with aspects of time, this technification of a synchronised rhythm reshaped most of modern life.

2.2 Time-Space

Not only did the possibilities for movement change, but also the perception and use of time entered a new dimension. In this section, the temporal aspect of repetition in relation to space will be discussed. It is structured as two sections: the first one discussing the practice of the rhythmisation of time in the sense of a countable system of repetition that serves as an orientation guide and the second discussing the relationship of time and space with a focus on the urban environment and the creation of such context. In this context, time geography is the discipline that offers concepts to deal with this problem and will be critically discussed. The city as a place is often described as simultaneous times with parallel realities. The space-time problem is widely discussed and subject to a number of attempts to conceptualise it, ranging from Lefebvre (2004) to Glennie and Thrift (2009). This discussion does not attempt to invent a time-space theory, but rather to summarise it, both from a theoretical perspective and also from the perspective of experience.

Science has occupied the debate and dominated the conception of time as linear for the most part since Galileo's pendulum. Only in the twentieth century have the discussions on the topic in various disciplines been taken on a different perspective. Where Newton strongly argued for an objective time, a single and true time concept, Einstein developed an alternative concept with the theory of relativity, allocating time a subjective role, with time passing at different speeds for different observers. This development took place in physics, but simultaneously the conception of time was being re-examined by philosophers, especially the French league with Bourdieu (1990), Lefebvre and Nicholson-Smith (1991), Lefebvre (2004), Serra, de Certeau (1984) and others. Here the debate between objective and subjective time continued with the phenomenologists entering the debate. However, as Elias (1992) points out in his *Time: an Essay*, that in their understanding of the concept that time is something given, a fixed instance, measurable. In this context the third concept of *social time* was developed. As Elias describes it:

> To perceive time requires focusing-units (humans) capable of forming a mental picture in which event A, B and C follow one after another, are present together and yet, at the same time, are seen clearly as not having happened together; is a synthesis only humans are capable of and learned and developed over generations (Elias 1992).

In the current debate most philosophers would agree with a concept of time that is nonlinear, and the concept of multiple times is accepted as a social aspect. However, in exact science, the linear singular *clock time*, as a single continuous count of units, continues to dominate the discussions, as well as the projects and work undertaken. This led to the increasing exclusion of the temporal aspect due to arising problems with the integration of linear time that does not fit with the complex systemic concepts any longer.

This could be about to change with the recent development in spatial research and the focus on mobility and location data. Aspects of temporality all of a sudden move into the spotlight. This of course goes hand in hand with a shift in the development and availability of specialised and capable technology delivering data in real time.

In visualisation terms, some interesting approaches have been developed. The time-space cube developed by Hägerstrand (1978) is one example that goes beyond a visualisation, having developed into a theory of time-space usage, but also into time-distance-based distortion maps or animated visuals. Also, the aspect of comparison between two instances is used quite often. Usually the same object is shown at two different stages to visualise the change that took place between the two points in time—implying the Newtonian time aspects apply.

Here, time is seen as an intellectual construction invented for structural purposes. Even though nowadays we regard time as *natural* and the watch as a technical device to measure this *natural* phenomenon, time is little more than a social convention (Glennie and Thrift 2009). It has entered our lives in all areas and has grown to be a constant companion. The history of time is very long, and humans have made many attempts to conceptualise the passage of time. Lynch describes the concept of time in *What Time is this Place* with:

> Rhythms, objects, and events exist; but time and space are triumphant human inventions. Past, present, and future are created anew by each individual (Lynch 1972, p. 119).

The observation of repetitive patterns has led to the idea of a continuous flow. Initially time was based on the observation of change in repetitive patterns that inscribed a sequence. Bodily functions, such as the heartbeat, provide a simple indicator of such a continuous sequence. Similarly, singing was used as a measure of time (Glennie and Thrift 2009, p. 2). On a larger time scale, the passage of the sun was an initiator of a time concept. The movement of specific stars or constellations gave a sense of time passing. Those are all natural rhythms that structure our environment by inheriting an *out of reach* timescale compared to human influence. However they still have a long-term cyclical character and therefore are not exactly accurate in the sense of modern daytime keeping. The improvement in precision over those 10,000 years is simply the successful story of technological progress. However, the fundamental concept of time as a structuring element has as much to do with social organisation than anything else (Reichholf 2008). In particular, the arrangement within growing communities (larger than families) demanded a structuring element applicable to any sort of arrangement. Amongst the first conventions were rites and religious/cultural activities. Priests were the first to be in charge of inventing calendars as a structural framework and tune the counting to ever more detailed units (Richards 1998, p. 5).

Each of the major religions has its own calendar, based on evolved conventions over time. As an example, the structural impact of the religion or faith practised by a group of people results in the persistence of religious calendars today. As Richards (1998) shows, for example, Friday is the day of rest for Islam, whilst Saturday is the day of worship and rest for Judaism and Sunday is the Lord's Day for Christians. Repetitive phenomena were used initially as counters. Each cycle was counted as the beat of time. Solar or lunar calendars are amongst the most popular ones. The Egyptians used both calendar systems in parallel, and still in modern Christianity, two calendars are used, the Gregorian calendar for secular purposes and the lunar calendar to set the date of Easter. The units of the year, the day and the month are loosely based on natural counters. The day, as the most fundamental unit consisting of day and night, is a feature of almost every calendar system. The month is based on the lunar cycle between two new moons, and the year is based on the seasons. All are repetitive patterns over different cycle lengths that can be observed without the need for special tools or instruments. However, the three units are incommensurate: the days vary in length over the year, the counting of the days in 1 year comes to 365 and the days in a lunation are 29–30, plus there are sometimes 12 and, at other times, 13 lunar cycles in 1 year. Today the three units are a rough guide and mathematics accounts for the gaps and misfits. It is cleverly constructed to fit most aspects. As seen before, most major systems agree on the same cycles. However, there is dispute over the start and end of such a unit. Does the day start at noon, midnight, dusk or dawn? Similarly, with the year, what marks the beginning of the cycle? In different cultures, these norms were different, and usually the starting point was connected to the method of observing. If the horizon was the reference, autumn or spring was the start; where the shadow is the reference, summer, with the shortest or no shadow, marked the start.

Synchronisation is a major topic in research on time. Many calendars are deeply rooted in the local culture and have been replaced successively by an international calendar. This could be called the colonisation of time. But still, synchronisation has not quite been reached, as the different time zones still require a different time according to place. Western culture has managed to dominate cultural values, but it is having difficulties overcoming the spatial component. The week is the missing link. There is a significant gap between the concept of the week and the concept of the month. Both share the day as a naturally defined unit. However, the week is a completely theoretical concept based on practice and culture, where as the month is loosely based on the cycle of the moon. It would be possible to add the hour as a third concept that is based on location. It too connects to the day as the naturally defined unit, as the week and the month do. The week does not have a corresponding *natural* phenomenon. Here it is solely based on convention that we create a week of, for example, 7 days. In the past, the week had any number of days between 3 and 15. The 7-day week unit was observed by the Babylonians, Jews and several nations in West Africa (Richards 1998, p. 266). The Jewish 7-day week is mentioned in the creation of the world as an act of 7 days in the book of Genesis:

> And on the seventh day, God ended his work which he had made; and he rested on the seventh day from all his work which he had made. And God blessed the seventh day, and sanctified it: because that in it he had rested from all his work which God created and made (Genesis 2:2-3 as quoted in Richards 1998, p. 267).

However, in *The seven day circle: the history and meaning of the week*, Zerubavel (1989) argues that the week can basically be seen as the concept of the market. With this, it goes further back in history than the religious Jewish or Christian references in the Bible. Examples of market weeks can be found in cultures across the world from a 3-day market week in New Guinea (as before) and Colombia, a 5-day cycle in ancient Mesoamerica and Indochina to a 10-day market week in Peru and a 12-day market cycle in ancient China (Zerubavel 1989, p. 45). The week in this sense makes for an interesting example of interconnection between the concept of time and space in the case of the city. The relationship between the city and the countryside, as described by Geddes and Thomson (1911) in his *Valley Section*, comes into focus as a region within which the different system elements play. The sustainability of the city is tightly interwoven with its hinterland and vice versa. The market week illustrates this relationship with the physical movement of farmers into the city to trade products; at the same time, the routine cycle sets a model for a time unit, similar to the observations of natural phenomena as discussed before. The major difference is that it is a construct, a logical sequence of dependencies, in itself. The cyclical migration of people points to a concept of space where space as distance, beyond the movement of the body, becomes important. The concept of the market place as a square fixes the location and the time of the market, structuring both conceptualisations, resulting in the dominating enforcement of power. This is a result visible throughout the history of urban morphology. To illustrate this connection, we can look at the Roman concept of the 8-day market week. The 8-day cycle was introduced by the Etruscan civilisation, a culture inhabiting Italy, the Romans inheriting the concept. The number eight here plays a dualistic role, as the Roman number concept was nine in eight, from the inclusive counting system, where the last and the first day counted twice, resulting in some sort of overlap. However, the eighth day was the market day where farmers came in to the city to trade products. At the same time, the citizens were allowed to host five guests overnight, as opposed to the limit of three during the rest of the week (Zerubavel 1989, p. 46). This illustrated the interconnection between the different locations and the concept of connecting them through a framework of time, as well as a concept of movement resulting in distance.

A completely different concept of time exists within the Hopi Indian culture. According to Whorf (1956), the Hopis use two terms to structure reality: the *manifested* (object) and the *manifesting* (subject). In them, time and space are the same. For example:

> Far away and long ago are the same: they are the manifest. Future and inner are also the same and lie in the manifesting (Lynch 1972, p. 131).

Tuan gives an example of the temporal concept:

> The Hopi do not abstract time from distance, and hence the question of simultaneity is to them an unreal problem (Tuan 1977, p. 121).

See Fig. 2.9 for an diagrammatic illustration of the concept. An event in a distant village can therefore not be at the same time, as it is far away and will be known in one's village only much later. The greater the distance, the greater the lapse of time. In comparison it becomes clear that the concept of time-space is nowadays perceived as *natural* ruling out simultaneity in its representation as an arrow. The same could apply for space. Our concept of space is built around the idea that two objects cannot be in the same place at the same time.

From this mainly time-based review, we have progressively made our way into the time and space connection. It seems increasingly difficult to actually think of the two separately, and in the early history discussed above, we have pointed out a link between both conceptions as conventions in connection to human settlements.

2.2 Time-Space

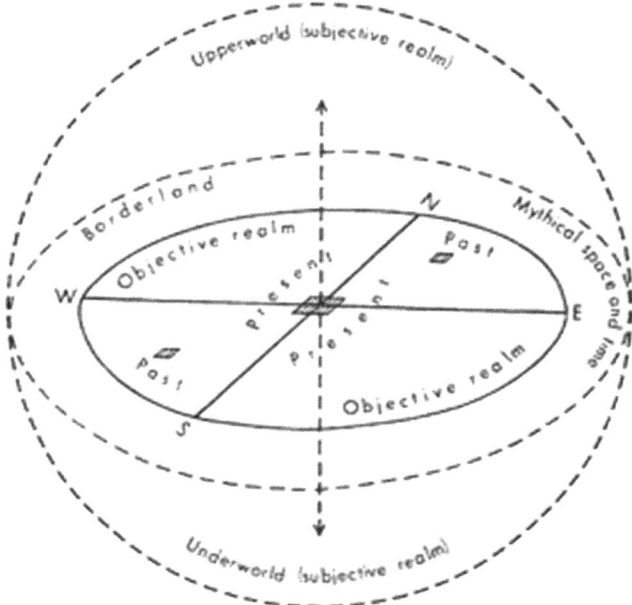

Fig. 2.9 Hopi space and time: subjective and objective realms. The objective realm is the horizontal space within the cardinal grid, but at the distant edges, it merges with the subjective realm as represented by the vertical axis (Tuan 1977, p. 121)

The field of time geography took on this huge topic in the 1960s and 1970s. The most influential figure in this academic field was Torsten Hägerstrand. He was a Swedish geographer based at the University of Lund. He worked with a large group of people on the subject of time geography, including Tommy Carlstein and Bo Lenntrop (Carlstein et al. 1978a,b). The research work they undertook focused on human movement in order to explore time and space. Hence the rule of time-space:

> The fact that movement in space is also movement in time (Hägerstrand quoted in Giddens 1986, p. 112).

To study these phenomena, a unique visualisation technique was developed, called the *time-space aquarium* (Kwan and Lee 2003). The trajectory of an individual is plotted in a 3D space, with x- and y-axes representing horizontally the spatial location and the z-axis plotting the time vertically. See Fig. 2.10 for illustration of the concept. This cube or aquarium maps the location of objects or individuals in space-time. In the 1970s, it was a completely new visualisation unknown to social scientists. Thrift states:

> Perhaps one way of looking at Hägerstrand's work is as a means of saying *hello* in a language many can understand: drawing as a kind of visual Esperanto (Thrift 2005).

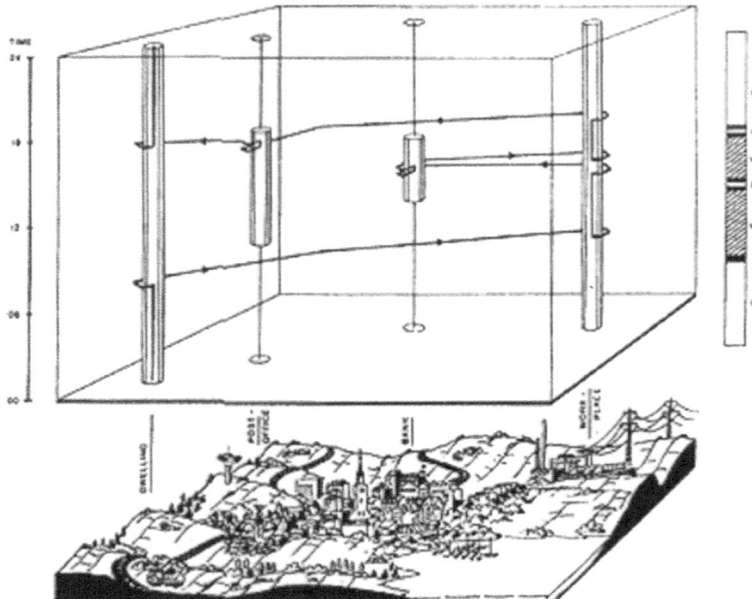

Fig. 2.10 The *time-space cube* or *aquarium* after time-space movement visualisation developed for time geography by Hägerstrand, visualisation taken from Carlstein et al. (1978b). The time dimension is shown vertically on the z-axis, where as the spatial dimension is laid out horizontally as x and y. In this example, a 24 h period of a *normal* workday is shown with emphasis on static locations where more time is spent

However this is a very descriptive form of visualisation, built upon the Euclidean idea of space by using the geographical referencing system. Essentially the cube represents the boxed space, where objects, goods and people can be placed and moved at random. However, Hägerstrand offers a detailed concept of influencing factors upon the random placement. After Giddens (1986), they are formulated as:

1. The indivisibility of the human body and of other living and inorganic entities in the milieu of human existence. Corporeality imposes strict limitations upon the capabilities of movement and perception of the human agent.
2. The finitude of the life span of the human agent as a *being towards death*. This essential element of the human condition gives rise to certain inescapable demographic parameters of interaction across time-space. For this reason, if no other, time is a scarce resource for the individual actor.
3. The limited capability of human beings to participate in more than one task at once, coupled with the fact that every task has a duration. Turn-taking exemplifies the implications of this sort of constraint.
4. The fact that movement in space is also movement in time (as referenced above).

2.2 Time-Space

5. The limited *packing capacity* of time-space. No two human bodies can occupy the same space at the same time; physical objects have the same characteristic. Therefore any zone of time-space can be analysed in terms of constraints over the two types of objects which can be accommodated within it.

It should be noted that these five points have a certain negative or restrictive aspect to them, almost as if they are a direct child of the modernist space view, glorifying the machine. There is a sense of the incompleteness of the human being. One might rather expect a more positive description of the human capacity. However, in general, time geography builds largely upon the concept of limitation. From there, the trajectories of individuals are constructed within the cube. Nevertheless, Hägerstrand (1982) assigns each individual the capacity of planning, to account for humans as intentional beings. This he calls *project*, and it represents what each individual seeks to realise according to Giddens (1986, p. 113). Hägerstrand (1970) also identified three categories of limitations to human movement, or *constraints*: *capability*, *coupling* and *authority*. Capability constraints refer to the limitations on human movement due to physical or biological factors. Thus, for example, a person cannot be in two places at one time. A person also cannot travel instantaneously from one location to another, which means that a certain trade-off must be made between space and time (Corbett ca. 2001). The coupling addresses the fact that, in certain situations, individuals are connected to one another, where they have to (in his words) undertake something jointly, resulting in a *bundled* path, as in the morning rush hour in the subway. The third aspect of *authority* is *domain*—the area that is controlled by a group or institution restricting access for other individuals or groups, such as a hospital, a military base or a private club. This describes a territoriality of human beings, transferring the time-space concept from the small-scale, individual level to the scale of nations and global dimensions. It has to be noted here that this also applies in reverse, meaning that nations do apply *authority* constraints to the individual, but in this concept, they do so to establish themselves in time-space. This setup with *constraints* produces a distinct view, that of time-space organisation forming a string of dependencies. As Giddens puts it:

> Hagerstrand made a particular effort to employ time-geography to grasp the seriality of the life *paths* or *life biographies* of individuals (Giddens 1986, p. 114).

This results in a clockwork-like description of activities interlocking in time. This is in line with the idea of the Urban Machine as an interlocking sequence of functions.

Back in the days when the time-space cube visualisation technique was developed, the computer software used to handle spatial data, such as a GIS system, was only in its infancy. Therefore, initial visualisations were limited to a few locations connected by trajectories, or highly abstract graphics representing the idea. In recent years a number of researchers have, with the use of new technologies and software, revisited the approach and fine-tuned the visualisation method. With GIS, specialised software such as Oculus's *GeoTime* (Kapler and Wright 2005), or even free software like Google Earth, has made these types of visualisation

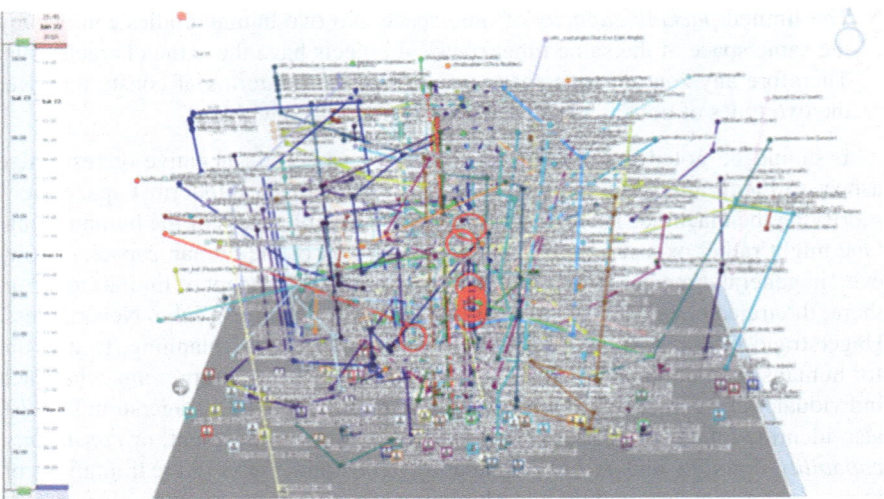

Fig. 2.11 A screenshot showing the visualisation in GeoTime of Twitter data collected for London over a period of a long weekend. GeoTime is an application developed by Oculus based on the time-space aquarium concept (Kapler and Wright 2005). The time dimension is represented vertically and the spatial location horizontally. Visualisation by the author (See http://urbantick.blogspot.co.uk/2010/02/urban-narrative-tracking-movement-via.html for a larger version)

quite simple to achieve; and they are capable of handling large data sets. See Fig. 2.11 for an illustration of data visualised in GeoTime. It is conceptually focused on the visualisation, and the process of analysis has not been extensively developed. In more recent works of, for example, Kwan and Lee (2003) or Miller and Ahmed (2007), the visualisation aspect is further developed with the help of new technologies and software. See, for example, Fig. 2.12.

As hinted above, there have been critiques of the time geography approach, especially from the social sciences, feminists and philosophers. Hägerstrand is widely regarded as the father of time geography (Thrift 2005). However, since he introduced the concept around 1970, the field has progressed and evolved into a multidimensional discipline with a large base and contributors. Martin Gren summarises the main points in his chapter *Time-Geography Matters*:

> It has also been criticised for being too *physicalistic*, too *reductionistic*, too *objectivistic*, too *masculinistic*, and for having a too *narrow or naive conception of power and human thought-and-action* (Gren 2001, p. 209).

Giddens specifies a few of these points: for instance, what the actual origin or desire of a *project* is left undefined. Furthermore, in general all elements are assumed to be given and therefore *natural*. Giddens (1986, p. 117) states the time geography operates with:

> ...uninterpreted processes of institutional formation and change.

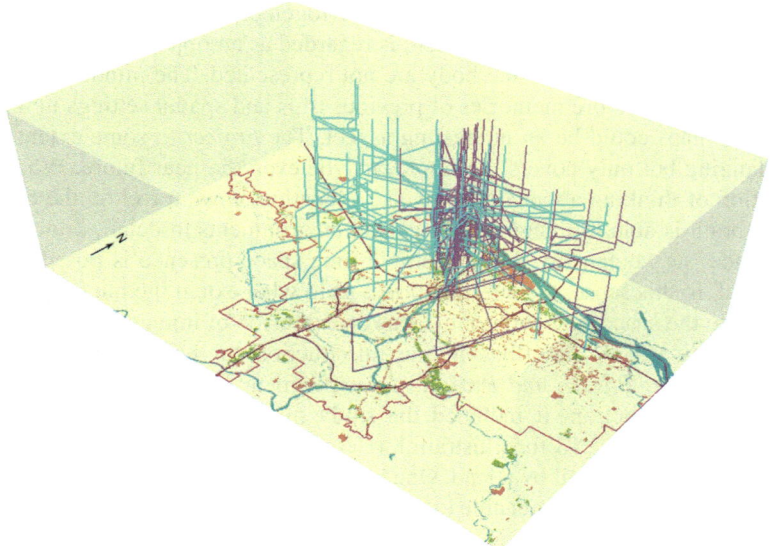

Fig. 2.12 Kwan (2004) *Geovisualization of Human Activity Patterns Using 3D GIS: A Time-Geographic Approach*. Space-time aquarium showing the space-time paths of African and Asian Americans in the sample

This results in a concept that excludes the essential transformational force of human action. Giddens also notes the weakly developed theory of power. As we saw earlier in the conceptualisation of time and space, power is an inevitable result shaped in the form of the urban settlement. Hägerstrand does refer to elements of power in his set of constraints as *authority constraints* but only as an *object*. In addition the concept lends itself to readily reproduced dualistic oppositions like *material-social, fiction-fact, reality-representation, sign-object, body-mind, culture-nature, inner-outer male-female* and so forth (Gren 2001, p. 210).

In its conception, time geography is directly concerned with the individual's movement in time-space. However, Gren points out that, regarding the visualisation, Hägerstrand has created a tricky dilemma. Whilst time passes, the individual track is rising vertically to the spatial reference plane. Effectively, this disconnects action and environment. It can no longer achieve what it claims to do—represent the physically embedded human body in its environment. There is, furthermore, a fundamental difference between the visualisation for an external observer and the experience of the individual. In this sense, the name "time-space aquarium" is appropriate. The software GeoTime tried to account for this problem by offering a vertically sliding ground plane. The currently active location is still the only one visualised in its actual context.

To a large extent, Hägerstrand's factors and constraints are based on the physical aspect of objects and the human body. To some extent, the concept even helped to bring objects and the human body closer together in terms of the theoretical approach. But in general, the body here is regarded as an object that moves in time and space. Other aspects of the body are not represented. The mind, for example, and its capacity to store memories of previous trips and spatial settings in the form of mental maps could be an interesting aspect. The *project* to some extent covers this planning but only covers the future, maybe even the near future. What about the option of thinking about someone else's location as we travel on the bus? The human brain is able to imagine settings and environments in detail, drawing from experience: so we can say that a multiple time-space presence is possible for an individual. In this sense, the question arises as to what extent the mind can connect the body to the space-time environment as an extension of immediate experience.

Spuybroek (2004) describes in the introduction to the book *Nox: machining architecture; Bauten und Projekte* an experiment undertaken by scientists to research brain function. It included the study of two kittens developing spatial experience. See Fig. 2.13 for illustration (Held and Hein 1963). Both kittens were attached to a circular trail by a lead. One kitten was able to walk around the course, whereas the other one was not able to move, sitting in a little box, but was dependent on the other cat's movement. Both kittens grew up with these imposed constraints and their spatial experience was based on this model. After being released, the kitten that was able to walk did not show any special behaviour, whereas the kitten that grew up dependent on the movement of the other subject was unable to navigate the environment, constantly bumping into objects. Scientist concluded that this cat had been unable to build up a memory stock of spatial implications through the body. To extend this observation to the human body, our spatial experience is essential to navigate and move the body in space. Through nerves and muscles, we have an

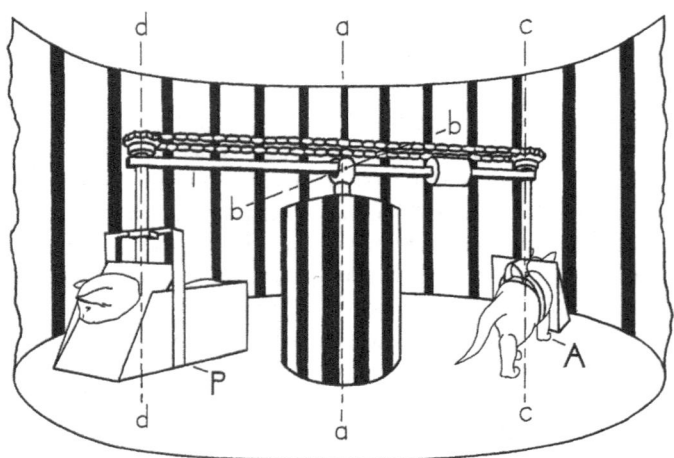

Fig. 2.13 Richard Held and Alan Hein, 1963. Apparatus for equating motion and consequent visual feedback for an actively moving (*A*) and a passively moved (*P*) subject

exact knowledge of the position of each element relative to one another and relative to other objects from our surroundings. This enables us, for example, to pick up a cup of tea without spilling it, or to pour water from a bottle into a glass. Furthermore this *knowledge* can be transferred onto objects and gadgets. If someone wears a hat with a feather on it, this person would automatically bend down their head whilst passing through the door, in the awareness of being too tall to go through the door without bumping the feather into the door frame. The same applies to driving a car. The mind simply extends the body and regards the car in which the body is sitting in as an extension, and we are able to drive the streets whilst normally not crashing into other objects or passers-by. To connect this to the above problem of spatial time-space perception of the individual, we will ask the question: to what extent can the human mind extend the body out into the city? Or to what extent is the city-space a creation of this extended body construct?

This leads us to the question of the relationship between the body and the city which will be discussed in Chap. 3.

Bibliography

Alexander C (1966) A city is not a tree. Design, vol 206. Council of Industrial Design, London, pp 58–62
Baxmann I (2011) The form of movement and life: modern ergonomics ad rhythmicising life. In: Zinsmeister A (ed) Gestalt der Bewegung Figure of Motion. Weisenhof edition. Jovis Verlag Gmbh, Berlin, pp 114–129
Bonnett A (2000) Buses. In: City A-Z. Routledge, London, pp 26–28
Borck C (2007) Communicating the modern body: Fritz kahn's popular images of human physiology as an industrialized world. Can J Commun 32(3):495
Borden I (2000) Hoardings. In: Thrift NJ, Pile S (eds) City A-Z. Routledge, London
Bourdieu P (1990) The logic of practice. Polity Press, Cambridge
Burdett R, Sudjic D (2008) The endless city. Phaidon Press, London
Carlstein T, Parkes D, Thrift NJ (eds) (1978a) Timing space and spacing time. Edward Arnold, London
Carlstein T, Parkes D, Thrift NJ (eds) (1978b) Making sense of time. Edward Arnold, London
Corbett J (ca. 2001) CSISS classics—Torsten Hägerstrand: time geography. http://www.csiss.org/classics/content/29
de Certeau M (1984) The practice of everyday life. University of California Press, Berkeley
Elias N (1992) Time: an essay. Blackwell, Oxford
Fischli P, Weiss D (1988) Der Lauf der Dinge. DVD Icarus Films
Geddes P, Thomson JA (1911) Evolution. Williams & Norgate, London
Giddens A (1986) The constitution of society: outline of the theory of structuration. Polity Press, Cambridge
Glennie P, Thrift NJ (2009) Shaping the day: a history of timekeeping in England and Wales, 1300–1800. Oxford University Press, Oxford
Gren M (2001) Time-geography matters. In: Thrift NJ, May J (eds) TimeSpace: geographies of temporality. Number 13 in Critical geographies. Routledge, London
Grosz E (1998) Bodies-cities. In: Nast HJ (ed) Places through the body. Routledge, London
Hägerstrand T (1970) What about people in regional science? Pap Reg Sci 24(1):7–24
Hägerstrand T (1978) Survival and arena. In: Carlstein T, Parkes D, Thrift NJ (eds) Timing space and spacing time. Volume 2: human activity and time geography. Edward Arnold, London

Hägerstrand T (1982) Diorama, path and project. Tijdschr Econ Soc Geogr 73(6):323–339
Held R, Hein A (1963) Movement-produced stimulation in the development of visually guided behavior. J Comp Physiol Psychol 56(5):872–876
Kapler T, Wright W (2005) GeoTime information visualization. Inf Vis 4(2):136–146
Kostof S (1991) The city shaped: urban patterns and meanings through history. Thames and Hudson, London
Kwan M-P (2004) GIS methods in time-geographic research: geocomputation and geovisualization of human activity patterns. Geogr Annal Ser A Phys Geogr 86B(4):267–280
Kwan M-P, Lee J (2003) Geovisualization of human activity patterns using 3D GIS: a time-geographic approach. In: Goodchild M, Janelle D (eds) Spatially integrated social science: examples in best practice. Oxford University Press, Oxford
Le Corbusier and Goodman J (2008) Toward an architecture, 1st frances lincoln ed edn. Frances Lincoln, London
Lefebvre H (2004) Rhythmanalysis: space, time and everyday life. Continuum, London
Lefebvre H, Nicholson-Smith D (1991) The production of space. Basil Blackwell, Oxford
Lynch K (1972) What time is this place? MIT, Cambridge
Lynch K (1981) A theory of good city form. MIT, Cambridge, MA
Marshall S (2009) Cities, design & evolution. Routledge, Abingdon
Marshall S, Batty M (2010) Geddes' grand theory: life, evolution, social union and the "great transition." CASA working paper, (162), pp 1–33
Massey D (2009) Tate Britain the status of difference: geographies of difference. Available at: http://www.tate.org.uk/britain/eventseducation/talks/19718.htm
Miller HJ, Ahmed N (2007) Time–space transformations of geographic space for exploring, analyzing and visualizing transportation systems. J Transp Geogr 15:2–17
Reichholf J (2008) Warum die Menschen sesshaft wurden: Das größte Rätsel unserer Geschichte, 2nd edn. Fischer, Frankfurt
Richards EG (1998) Mapping time: the calendar and its history. Oxford University Press, New York
Scott R (1982) Blade runner. DVD Warner
Shane DG (2005) Recombinant urbanism: conceptual Modeling in architecture, urban design and city theory. Wiley, Chichester
Spuybroek L (2004) Nox: machining architecture; Bauten und Projekte. Thames and Hudson, London
Strogatz SH (2004) Sync: the emerging science of spontaneous order. Penguin Books, London
Sullivan LH (1896) The tall office building artistically considered. Lippincott's Mag 57(March):403–409
Thrift NJ (2005) Torsten Hägerstrand and social theory. Prog Hum Geogr 29(3):337–340
Tuan Y-F (1977) Space and place: the perspective of experience. University of Minnesota Press, Minneapolis
UN News Centre (2009) Half of global population will live in cities by end of this year predicts UN. http://www.un.org/apps/news/story.asp?NewsID=25762
Whorf BL (1956) The Hopi language. University of Chicago Library, Chicago
Zerubavel E (1989) The seven day circle. University of Chicago Press, Chigago

Chapter 3
Body Space and Spatial Narrative

Abstract In this second part of the literature review, section "Body Space" discusses the rhythmic structure as a function of the human body as well as investigating the body-city relationship. The body is discussed as the ultimate point of reference to actually measure any repetitive pattern. Very much in the sense of Lefebvre's Rhythmanalysis, this establishes the human capacity to perceive cycles and relate to them. The second part of this section discusses the connection to cycles external to the body and focuses especially on the relationship of the body and the city. It examines the relationship overall as a model but also specifically related to individual or a group of cycles, discussing the possibility of defining space in relation to the body or body extension.

The section "Urban Narrative" moves on from the concept of body space towards outlining a concept of structure as a sequence of events constituting an overall story: a story unfolding as a course of activity in the context of the everyday. The narrative is discussed as an element of structure to describe and capture the nature of ongoing processes. It is utilised to provide a framework that can actually integrate the numerous different aspects examined as part of everyday experience.

In this second part of the literature review, Sect. 3.1 discusses the rhythmic structure as a function of the human body as well as investigating the body-city relationship. The body is discussed as the ultimate point of reference to actually *measure* any repetitive pattern. Very much in the sense of Lefebvre's *Rhythmanalysis*, this establishes the human capacity to perceive cycles and relate to them. The second part of this section discusses the connection to cycles external to the body and focuses especially on the relationship of the body and the city. It examines the relationship overall as a model but also specifically related to individual or a group of cycles, discussing the possibility of defining space in relation to the body or body extension.

Section 3.2 moves on from the concept of body space towards outlining a concept of structure as a sequence of events constituting an overall story: a story unfolding as a course of activity in the context of the everyday. The narrative is discussed as an element of structure to describe and capture the nature of ongoing processes. It is utilised to provide a framework that can actually integrate the numerous different aspects examined as part of everyday experience.

3.1 Body Space

In this chapter the relationship of body and city is discussed under two main headings. One topic is the natural, biological and functional aspects of rhythms and patterns directly related to the human body, both internal and external. This is drawing direct lines to natural *constraints* resulting from the way we as humans experience the environment. Secondly this chapter explores the experience, meaning and creation of space through these functions and the body as a whole. The aspect of the body as a physical object in space is investigated in relation to how this is the main basis of space creation. This draws direct lines to the Urban Diary tracking project where the movements of individual *bodies* are recorded, resulting in a spatial inscription on the city. As a result of these investigations, a concept of space, termed *Body Space*, is outlined as a conclusion.

In nature, many functions take the form of cycles. This derives from the movement of the planets, including the sun, the moon and the earth itself. This movement results in repetitive patterns to structure all life on our planet, with the most visible example being the day and night pattern. Continuing on smaller scales, cycles appear also as life cycles and processes of regeneration as well as the most basic body functions down to the level of biological cell operations.

Early cultures based most of their time concepts on the observation of these phenomena. In Egyptian culture, for example, sunrise and sunset were both integrated in an understanding of the life cycle and interpreted respectively as the birth and death of the sun on a daily basis (Tuan 1990). This very strong daily rhythm of the sunrise and sunset influences, or even directs, the palette of body functions. In *Rhythmanalysis*, Lefebvre (2004) has developed a conceptualisation of rhythms based on the body, and even though he proposes to approach the discussion from a theoretical perspective, he does in fact initiate it from the experience of the body. In his portrait of the Rhythmanalyst, he writes:

> He listens—and first to his body; he learns rhythms from it, in order consequently to appreciate external rhythms. His body serves him as a metronome. (Lefebvre 2004, p. 19)

In biological research the concept of cyclical patterns has gained acceptance. This was initially initiated by sleep research, but it has since influenced concepts in all areas of biological research and is today known as *chronobiology*. The repetitive patterns studied by chronobiology are summarised in the term Circadian Rhythm, which refers to repetitive patterns within the 24 h cycle of the day. It was coined by Franz Halberg in the 1950s and described in his book *Introduction to Chronobiology – Variability: From Foe to Friend, of Mice and Men* (Halberg et al. 1994).

The philosophical discussion around the body goes as far as to say that we actually live in our body, as that we own our body (Nast and Pile 1998, p. 1). This description is separating the mind, as an abstract idea, and the body as a physical element. In Bodies-Cities, Elizabeth Grosz reflects on this usual understanding of the body, mind and body, inside and outside, experience and social context, subject and object and self and other, all as the underlying opposition of male and female

(Grosz 1998, p. 42). However, any experience, communication or movement can only take place through the physical element. Therefore the relationship might not be that simple as to say that we consist of two elements. Touching something with the hand is a complex task. Furthermore, processing information cannot be transformed into a single ultimate knowledge. It depends on a number of additional factors too. These can be skin condition, e.g. wet, sweaty and cold, but also mood or external elements such as the atmosphere in the room, the smell, temperature, etc. What we know and what we feel is what our body lets us – and therefore we are the body. It is not about two contradictory elements. So nothing that we know, can think of or can plan can be disconnected from our body.

By looking at the early development of a human baby, these aspects become clear. For a small child, life is all about bodily experience, to learn, to understand. This includes touching and feeling objects with the mouth, to physically learn how to stack them, to finally be able to move in relation to them, building up a memory of spatial experience that is the basis for anything to follow. Spuybroek (2004) describes this in the introduction to the book *Nox: Machining Architecture; Bauten und Projekte* as described in the previous Chap. 2 with the example of the cats.

The human body is not just a static construct that allows us to do a few things in space. The body is a very complex system of processes. Ultimately they are cycles and based on rhythms, as the body ultimately is a natural artefact that needs constant reinventing in order to function. From energy that needs topping up to activity and rest, to breathing and the heartbeat, it is based on repetitive patterns. This applies to different scales of the body, down to the renewal of cells. The different processes have different cycles in length. A lot of them are timed to be in sync with the natural day and night rhythm. This pattern is based on a 24.7 h duration and therefore does not exactly match the 24 h calendar structure. This means that we have to adjust our body clock on a daily basis. However, experiments have shown that the body is able to maintain a rhythm over a prolonged period of not syncing, slowly drifting away from the naturally imposed rhythm (Strogatz 2004).

The Circadian Rhythm in the 1980s was mainly studied in relation to sleep and sleep disorder. The extensive study of sleep patterns of people who suffer sleep disorder has resulted in conclusions drawing on rhythms as the main factors. Scientists were able to make sense of how newborn babies need time to grow into the grown-up cycle of sleeping at night and being awake during the day, or why teens stay up late and have difficulties getting up in the morning, and why elderly people often wake up when it is still pitch black outside but cannot go back to sleep. Extended research, including experiments with people spending weeks in the dark, has shown that daylight plays a big part in normal sleep patterns. The human body seems to be capable of synchronising with the light-dark rhythm of our planet. The suprachiasmatic nuclei (SCN), a bundle of nerves located in the brain's hypothalamus (Kiser 2005), are responsible for keeping track of the time. This region does not tell the time; it simply keeps track of it. The clock is not centralised as one organ or feature, but distributed and inherent in all cells and regulated to stay in sync. Strogatz (2004) describes in his book *Sync* three different levels of synchronisation related to the human body. The first is on the level of cells that are mutually synchronised. The next level is organs that stay in sync. This does not mean that they are all active at the same time, but they each keep their allocated rhythm within the larger context of the system. The third level Strogatz describes as the synchronisation between the body and the environment around us. On this third level, he does not go into detail why this might be and how this might manifest. But logically it must have real-life consequences, not only in social space but also in physical space.

A gene for the biological clock of mice was identified and cloned in 1997, the first such gene to be identified at the molecular level in a mammal. New research on the Circadian clock's role in the organism suggests that the process controls almost all behaviours and physiology. In a surprising revelation, a new study (Sato 2007) suggests that the function of all genes in mammals is based on Circadian Rhythms. Up to now scientists believed that only about 10 % of genes are influenced by the body clock. The importance of the daily rhythm has only recently been uncovered.

Scientists believe that the main synchronisation to orchestrate the vast number of independent elements that follow this rhythm is the daylight cycle. A number of studies have shown that if not exposed to the cycle of day and night, e.g. by staying in the dark for a longer period of time, the sync slowly drifts off. It will automatically re-establish itself once back to exposure (Sato 2007).

New research has now also tried to explain the differences in lifespan in connection to the Circadian Rhythm. Until recently, lifespan was thought to be roughly related to the frequency of the heartbeat. As reported by physorg.com (2008), NYU dental professor Dr. Timothy Bromage was carrying out research on the growth of tooth enamel when he discovered cycles or intervals of tooth and bone growth. The rhythm seems to vary from organism to organism and seems to have a direct impact on lifespan. For example, rats have 1-day interval, chimpanzees 6, and humans 8. During the 37th Annual Meeting of the American Association for Dental Research, Bromage said:

> The same biological rhythm that controls incremental tooth and bone growth also affects bone and body size and many metabolic processes, including heart and respiration rates. In fact, the rhythm affects an organism's overall pace of life, and its life span. So, a rat that grows teeth and bone in one-eighth the time of a human also lives faster and dies younger. (physorg.com 2008)

The human body is controlled into a number of cycles, as we have seen in this broad field discussed above. Ultimately these patterns have a direct impact on any activities and social relations coming from the body and thereby also the shaping of the immediate environment. Furthermore this highlights the importance of looking at repetitive patterns as elements of the society in an overall sense.

The relationship and interaction of body and city is our subject of direct investigation in the Urban Diary project. Through the GPS tracking of individuals in the city, body movement is recorded on a city scale and visualised as the extension of the interaction of the body with the urban morphology. Quite literally, the record can be visualised as the body's physical inscription onto the urban form. With the rhythmic constitution of the body in mind, this space *creation* of the physical body is investigated in the following discussion.

3.1 Body Space

In the more theoretical conceptualisation of the body, there is a great emphasis on issues of gender and sex. Although these are important aspects, they are too wide to integrate into this thesis. The focus will therefore be on a more physical and experiential conceptualisation, whilst being aware of the gender issues created.

If the body is read in connection with the city, the importance of the external is emphasised. The body shapes the city in a literal sense if the city is understood as a human artefact. They do actually stand in a two-way relationship, meaning that they directly influence one another, resulting in the city shaping the body. Taken from the dictionary, body does form part of the Greek word *polis*. Polis is a city, a city-state and also citizenship and a body of citizens. When used to describe classical Athens and its contemporaries, polis is often translated as *city-state* (Britannica Online 2012). As this shows, the human body played an important part in the early conceptualisation of the city through the *body of citizens* as a metaphor.

In the current debate about the relationship between the body and the city, two main models can be identified. The first one is a cause and effect model. The body here dominates the city as a structure, mainly derived from physical strength to actually build. In this sense, the body is projected onto the city. However, as Grosz points out in *Bodies-Cities* (Grosz 1998), recently an inverted view of this relationship has emerged. The urban environment is labelled as alienating and cities do not allow the body a *natural* context. It all fits into a specific view of humanity. The human subject, characterised as an independent agent individually and collectively, is responsible for the creation of culture, socially and historically. Going as far as denying any contextual influence, in the case of cities, this means humans make cities; overall, humans rule the world. From this viewpoint, new light is shed on the current debate around the overcrowded city as artefacts. Much of the current debate in urban planning is directed by this understanding.

With the UN's announcement in 2008 that now, for the first time in the history of the earth, more people live in cities than in rural areas, a huge wave of debate has passed through the professions working in related subjects, as reported, for example, in The Endless City (Burdett and Sudjic 2008). The above human-centred approach was applied to deal with rising predictions of city population. In a one-way relationship where the body is the cause and the city the effect, the solution is simple. In this debate it is presented as a question of cleverness to solve this *new* problem to regain dominance on a human creation. The concept of the city needs updating. Part of this problem is related to the disconnection of the body and the city. Our understanding has moved a long way from the meaning of *polis* as introduced above. The relationship between the two terms shifted from a dependency to a rivalry. However, there are signs of the development of a new concept. For example, the title of the publication *You are the City*, by Kempf (2009), already suggests a dramatic change in understanding the city.

The second approach is the direct modelling of the city on the state of the body. There are a number of concepts to transform the body, or use the body as an example in the attempt to model larger structures. Machines are amenable to such comparison, but also cities and even political systems have been subject to this sort of function/meaning transfer. The political model was mainly coined by Thomas Hobbes and developed in his book Leviathan (in 1651). He directly modelled his proposition of the ideal state on the human body, the head being the king, the nerves the law, the arms the military and so forth. A similar literal translation was undertaken by Francesco di Giorgio Martini in 1470 from the body to the urban form of cities. In his explanations accompanying the sketch, he said:

> One should shape the city, fortress, and castle in the form of a human body, that the head with the attached members have a proportioned correspondence and that the head be the rocca, the arms its recessed walls that, circling around, link the rest of the whole body, the vast city. And thus it should be considered that just as the body has all its members and parts in perfect measurements and proportions, in the composition of temples, cities, rocche, and castles the same principles should be observed. (Francesco di Giorgio Martini as quoted in Nesbitt 1996, p. 548)

Similarly, Le Corbusier is reported to have used similar references during a planning meeting for the city of Chandigarh, his only built city project. In *Cities of Tomorrow*, Peter Hall reports this monologue:

> Corbusier has the crayon and was in his element. 'Voilà la gare' he said 'et voici la rue commercial', and he drew the first road on the new plan of Chandigarh. 'Voici la tête', he went on, indicating with a smudge the higher ground ... 'Et voilà l'estomac, le cité-centre'. Then he delineated the massive sectors measuring each half by three quarters of a mile and filling out the extent of the plain between the river valleys, with extension to the south. (Hall 1988, p. 212)

The relationship here, between the body and the city, is a kind of parallelism. The two are understood as congruent counterparts with features and organisation mirrored in one another (Grosz 1998). The implication of such a relationship is not only the clear male dominance of the body over the city, but the resulting implied opposition between nature and culture. This is also based on a hierarchical structure, both for nature and culture.

A further connection between the city and the body can be established by applying the same usage, for example, through the use of advertisement. Towards the end of the twentieth century, the city has increasingly become a giant billboard, not only being plastered with images but also promoting itself as a brand. The urban environment changes itself constantly whilst being in use and being refurbished or rebuilt. The advertisement campaigns do visualise this aspect in a new short-term version. Typically a large-scale billboard campaign lasts about two weeks before it is replaced with a new product, slogan or label. Rotating billboards or screens can reduce the rate of change to a couple of seconds. The human body has increasingly become a space for advertisement too. Brands place very visible labels on products: glasses, hats, schools, gadgets and bags. Even smells, hairstyles or cosmetics may have a label attached. These are statements of class and cultural status. Iain Borden points out that:

> ...fitness is now a social issue. To be slim and trim is to be not just attractive but thinking, moneyed, urbane. My body is not just my temple, it is an advertisement for myself. (Borden 2000, p. 106)

3.1 Body Space   43

With these explorations, the relationship between the city and the body can be explained as a dialectic relation of one informing the other. Grosz names this relationship in her description:

> two-way linkage which could be defined as an interface, perhaps even a co-building. (Grosz 1998, p. 47)

The city is built by humans, but more importantly the meaning of its shape is as a product of social and cultural configuration (Hillier and Hanson 1984). We have also seen that the body is a product of repetitive functions, internally and externally existing in tight connection to the context of the earth. This in turn leads to the conclusion that these cycles transform onto the city, the urban environment and the morphology.

However, one aspect has been left out so far. Not only the aspect of movement of the body in space but also time does play an additional role in the body-city relationship. The cycle in itself already implies the passage of time, but the movement as an activity or task results directly in the creation of space as a moving object in time. The visual aspect of body movement was extensively documented and researched by the photographer Edward J. Muybridge in the late nineteenth century. He is most famous for his photographs of moving subjects such as galloping horses, as shown in Fig. 3.1. With the use of photography, he tried to capture the

Fig. 3.1 Eadweard Muybridge, 1878. The horse in motion

motion with a series of still images. The legend has it that it all started to settle a bet regarding the movement of horses. The question was whether the horse lifts all four legs at once. He was eventually able to demonstrate and settle the bet by producing a photo that showed the horse indeed lifted all four hooves at once. However people were divided. The sculptor Auguste Rodin thought it an abomination:

> It is the artist who is truthful and it is photography which lies, for in reality time does not stop. (Auguste Rodin in Muybridge 1979)

Muybridge not only invented the field of motion photography, later to become the cinema, but also the technology needed. This ranged from electric triggers to chemical formulas for processing, to what we now know as the movie theatre. It was invented for his lecture series entitled the Science of Animal Locomotion in the Zoopraxographical Hall, Chicago, 1893. Beyond the physical creation of the city, the more abstract creation of space through the body movement is a direct aspect of the relationship. It is described by Tuan as:

> Furthermore, if we think of space as that allows movement, then place is pause; each pause in movement makes it possible for location to be transformed into place. (Tuan 1977, p. 6)

This statement sets out a concept of space based on two things, the body as a tool but also an object and time as an entity. Together this forms the narrative. We will come back to this concept in the following Sect. 3.2 on *Urban Narrative*. A concept of space creation would mean looking beyond the geographic definition of space as a Cartesian box with a pretended objective single fixed viewpoint.

This conceptualisation of space is based on a general system that is designed to incorporate every possible configuration. In the Cartesian grid, every possible location on earth space can be objectively described not only as one point but any at the same time, not only once but forever and always. The box space concept was, for example, discussed in detail by Albert Einstein Einstein in his article *Relativity and the Problem of Space*.

Space is conceptualised as a box or a container, just like a room. The simplicity and experience-based origin o the box space concept is probably also what makes up the attraction of Google Earth. It shows the earth in 3D, but again visualises the space as a box, everything at the same time. It is so attractive because it seems to finally capture the space, the entire planet earth, completely.

In architecture, the space as a box concept for the creation of form was overcome in the legendary design by Mies van der Rohe for the Barcelona Pavilion (see Fig. 3.2 for illustration). It contradicted every concept of space in architecture up until it was created. Even though it is a rather small temporary structure for the German pavilion of the World Exhibition in 1929 in Barcelona, it had a big impact on architectural design thereafter.

> Looking at the visual experience of the Pavilion, it is suggested that it had no overriding system based on axial symmetry. (Psarra 2009, p. 43)

If we take the human body into account and move away from spatial descriptions in proper mapping terms such as north, south, left and right towards the earlier explained fundamental body experience of a very young child, understanding

Fig. 3.2 The German Pavilion for the World Exhibition in 1929 in Barcelona designed by Mies van der Rohe. Here shown a reconstruction of the original building

becomes different. If instead we use experience-based expressions such as in front of, on top, on the back, standing up (vertical), lying down (horizontal), in the inside and on the outside, the spatial understanding is definitely going to change – not to mention the large group of people that might suddenly understand spatial descriptions.

This will then allow us to talk about the creation of space as a personal temporal experience. There is not one centre 0/0/0, but many or even endless centres. It is an impossible proposition, but interesting in terms of the result. This results in the city no longer being in one piece, but a group of things. Many things now no longer happen in the same space but at the same time. Further, it becomes possible to describe the city not as a construct that is built once and for all to be used by the people; rather, it is a constant, collective effort, in which the city is created through activity and body movement.

3.2 Spatial Narrative

One could argue that the combination of inherited body function, rhythm and routine behaviour in the sequence of tasks and activities results in a daily story of movement and activities, both individually and also in the city as a whole. In this view, the city is at the same time the stage and a result of the opera. This sequencing of movement and action is here termed *spatial narrative*. The individual similarly has his or her identity created through and from this Urban Narrative. This draws from both action (in the present) and memory (from the past). The body, being the vehicle to receive, process and deliver information, must be regarded as the central element for this choreography to take place, even if it is extended by machines or absent from memory or in virtual worlds (Hudson-Smith 2003). It is the processing of the fundamental experience that lets us visualise the essential information.

Aspects of mobility are important in the preliminary conception of Urban Narrative as a succession. Graham Shane points out that Foucault identified the ship as the heterotopia par excellence, mainly because of its quality of mobility and time (Shane 2005). Shane introduces the narrative as:

> Because of the increasing speed of travel and communications, the Picturesque landscape entered into the narrative of the journey and city. (Shane 2005, p. 252)

A series of projects and investigations fit into this approach of the narrative. For example, this is Jackson (2000) describing, in *The Stranger's Path*, the town from the perspective of an arriving stranger (male) and how the town is read as a sequence of elements resulting in an aggregated narrative. Similarly the investigation by Venturi et al. (1972) of a similar context in *Learning from Las Vegas* can be read, which adds a different perspective, from behind the wheel of a car. Similarly, before Venturi and Scott Brown, Appleyard et al. (1964) focused on a similar narrative in *The View from the Road*. See Fig. 3.3 for an illustration of the car journey sketches. These authors all document the scenography and choreography of movement and flows within the city or town, beyond and into the landscape. This can be termed the narrative of the machine, in reference to the previous Sect. 2.1 and the functional city.

A further example is *AS in DS: An Eye on the Road* by Smithson (2001). In this book, similar to *The View from the Road*, Alison Smithson documents a car journey in her Citroen DS. It is a diary where she describes:

> ...a passenger's eye on English roads in the 1970's - a sensibility to car movement has its beginnings. (Smithson 2001, blurb)

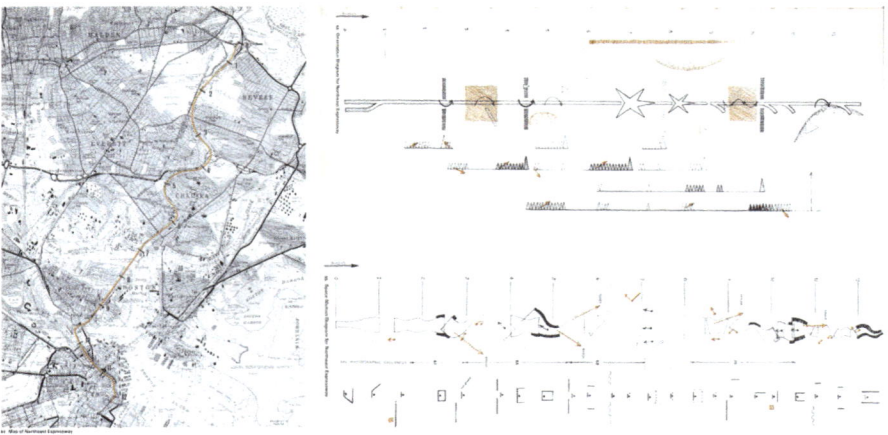

Fig. 3.3 Appleyard et al. (1964) in *The View from the Road*. Visualising the characteristics along a road by coding the features linearly as they appear in sequence

3.2 Spatial Narrative

The journey is narrated in linear fashion as notes, sketches, maps and the occasional black and white photograph. The motivation for the diary was a *new found freedom* through the car, the joy of riding and, as already highlighted in the work by Appleyard et al. (1964), a new, now consciously experienced, view from the front seat of a car, in this case the passenger seat.

The publication form of the book enforces a linearity in the narrative and in doing so underlines the linearity of the movement. It is a very literal description of the *path* and, as such, a brilliant example of the experience on the move. Whilst the text is sketchy and more in a noting style hurtling along, catching up with the progress and the experience is ultimately translated for the reader. On the other hand, the sketches and map sections added later on act as markers and points of orientation. Someone familiar with the landscape might easily recognise these guides.

The text is illustrative and detailed, at times poetic, to the extent that the reader asks himself "what else was there", "what else can be seen" at this moment or "what is on the left" if something on the right is described. It is after all a highly selective recounting of the situation, coloured by a personal perspective. However, the narrative is captivating, pulling forward and dragging the reader along, intertwining her or his own memories of past *path* experiences.

However, the Urban Narrative is not, in the machine sense, a linear sequence of events. It is far more of a framework to integrate:

> On the one hand, architecture as a thing of the mind, a dematerialised or conceptual discipline with its typological and morphological variations, and the other, architecture as an empirical event that concentrates on the sense, on the experience of space. (Tschumi 1996, p. 83)

Sophia Psarra notes, in the introduction to architectural narratives, that these two elements of conceptual and perceptual aspects of space, as mentioned before by Tschumi, are portrayed in opposition in recent architectural history (Psarra 2009, p. 3). These two concepts were discussed earlier in Sect. 3.1, and in terms of the body, the separation of the two aspects results in a distorted definition of space as opposed to the combination of body and mind. Here the concept of the narrative is employed to overcome these binary conceptions and open the framework to allow multiple combinations. As Psarra points out, these two terms of perception and conception could become crucial in analysing structure and morphology towards its narrative qualities, but as specific viewpoints, not necessarily in opposition.

As the records of the Urban Diary project will show in the following chapters, the spatial narrative of these activities taking place over a period of time has an individual characteristic spatial shape. Each individual produces, via the body, movement in the environment *trajectories*. Recorded with a device such as a GPS, over time, a drawing emerges as a result of the movement. This forms part of what is here termed the habitarium, the extent both in time and space of the everyday routines of individual citizens. The idea, for example, *What Shape Are You?* (see Fig. 3.4 for illustration), is rooted in much earlier formulations of similar concept (Neuhaus 2010). Such a concept of sequential visualisation of locations was used, for example, by Hägerstrand (1978) as a *time-space cube*. Later the term

Fig. 3.4 Visualisation generated from the Urban Diary data set, showing the extent of the spatial movement of six participants (See http://www.flickr.com/photos/40984848@N04/8701152303/ for an online version)

"*aquarium*" was used by Kwan (2004). The term habitarium is intended to expand upon the mainly linear meaning of the cube and include the wider aspects of the habitus as discussed earlier in the Introduction Chap. 1. This, in many ways, is the narrative of the everyday routine that plays an important role to describe and characterise a habitarium.

An individual characterisation of the drawing of a human body imprints on the city is the path travelled, which describes the physical extent of a specific habitarium. Its characteristics depend on a number of factors, including the location of home, the location of key destinations, frequency of travel and mode of transport. However, the elementary aspects across different individuals remain the same despite the influence of different factors and configurations. There is a home location and a handful of additional key destinations towards which movement is directed, while always returning home afterwards. Each location has a neighbourhood area around it, the size of which depends on the intensity of visits. These *basic units* could be called the individual elements of the city space. The question then is how do they link up and fit together; indeed, do they at all?

The narrative is in this perspective inscribed as a result. It partly is because of our reading of the result where our experience lets us interpret the drawing as a container of activities. We remember our odyssey through the city and can clearly visualise the numerous little anecdotes taking place at each bend of the trajectory. However, this in turn stresses the importance of the body-city relationship. To some extent the machinification of the body can be seen as a response to the city as a machine, in order to sustain the body as an active player in the city. Moreover the body as a machine raises the body over the city as a sort of *uebermachine* to steer the black box. The resulting construct is not to be used to solve the conflict of power between them, nor will it be from any such technical perspective. One way to escape this problem could be formulated by describing the two, body and city, as a connected pair with a two-way impact and information exchange link.

In terms of the time-space problem and the invention of the two conventions *time* and *space* as a way for humans to live in larger groups and to form cities, the body as a manifestation might serve as a valuable model. It represents space as such. The subject is in the flesh as a metaphor, but at the same time the body functions clearly represent time. This point of view was addressed in early time geography, and the *time-space aquarium* was used to visualise this duality of the human body as an *actor* in the environment (Hägerstrand 1978; Pred 1981; Carlstein et al. 1978). Here again, the narrative is inherent in the visualisation, and unfortunately it is at the same time part of the criticism made of this subject. Part of the success of the visualisation is the ease with which one can relate to what can be seen and linked to personal memory of real-life experience. The trajectory relates to how we just got here. But, it is overlooked that actually this is not the way we got here. This is the way some supernatural observer could see how we got here, in abstract form. The body experience tells us less of a continuous walk down the road than of a step by step repetition of the same process, constantly adjusting slightly to the unevenness of the paving slabs laid out on the footpath. Of course, if this perspective is taken further, the view of individual movement gets overcomplicated, but similarly the trajectory is oversimplifying what actually happens. The idea of the *project* is introduced by Hägerstrand (1982) as part of the concept to actually allow for partial individual freedom, but it does not feature in the visualisation nor does it implement any of the crucial production capacity it could inherit.

As production capacity we summarise the aspects of creation that the narrative is a direct vehicle for. This capacity is generally discussed as *agency* and consists of the capacity of human beings to make choices and act upon them (Barker 2000; Scott and Marshall 2009). On the one hand, this has a cultural aspect, focusing on the individual force to contribute and create to the culture, as well as a social aspect, and to play a role in society, taking care of someone else and creating a community. These are decisions the individual can influence, resulting in relative randomness on the level of the individual. The apparent randomness appears if numerous individual narratives are being summarised and agency is not distinguished from *free will*. A task that is, because of the creative force each individual uses, nearly impossible on the level of the individual decision. Creating the everyday narratives in great detail is however something that happens within the habitus. Agency and its decision-making aspect includes a process which numerous factors influence to various degrees; hence agency is not the same as free will. The habitus, as well as various other cultural, social and material limitations, act as guidelines, and the apparent randomness of individual decision making is no longer random. Hägerstrand (1970) summarised the guiding aspect as *constraints* that direct the individual activity, but which then over-instrumentalise these constraints to the extent that all creation or agency capacity is eliminated. This was the source of much criticism of the concept of time geography, especially from philosophers and feminists (see, e.g. Giddens 1986 and also Rose 1993).

Agency in many ways is an observer's term. With the spatial narrative, the agency is described from the perspective of the individual as it happens. It also has a poetic side to it, as well as the influence of a play as a distance reference in the sense of identity (Ferguson 2009; Goffman 1959). This also includes the geographical identity, described, for example, by Pile (1993), as a concept that does not rely on the *agency-structure* dichotomy.

If time and space are conventions to allow large groups of humans to live in close proximity, their production cannot be taken as given and constant. Which of the two, space or time, came first is a sort of chicken-and-egg question. The road is the most convenient way to move about the city. But this does not mean that the road came first and then people started moving along roads. It could equally be that movement created the road, probably creating a path first which turned into a road. In many old cities, this can actually still be observed and archaeologists have reconstructed numerous such stories. Furthermore great cities have been built on the basis of a path location, for example, along trade routes where cities like Jerusalem rose to great fame and a relatively large size. The path, however, is still an important element and demonstrates by its existence the fundamental experience of moving from one location to the other. It features as one of the five characteristic elements used to describe the city in a mental ma Lynch (1960) in *The Image of the City* defined them as *Path, Edge, District, Node and Landmark*.

This concept to a large extent was developed against the background of modernist city concepts. Maybe this is why Lynch's work has been so influential. After the reign of the plan and the functionalisation of space, such a new point of view from an experiential and an individual perspective offered something fresh, and still does so today, probably because it is fundamentally anti-machine. It is the city defined by inhabitants as opposed to the city defined for inhabitants.

Maintaining the machine requires a lot of effort as well as regulations on the user side. Do not touch this, please walk on the left, and stand on the right. In addition to the concept of space and time, the machine has invented detailed regulation for their use. They deal in part with everyday life in the city and allow the regulation of people living very close, but in part they also enable the distinction between the city and its hinterland as two entities. In this conception, the desire to establish clear boundaries serves the implementation of power: not necessarily in favour of the "user", but mainly serving the idea and the concept of the plan. To some extent, *constraint* of the *authority* as introduced by Hägerstrand (1970) could be a direct result of this installed power, as it refers to defined limitations and exclusivity of large parts of the city. This implementation of power obviously serves also to disconnect space and the individual.

Lynch's approach, to some extent, offered new possibilities, as the installed areas of power were at least described from the perspective of the individual

(Lynch 1960). If we now take the description to the level of agency, the power concept is undermined. Creating a concept of space from the perspective of individual actions will completely change this. The user no longer follows an order, but instead the instruction is part of the temporal creation of space. This would result in the standpoint that we are not all "users" of the space provided by "the city", but we are actually creating the city through individual narratives. This is not to say the city does not need regulation any longer, but to switch the way the elements are read.

In this respect, the Urban Diary project is expanding Kevin Lynch's *The Image of the City* study with a definition of space creation to *the use of the city*. Essentially the aspect of time is added as an equivalent dimension to that of space.

Urban spaces are not designed once to then remain static thereafter. The activity within the space constantly creates it anew. This constant urban mutation allows the city to remain relatively stable over time while adapting to new challenges and changing needs. The changes are multilayered and might apply only to some of the challenges. Over time, this builds up a history of place. This history is what we have as our memory of place. The memory is also what feeds and influences how we plan for the future. However, history is not fixed. The more time lies between the event and the present, the more the history depends on memory, and this in turn very much depends on the present. This can be the cultural context, aspects of technology or (for an individual) experience or mood. History is, so to speak, a construct or a tale, a projection into the future very similar to a proposed project.

The narrative is approached as a form of organisation: a form of structuring information and providing a sense of direction with a beginning and a progress and often an end. It can take the shape of a map, an atlas, an A to Z guide or a database or any form of organised information. Through the act of narrating, the information can be transformed into knowledge. At the same time, the narrative is also a form-giving transformation. Through the organisation, it shapes something new which, even though it is based on something else, often has the power to stand on its own. It can become an independent artefact. Like a map of the world, this is a map in the context of geography. It can develop its own powers and areas of influence.

The only rule in narrative is its inner, self-referential coherence. There is no truth, but only ideas about *truth*. Narrative is about what we believe is the truth, or what we want others to believe. The urban condition as narrative battlefield could therefore be contradictory, or tell different *stories* at the same time.

In this context, the urban environment is understood as a continuous story of everyday experience. Through activity within the physical form, a spatial narrative is created, a process that involves individuals shaping the collective identity, which results in a culture and in turn influences the built form.

However, it is not the intention to propose the narrative as a design strategy as such. Rather it is the idea to offer an open platform for a variety of approaches to design. The story is just the vehicle for the delivery of a proposal where the designer is asked to contribute and develop their own individual approach.

The idea of place is also part of the discussion around the Urban Narrative. It is a construct, a form of organisation of time and space. One could argue that through the organisation, it creates both the space or a place and the time. Doreen Massey dedicates a lot of her work to the definition of place, and her position is quite clear: place does exist and it is a fundamental element of the social and political organisation (Massey 2005). However, the concept of place has been critiqued, for example, by Hardt and Negri (2001). Regarding the explorations in the Body Space section in this chapter, we would agree with Massey on the concept of space purely because of the fact that the human body itself is already the definition of space and time and therefore place. Place does require a boundary and by naming something we are automatically excluding something else, effectively making it distinct and therefore creating a boundary for its definition. However, this does in one way conflict with Tuan's description of place as rest (Tuan 1977). His definition exists together with the partner definition of space as motion. Place in this sense is somehow a location on the globe, and space a trajectory on the globe. Both these definitions exist in a contrary conceptualisation and also require the definition of time as a separate entity that exists outside the two. However, the notion of place in Massey's work can be understood more in the sense of place as unifying space and time into one as a definition of the term. In a similar sense, the narrative is understood as presented here.

If this description is now applied to the time-space aquarium of the time geographers in the 1970s, what do we get? Tuan's model can be seen as a description of exactly this visualisation. Either an individual moves or it does not and the time passes anyway, and the space continues to exist around it. Tuan's places would sit on the trajectory somewhere, where his space would be defined by the motion. Massey, however, would argue that the time-space aquarium, the habitarium as such, is place. And this view results in the city as a multiplicity of places. And this is where the idea of the narrative has come from, as a description of all the aspects in the habitarium, while allowing for a multiplicity of narratives simultaneously.

In turn, the aspect of rhythm and cycle could help to structure the multiplicity of places in the city. As the Urban Diary project shows, there are strong routine patterns in individual travel over time. Restrictions apply according to the spatial narrative. The collection of the narratives will make up the individual body of the city. Because of the dual relationship of city and body, the same entity can also be described as an extended body, constituted through both experience and conception. Cognitive maps collected as part of the Urban Diary project have clearly shown that this narrative, or at least the spatial dimension of it, can be visualised. The remaining question is how to comprehensively visualise the time part of it. It is partly integrated with the narrative, but not explicitly. As a form of visualisation, this has to be explored further.

Bibliography

Appleyard D, Lynch K, Myer JR (1964) The view from the road. MIT Press for the Joint Center for Urban Studies of MIT and Harvard University, Cambridge
Barker C (2000) Cultural studies: theory and practice. SAGE, London
Borden I (2000) Hoardings. In: Thrift NJ, Pile S (eds) City A-Z. Routledge, London

Bibliography

Britannica Online (2012) Polis. http://www.britannica.com/EBchecked/topic/467403/polis
Burdett R, Sudjic D (2008) The endless city. Phaidon Press, London
Carlstein T, Parkes D, Thrift NJ (eds) (1978) Making sense of time. Edward Arnold, London
Einstein A (1954) Relativity and the problem of space. Relat Lond Methuen 4(2):135–157
Ferguson H (2009) Self-identity and everyday life. The new sociology. Routledge, London
Giddens A (1986) The constitution of society: outline of the theory of structuration. Polity Press, Cambridge
Goffman E (1959) Presentation of self in everyday life. Doubleday Anchor Books, Garden City
Grosz E (1998) Bodies-cities. In: Nast HJ (ed) Places through the body. Routledge, London
Hägerstrand T (1970) What about people in regional science? Pap Reg Sci 24(1):7–24
Hägerstrand T (1978) Survival and arena. In: Carlstein T, Parkes D, Thrift NJ (eds) Timing space and spacing time. Human activity and time geography, vol 2. Edward Arnold, London
Hägerstrand T (1982) Diorama, path and project. Tijdschrift voor economische en sociale geografie 73(6):323–339
Halberg F, Cornélissen G, Carandente A (1994) Introduction to chronobiology – variability: from Foe to Friend, of Mice and Men. Chronobiology seminar, vol 7. Medtronic, Minesota
Hall P (1988) Cities of tomorrow: an intellectual history of urban planning and design in the twentieth century. Basil Blackwell, Oxford
Hardt M, Negri A (2001) Empire. Harvard University Press, Cambridge
Hillier B, Hanson J (1984) The social logic of space. Cambridge University Press, Cambridge
Hudson-Smith A Digitally distributed urban environments: the prospects for online planning. PhD thesis, University College London, the Bartlett School of Architecture
Jackson JB (2000) The stranger's path. In: Landscape in sight. Yale University Press, London
Kempf P (2009) You are the city. Lars Müller Publishers, Baden
Kiser K (2005) Father time. Minn Med 88(11):26–28
Kwan M-P (2004) GIS methods in time-geographic research: geocomputation and geovisualization of human activity patterns. Geografiska Annaler Series A-Physical Geography 86 B(4): 267–280
Lefebvre H (2004) Rhythmanalysis: space, time and everyday life. Continuum, London
Lynch K (1960) The image of the city. MIT, Cambridge, MA
Massey D (2005) For space. SAGE, London
Muybridge E (1979) Muybridge's complete human and animal locomotion. Dover, New York. ISBN:0486237923, 9780486237923
Nast HJ, Pile S (eds) (1998) Places through the body. Routledge, London
Nesbitt K (1996) Theorizing a new agenda for architecture. Princeton Architectural Press, New York
Neuhaus F (2010) UrbanDiary – a tracking project capturing the beat and rhythm of the city: using GPS devices to visualise individual and collective routines within central london. J Space Syntax 1(2):315–336
physorg.com (2008) NYU dental professor discovers biological clock. http://www.physorg.com/news126712217.html
Pile S (1993) Human agency and human geography revisited: a critique of 'new models' of the self. Trans Inst Br Geogr 18(1):122–139
Pred A (1981) Social reproduction and the time-geography of everyday life. Geografiska Annaler. Series B, Human Geography 63(1):5–22
Psarra S (2009) Architecture and narrative: the formation of space and cultural meaning. Routledge, Abingdon
Rose G (1993) Feminism & geography: the limits of geographical knowledge. University of Minnesota Press, Minneapolis, MN
Sato R (2007) Nature's clock – The rhythm of life. http://www.dailygalaxy.com/my_weblog/2007/06/circadian_clock.html
Scott J, Marshall G (2009) A dictionary of sociology. Oxford paperback reference, 3rd edn. Oxford University Press, Oxford. Rev edn

Shane DG (2005) Recombinant urbanism: conceptual modeling in architecture, urban design and city theory. John Wiley & Sons, Chichester
Smithson A (2001) As in DS: an eye on the road. Lars Müller Publishers, Baden. First Published by Delft University Press, Delft 1983
Spuybroek L (2004) Nox: machining architecture; Bauten und Projekte. Thames and Hudson, London
Strogatz SH (2004) Sync: the emerging science of spontaneous order. Penguin Books, London
Tschumi B (1996) Architecture and disjunction. MIT, Cambridge, MA
Tuan Y-F (1977) Space and place: the perspective of experience. University of Minnesota Press, Minneapolis
Tuan Y-F (1990) Topophilia: a study of environmental perception, attitudes, and values, Morningside edn. Columbia University Press, New York
Venturi R, Scott Brown D, Izenour S (1972) Learning from Las Vegas. MIT, Cambridge, MA/London

Chapter 4
Urban Diary

Abstract This chapter presents fieldwork undertaken in the Urban Diary project. This is a study to track individual participants, recording the spatial extension of their everyday lives. It includes two locations—London, UK, and Basel, CH—each with 20 participants. Also covered in this section are method and technology employed in the fieldwork. This included GPS tracking but also mental or cognitive maps and interviews.

In separate sections, the study setting, the tracking, the technology, the cognitive maps and the interviews are presented and discussed. In the introduction to GPS technology, the possibilities and the problems associated with it are highlighted. Also examined are best practice and different options to validate the data.

This chapter presents fieldwork undertaken in the *Urban Diary* project. This is a study to track individual participants, recording the spatial extension of their everyday lives. It includes two locations—London, UK, and Basel, CH—each with 20 participants. Also covered in this section are method and technology employed in the fieldwork. This included GPS tracking but also mental or cognitive maps and interviews.

In separate sections, the study setting, the tracking, the technology, the cognitive maps and the interviews are presented and discussed. In the introduction to GPS technology, the possibilities and the problems associated with it are highlighted. Also examined are best practice and different options to validate the data.

4.1 Introduction

Day and night, the rush hour, weekends, train timetables, paydays or annual celebrations are examples of repetitive patterns occurring in the city. Such patterns could theoretically be the result of spatial and social configurations; here they are seen as based around the organisation of the urban environment. As we have explained, in *The Social Logic of Space*, Hillier and Hanson (1984) provide numerous examples demonstrating the connection between social configuration and morphology of the built form in a static setting. Through specific fieldwork study, these aspects are being researched in the everyday setting of individual citizens.

4.1.1 Discussion of Space-Time

As discussed in Chaps. 2 and 3, the fieldwork aims to address the points raised about perception, orientation, general use and interaction of and with the urban environment. This fieldwork creates a basis to test the claims and provides data as a practical example to illustrate the points about rhythms of the everyday in time and space. Furthermore, it acts as a basis for the development of a reinterpreted view on those theories mentioned by Hägerstrand (1978) on space-time, and *The Image of the City*, presented by Lynch (1960) using mental maps.

Traditionally the city is mapped as a network of streets, buildings and blocks that form the space within, where this space is generally taken as given. The universal and objective view of space is described by Giddens (1986) in *The Constitution of Society: Outline of the Theory of Structuration*. Giddens describes space as given, a matter of fact, completely surrounding and rooted in place. Within this static construction of space, ephemeral elements such as movement and change are treated as placed attributes, unlike the physical elements of the city. Such ephemeral elements have not one state but many. As Loew (2001) outlines in *Raumsoziologie* (A Sociology of Space), such a static space definition misses the chance to structure space and time as a sociological term, taking account of the human capacity of activity and decision making, referred to here as *agency*, as well as the shifting of perception through experience. The Urban Diary (UD) fieldwork study aims to investigate the problem of *many states at the same time* by examining techniques and methodologies to observe and map the change of movement and time directly. The viewpoint adopted is from within the given and generally understood concept of space, insofar as this study combines aspects of process into the overall description of urban space by tracking activities. It aims to generate new perspectives on how to define and interpret the city as a collective product of patterns in time. Besides this, the project also clearly focuses on the social aspect of the production of space, as described, for example, by Hamm (1982) as a result of many factors related to the individual activity, aim and method.

Transport planning and transport studies have traditionally focused on the representative day with the representative peak hour (Ortúzar and Willumsen 1994; Yáñez et al. 2010). Furthermore, traffic and movement surveillance often deal solely with data based on the spatial locations of activities, such as travel diaries or questionnaires. This results in only some specific points being represented and interpolated. In the context of a detailed social production of activity space, this is again not a representation of individual parameters leading to time-space-specific decisions which are task related. The motivation, as a result of the lack of detail, tends to be reduced to pure optimisation, best being short and quick. This is fuelling

4.1 Introduction

the legend of the city as a machine as discussed earlier in the Chap. 2 resulting from modernist city programming. The task of travelling in this setting is to try and trick the branded *system*, beating it in both time and space. By letting the actions and moves of an individual tell the *narrative* of what happened spatially and how time structured their planning, the Urban Diary study aims to challenge this preoccupied perspective, opening up to individual reasons and decision making in the context of the spatial and temporal situation, as opposed to an urban-wide standard model.

4.1.2 Setting

This study extends this perspective, both in time and space, by specifically looking at the route chosen between locations and the pattern of repetition occurring through rhythmic schedules. To do this, a number of volunteer individuals were equipped with a GPS device that allowed their journeys through the city to be tracked. Over a period of 2 months, 48 participants recorded their personal spatial diary, mapping the extension of their personal everyday life in space and time. The perception and experience side of the data was recorded using in-depth interviews focusing on contextual information and individual routines and habits. The expressions *everyday*, *everyday life* and *routine* in this context are used in the sense of de Certeau (1984). The data collected and the routines used are very much on an individual level, but combined as a collective map, they represent a spatial diary of events taking place in the urban environment and ultimately capture aspects of the rhythm of the city that we seek to explore.

For this study, the participants wore a watch-like GPS device on their wrist, illustrated in Fig. 4.1. The specific technology and accuracy of GPS systems is explored later on in this chapter in the Sect. 4.3 on page 61. Wearing the device on a daily basis ensured that personal routines were captured at the required level of detail over the duration of the study. The study lasted 2 months to ensure that monthly and weekly, as well as daily, patterns were included. The data record contains location and time information, which can be mapped using a variety

Fig. 4.1 Garmin Forerunner 405: The gadget used for some of the test tracking

of methods and tools. Participants were not specifically selected, but they were all adults of different ages, with a mix of female and male candidates from different backgrounds, family and work status. Each participant was met weekly or biweekly to download the data. This ensured a close contact between researcher and participant and allowed for informal discussions about the data collected and personal routines. Towards the end of the 2-month recording period, a formal interview was carried out, allowing the presence of routines or habits to be recorded that may not have been identifiable in the GPS data.

The study was set in two locations. One is the Greater London area in the UK and the other the city of Basel, Switzerland. The two locations were chosen because of the existing detailed local knowledge of the researcher. London is a metropolitan area with approximately eight million people. It has an extensive public transport network, and its morphology has grown over about 2,000 years. Such a layered structure leaves traces and patterns of previous times inscribed in today's appearance, creating unique and *organic* structures. London is the main centre of the UK but is also one of the top ten global cities. Basel, on the other hand, is the third largest city in Switzerland with about 170,000 inhabitants. Basel too has a medieval core with two wall extensions, as well as later extensions especially during industrialisation. Here the morphology is not as much a layering as it is in London but is based on extending and adding. Basel has a very tightly knit public transport network, reaching out into surrounding areas. The city is an important centre for the northern part of Switzerland and the local tri-national region of France, Germany and Switzerland. Both studies have the same settings with participants being tracked using GPS technology over a longer period of time.

Some additional data collection has also been undertaken in Moscow, Russia and Plymouth, UK, with smaller samples. These two were intended as reference collections to get an impression of routines in a different cultural context—as in Moscow—and to get a feeling for lesser pulsing and more rural locations, as in Plymouth. In addition to the GPS tracking, cognitive maps and interviews are being used to collect detailed information on spatial activities, including both details of trips falling into the recording period as well as details on more general routines related to spatial activities.

4.1.3 Objectives

The hypothesis we are exploring suggests that individuals maintain a personal network of nodes that are spatially connected and individually loaded with identities built from experiences. In order to investigate the hypothesis, we are looking at the temporal production of space within a diary framework, as well as investigating how external circumstances lead to the constraints as discussed by Hägerstrand (1978) in terms of space-time defining elements, which are primarily external to the subject. As the hypothesis of this research, cycles as repetitive activities are believed to be a third dynamic element within the system of objects, the first element, and their interrelationships, the second element. As Barry Curtis puts it:

> In any urban setting, these repetitive patterns are the main source of identity, provide orientation, and are a main creator of memory. (quoted in Borden 2001, p. 63)

and regards memory as:

> one of the key ingredients in the creation of place.

He reflects on this by adding:

> Memory is rarely without contradictions, and it must be compromised in order to function.

The Urban Diary is an explorative project, led by two main research questions: Firstly, do cycles participate in the shaping of cities, especially on the level of urban form and, secondly, how can cycles be incorporated as a tool in the urban design and planning process? Distinguishing between three main groups of cycles, natural, activity and material cycles, the focus will lie on the activity group, daily rhythms and routines of individuals, living in the city and how these habits are manifested in space (see, e.g. Neuhaus (2010) in *Cycles in Urban Environments: Investigating Temporal Rhythms*). With both a theoretical and practical context, the Urban Diary project examines the spatial extension of an individual's routines in urban environments.

4.2 Lines of the Everyday: Tracking

Tracking summarises the different methods used to record a sequence over time (Golledge and Zhou 2001). This method was used in different forms for several aspects of the field work for the present research work. Tracking can apply to many different activities and is an overall term describing a method used to record activity and actions. The characteristic of tracking is that the two elements, space and time, are both recorded at the same time and in sequence. For this research, broadly speaking, the idea is to record a person's whereabouts in both space and time in relation to the city as a larger reference point, allowing for analysis in different contexts with changing reference data.

Capturing something as trivial and ephemeral as passing by is difficult. Normally there are hints through leftovers to tell the story of such an event after they are long gone. Someone dropped a coin or ticket, or on the table remains an empty coffee cup. These objects can be used to interpret possible activities using the personal experience. Using GPS technology, spatial movements are tracked by the individual, recording a series of locations along the way, tracing their steps and thus capturing the trivial and ephemeral actions of passing by. Particularly of interest is how the everyday routine creates a spatial *habitus*.

After a relatively quiet period following the explorations in mapping time geographies in the 1960s and 1970s (Steinitz 1968; Hägerstrand 1978; Carlstein et al. 1978; Hanson and Burnett 1982), a number of recent projects have picked up on this interpretation practice again. There is a renewed interest, on the one hand, in the practical methods of investigation but also, on the other hand, in the theoretical framework of time-space and movement. This also relates to the availability of new technology, enabling large-scale accurate tracking but also the emergence of Geographic Information Systems (GIS). Different technologies are being experimented with, such as GPS (Kwan 1999; Chen et al. 2010), but travel diaries and questionnaires are still very much in use. The MIT's Senseable City lab researchers have successfully used mobile phone technology for tracking studies (Ratti et al. 2006; Reades et al. 2009). There have also been tracking studies at ETH, for example, using the MobiDrive (1999) data (Axhausen et al. 2002). Regarding the theoretical side of the time-space development, Miller (2004, 2005a,b, 2006a,b) and Miller and Ahmed (2007) has written extensively about the concepts developed by Hagerstrand at the Lund School and updated some aspects in regard to the newly available technology. Similarly Sheller and Urry (2006) and Larsen et al. (2006) have written about the theoretical aspects of the path in a time geography sense and newly available mobile technologies.

With the focus on routines and habits, we are not looking at a single event, but at an overall practice that only becomes visible in the long term over a period of time. The individual action is of little importance in this particular context; the reference point is the overall or average practice within a long sequence. The tracking captures individual actions over a long period making the long-term pattern and trends visible. To capture the practice in more detail, tracking is undertaken using a range of different methods and technologies. Each of them will be discussed separately in the following sections. Sequential recording, especially in a technological context, implies the linear recoding of time. The intention is to break out of this constraining framework of one-dimensional time definition. Nevertheless, technical methods play an important reference role. The two perspectives will be used to inform one another (Fig. 4.2).

The two tracking aspects are time and location. These are two far-reaching terms, but in combination, they hold the data for the narrative invented by individuals. Time in this collection of sequential activities is more than merely the cataloguing indicator. Time is seen here as the primary element guiding the repetition and providing the framework for the spatial practice. As such, time is not a primary element, and it is recorded via the repetitive patterns. The units can vary depending on the scale of practice and the tasks. As a reference, however, the units relate to *Clock Time* as a social concept, ensuring that the practice can be understood in a wider context. Even if the qualitative approach will not allow for a larger generalisation, this connection to larger social practice will ensure that the results can be read in different contexts and the wider field. It is important for Clock Time to use the same basic units in order to fit into the same organisation method as a collective agreement (Glennie and Thrift 2009). Furthermore, the concept of Clock Time is based on the idea of repetition structuring every day anew using the same

4.3 Between Object and Subject: Technology

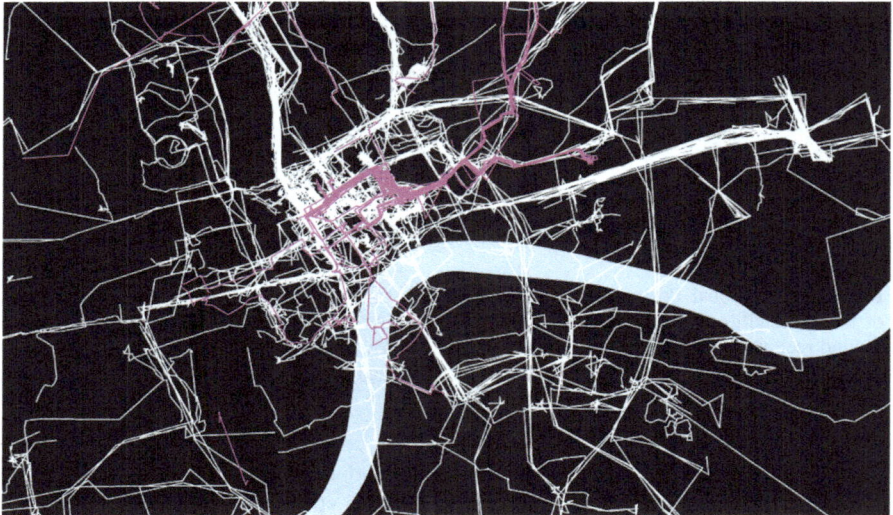

Fig. 4.2 Zoom into the centre of London with movement tracks from all participants of the Urban Diary study. One individual track is highlighted in *purple*

pattern. This of course encapsulates the idea of rhythm. The temporal aspect of tracking extends this by taking the concept of collective agreement and extending it to the point where it serves to record patterns in everyday practice.

Space covers the part of tracking where activities extend over different locations or connect different locations. This is how practice interacts with the urban context it sits in. Through action in space, the individual shapes an individual experience of the city. Based on the rhythm, this spatial routine can grow into different intensities, highlighting the importance of each element. The spatial element is incorporated in the different practices of place and place making (Tuan 1977; Massey 2005) but also in the production of space as a social practice (Hillier 1996; Lefebvre 2004). Using such a detailed spatial tracking approach enables the recording of spatial interaction and in-depth analysis of the importance of urban concepts.

4.3 Between Object and Subject: Technology

For the study, GPS was used to track participants as they move around the city going about their everyday business. In this section, the technology behind this approach will be discussed in more detail, to create a context for the data collected and the nature of the findings. GPS stands for Global Positioning System and is a global navigation satellite system. It was developed around 1973 and is created and realised

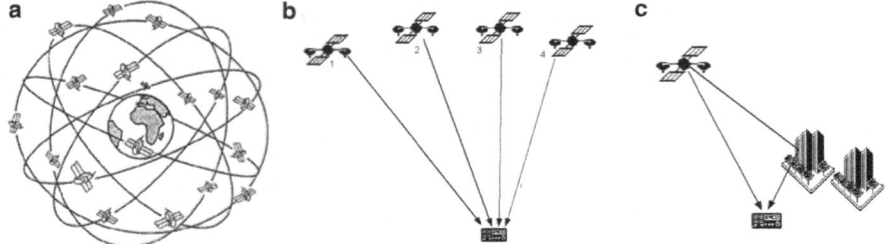

Fig. 4.3 The Global Positioning System explained. Figure (**a**) shows an abstract constellation of the 24 satellites orbiting the earth. Signals from four different satellites are required for the ground receiver to calculate an accurate position, shown in figure **b**. Numerous factors influence the accuracy of the signal, including reflection of building facades, as shown in figure **c** (Illustration after Kilroy (1998); Chadha and Osthimer (2010))

by the US Department of Defense, being fully operational since 1994. Based on a signal sent from satellites orbiting the earth, a specialised receiver device can accurately define its location in the framework of the Cartesian system.

> Current generation navigation systems ... determine the user terminal position through the time of arrival. In general, this kind of ranging technique is based on the measurement of the time interval employed by a signal transmitted by an emitter (e.g. satellite, radio beacon) at a known location to arrive at the user receiver (Prasad and Ruggieri 2005, p. 15).

For pinpointing a location, theoretically three satellites (reference points) would be needed to triangulate the position. See Fig. 4.3 for illustration. However, due to a constant unknown bias, usually differences in clock time, for an accurate location, the signal from at least four satellites is necessary (Tsui 2005, p. 9), where the fourth one is used for time correction. Accuracy can be up to a few metres depending on the quality of the satellite signal. Each location is determined as a latitude-longitude pair of coordinates and, together with the time information, is stored as a location point on the device's internal memory. With time information, this data can, in sequence, produce a track, or indeed *path*, as a movement line.

The signal strength of GPS devices is dependent on a number of environmental factors, such as the weather and the nature of the physical environment, whilst errors can also be on the satellite side, such as *ephemeris or orbital error* related to incorrectly transmitted position or time. Even though each satellite is based on four atomic clocks, errors occur due to instability, and a deviation of 10–8 s results in around a 3.5 m error on the ground (Parkinson 1996b as cited in Spencer 2003, p. 52). The atmosphere can also introduce errors as the signal passes through, resulting in a 2–6 m displacement from the exact location. The largest impact on errors, however, is the *position dilution of precision* (PDP) and describes the signal quality as a result of satellite positions, relative to one another. A good signal is received if the four necessary satellites are distributed at the same height in the sky, whereas a low-quality signal results from clustered satellite positions or satellites being very low on the horizon (Spencer 2003, p. 28). With the help of the corrected

4.3 Between Object and Subject: Technology 63

satellite position, published sometime after the event by the GPS Master Control Stations (MCS), a correction filter for the data can be calculated. Similarly, the location for a survey can be prechecked, if the time is known, regarding the satellite signal, and especially if the PDP takes into account the quality of signal that can be expected.

On the ground, radio signals can be reflected by hard surfaces, resulting in multipath interference. Buildings or trees can, through this, have a significant impact on signal quality. The resulting errors can vary between 2 and 15 m from large, highly reflective surfaces such as bodies of water. This is significant in the urban setting of the UD project, as it is located largely in a dense urban environment. The combination of narrow streets and high buildings, plus a large amount of street furniture and signage, can make it difficult for the receiver to establish and maintain the satellite signal. An additional implication for the quality of the satellite signal is the mode of transport. In the context of London, the underground and the bus play a significant role in citizens' daily commuting. Underground on the tube, no satellite signal can be recorded. Similarly on the bus or in the train, it can be difficult for the device to register an accurate signal. A window seat is notably better than an aisle seat, and the further away the device is from the building facades towards the middle of the road, the better the signal quality.

Initially developed for military use, the technology has, in the last few years, become popular in everyday culture. Today a large variety of digital gadgets are equipped with a GPS receiver, ranging from in-car navigation systems to mobile phones and cameras. This was initiated by the former President Bill Clinton's decision to lift the imposed selective availability (SA) restriction in 2000 (Prasad and Ruggieri 2005, p. 07). The SA was initially imposed to prevent enemies from using the system in military action against the USA. Following the SA removal, civil and commercial GPS accuracy increased from around 100 m to somewhere between 3 and 15 m (Pendleton 2002 as cited in Spencer 2003, p. 56).

Besides the use in handheld devices for tracking and navigating, GPS is now built-in to a range of products from cameras, mobile phones to in-car navigation. Geocaching or geotagging of images are just two examples of the use of GPS in everyday life. A key player in the market is Google, with its location-based information services developed from Google Maps delivering background information to visualise the position in context. It is not only about knowing where one is but in which direction one is going, towards where and what one has to expect. The use is generally very simple with a direct feedback loop.

The main technological issue for the GPS devices is battery life. Most mobile phones are unable to support the energy-consuming GPS receiver together with the communication to mobile phone masts over a long period of time on a single battery charge. Specialised GPS devices currently perform better, lasting for days depending on the settings. The latest handheld devices can be the size of a watch. For these reasons, a specialised GPS device was used for this study.

In the UD study, we used two tracker models (see Fig. 4.4). The Garmin Foretrex 201 was used for the London study, and an iGotU GT-120 was used for the Basel

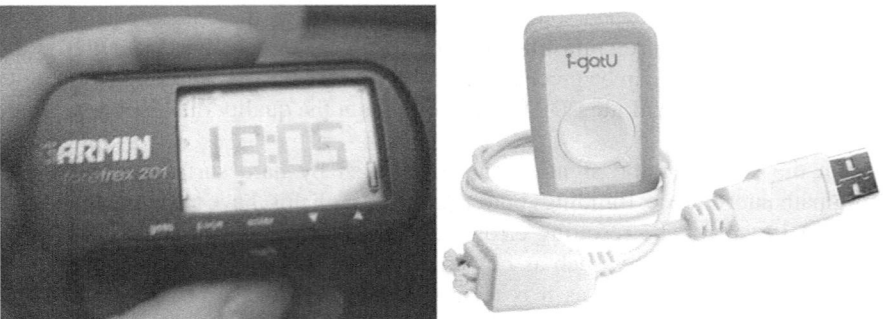

Fig. 4.4 Two GPS tracker models were used for the Urban Diary study. On the *left*, the Garmin Foretrex 201 and on the *right*, the iGotU GT-120

study. Both are simple handheld GPS devices. The 201 can be worn around the wrist like a watch; the GT-120 is a little box the size of a matchbox and can be either worn with a wristband or carried in a pocket. When the device is turned on, it starts automatically to search for satellite signals and, when its position is determined, it starts recording. Approximately every 2 days, the participants are required to recharge the device overnight. Reception of satellite signals on this device is quite good, but it can be affected by errors as previously described, due to the setting. Nevertheless, results have been determined as satisfactory so far. The handling of the devices is slightly different. The GT-120 is completely automatic. It is preset to continuously record and log a location every 7 s. With the 201, the user is required to turn it on and off since automatic programming is not possible. Therefore, the participants are instructed to turn the device on as they leave a building and turn it off as they enter the building. The recording time on the 201 is set to automatic with varying times between points as to optimise both storage space and battery life.

The data collected by the participant is stored locally on the device and downloaded manually by the researcher, usually on a biweekly basis. For both models, some data cleaning is required due to random location points saved when the device either loses signal or has a weak signal. This data cleaning is done by filtering out location changes at high speed, e.g. above 120 km/h for nonconsecutive occurrences. Both devices tend to occasionally record outlier points on tracks. To smooth out the edges, an averaging algorithm is used to smooth each point with a percentage of the previous four and the consecutive four recorded points. This averaging is based on a fading percentage of influence and has a filtering effect, as well as achieving a blending in. This technique also works well for clustering locations, where the participants stayed over a longer period of time.

A database stores the location points together with contextual and anonymised personal information. For visualisation purposes, the data can be made available in various table formats and processed further, for example, as a collective diary map for London (Fig. 4.5) or as an individual diary map (Fig. 4.6).

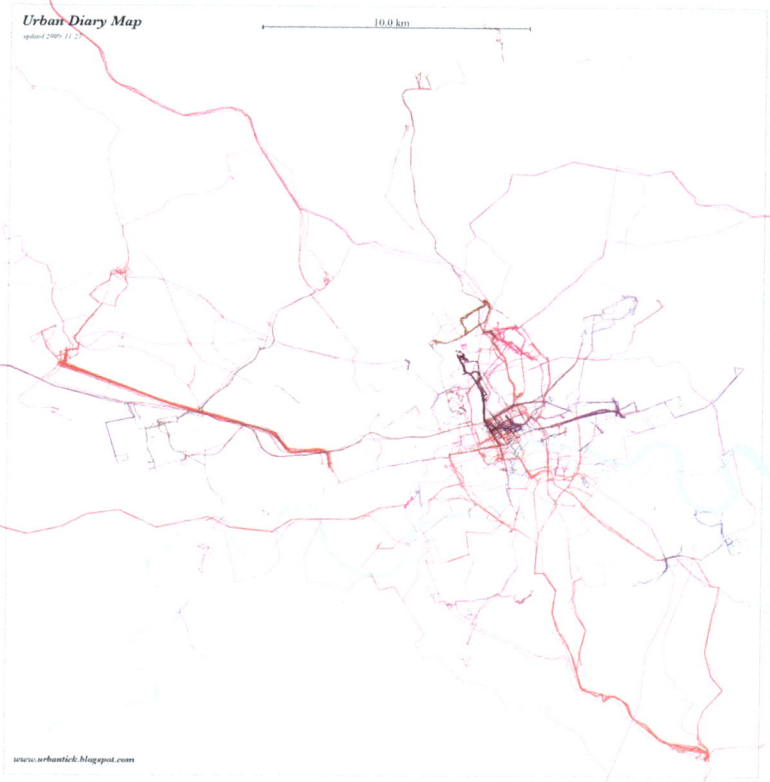

Fig. 4.5 Urban Diary map showing 20 participants tracked over a period of 2 months in the urban area of London. Each individual is assigned a different colour (See http://www.flickr.com/photos/40984848@N04/8701221129/ for an online version)

4.4 The Image of the City as Routine: Mental Maps

One of the most cited studies using mental maps is *The Image of the City* by Kevin Lynch. The study was carried out over 5 years and summarised in his 1960 book. Lynch (1960, p. 1) states:

> Every citizen has had long associations with some parts of his city, and his image is soaked in memories and meanings.

This is a fairly romantic description, with notable implicit hints into how environments may be understood socially. It also expresses the view that there is knowledge and meaning in each one of us about the environment we live in and navigate through. This is not about orientation, exact distance measurements or overarching, objective descriptions. Rather, it is about personal experience,

Fig. 4.6 Urban Diary map showing the records of an individual participant in the urban area of London (See http://www.flickr.com/photos/40984848@N04/8702343076/ for an online version)

judgment and what is physically and psychically important to the subject. As Lynch (1960, p. 2) puts it:

> Most often our perception of the city is not sustained, but rather partial, fragmentary, mixed with other concerns. Nearly every sense is in operation, and the image is the composite of them all.

Lynch was not the first to express these views. As noted by Gould and White (1974, p. 28), as early as 1913, the pioneer of mental maps, Charles Trowbridge, commented on how people have different senses of orientation. He concluded that there were two groups of navigators. Some people have imaginary maps in their heads centred upon the location of their homes. They are able to navigate a certain distance on familiar ground, but they would lose orientation in unfamiliar ground. The other group was described as more *egocentric* and orientated to their own position at the moment, with a better ability to navigate in unfamiliar territory.

4.4 The Image of the City as Routine: Mental Maps

Portugali (2004) uses the term *cognitive map* for mental map, essentially meaning the same. The term cognitive map goes, according to Portugali (2005), back to Tolman (1948) who, in his experiments in the 1930s and 1940s, demonstrated how animals and humans are capable of constructing representations in their minds about the external environment they have experienced. Portugali discusses cognitive maps in his book *The Construction of Cognitive Maps* Portugali (1996). He describes the cognitive map as:

> ...an internal representation - of a large object, one so large that it cannot be grasped in its entirety by means of a single cognitive act. Every cognitive map is therefore a mental construct, a cumulative structure of many small pieces of information. The component *map* in the notion *cognitive map* is, of course a metaphor. (Portugali 2004, p. 16)

The map is just one form of expression of these personal memories and descriptions. Although it is called a map, it has two fundamental differences from a conventional cartographer's map. It has no scale and no objective direction assigned to it, the drawing of its elements may only relate to this personal context, e.g. there is no assumed north point unless the author of the map assigns it. Nevertheless, some features of a map may be borrowed from conventional maps by the participant, such as a top-down view, symbols and so on. Other methods can be a description in words, both as a text or an interview.

The business building block system Lego Serious Play is another creative way of expressing memories and perceptions in a hands-on sort of way. David Gauntlett is a researcher working with this method. He explains in his presentation clip *Representing Identities* on YouTube (Gauntlett 2008), using these creative methods will encourage the brain to work in a different way. See Fig. 4.7 for an illustration of a Lego creation. He argues that individuals each have an embodied experience and that this experience is more easily accessible through body gestures.

The Urban Diary project participants' instructions to draw a mental map are intentionally simple. The focus lies on the content and not the beauty of the sketch; there is no right or wrong. The key is that the sketch is not copied from a map or image, but rather it is drawn from memory. Lynch introduces the mental map to the participants as follows:

> We would like you to make a quick map of .... Make it just as if you were making a rapid description of the city to a stranger, covering all the main features. We don't expect an accurate drawing - just a rough sketch. (Lynch 1960, p. 141)

It is a quick exercise that does not require a lot of planning and thinking. However, based on the experience of using them on this project, there appear to be three distinct phases to the creation of a mental map. The first is the skeleton phase, during which most of the important information, objects, direction, names and paths are set down. The second phase increases detail by linking between memories with information and description. This will often trigger some more memories and make the map rich and representative. The third and last phase is the beautification process, where no more important information is added, but rather the sketch is adjusted and critiqued.

Fig. 4.7 Screen shot taken from *Representing Identities* documenting a workshop by David Gauntlett using Lego as a creative material. Image showing one structure as presented in relation to identity by one participant of this workshop

Mental maps have been used in a variety of spatial research. On the one hand, there are studies such as Lynch's which focus on the physical environment. On the other hand, there are studies that focus on the quality of the perceived environment, as recorded through feelings like desire, stress, fear or happiness. Such a study has been carried out by David Ley in Philadelphia and presented in *The Black Inner City as Frontier Outpost: Images and Behaviour of a Philadelphia Neighbourhood* (Ley 1972), where the participants' responses have been processed to create an intensity topography. Matei (2003) has undertaken a similar project on fear in Los Angeles.[1] From the participants' responses, he was able to create a three-dimensional digital surface to represent the amount of fear in the Los Angeles region. The colours red and green are used to highlight areas of lesser or greater amounts of fear. For an illustration, see Fig. 4.8.

Gould and White (1974) summarise an investigation into people's desires using mental maps in the book of the same name. They posed the following research question:

> Suppose you were suddenly given the chance to choose where you would like to live - an entirely free choice that you could make quite independently of the usual constraints of income or job availability. Where would you choose to go? (Gould and White 1974, p. 15)

---

[1] The project can be accessed online on www.mentalmaps.info

4.4 The Image of the City as Routine: Mental Maps

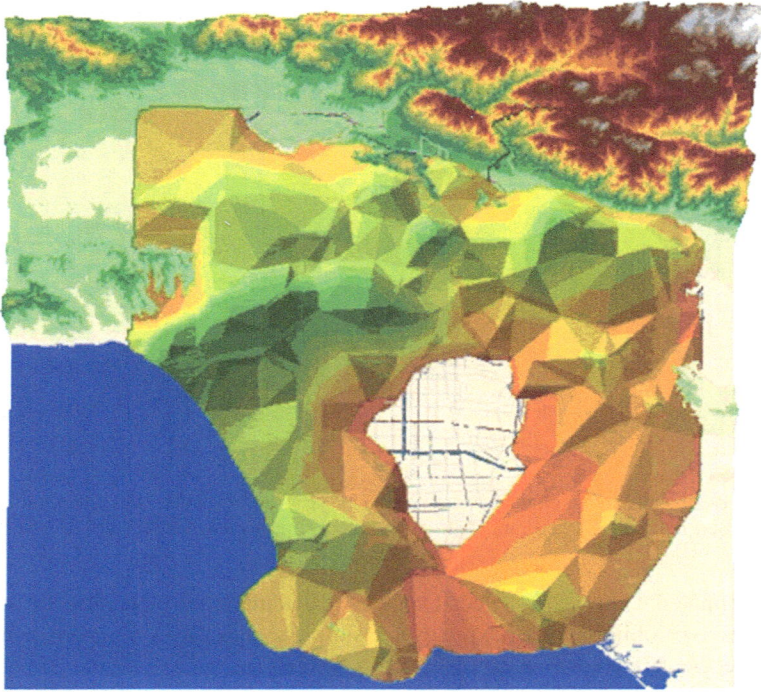

Fig. 4.8 A mental map of fear in Los Angeles visualising the results in 3D by colour. *Green* stands for very low fear and *red* for high numbers of the mentioning of fear. The *shades of yellow* and *orange* in between allow for the registration of shades (Project by Matei (2003))

From the responses, they generated a surface of desire for different areas in the world. In these early days, surfaces were visualised through contour maps, with each contour representing a change in value. In a very early 3D computer rendering of the data, they detail the UK by plotting the desired location of residents. Unsurprisingly, the taller peaks are in the southern part of the island. These approaches do not actually work with mental maps as sketched by the participants, but they use participants' responses to specific questions to generate them into a mental map that could be called collective.

Whilst working with children, mental maps are often used as a method of expression, as, for example, in *Environmental Fears and Dislikes of Children in Berlin and Paris* by Besten (2008); see Fig. 4.9 for illustration. Besten's paper examines the absence of children in today's cities and investigates the highly specialised urban environment from a child's perspective of safety, fear and joy. Drawing is often associated with something for children or something that one does at school. For research with children, the method seems appropriate but why not for adults? As Gauntlett (2008) demonstrated in his workshops, there is notable potential via creative methods. As part of this, the aspect of drawing should not be

Fig. 4.9 Two mental maps, on the *left* drawn by a 10-year-old boy from Paris, and on the *right*, a girl from Berlin, laying out her immediate environment (Images taken from (Besten 2008))

underestimated. In our experience, it appears that adults often have more difficulties, compared with children, in drawing even a simple sketch. Drawing is not something adults necessarily do very often.

During the UD interview, the participants were asked to sketch a mental map to allow participants to express how they navigate the space of the city. In addition to the technical GPS record, this personal view focuses on the participants' perceptions of space based on memory, experience, circumstances and current concerns. Through the comparison of the two different maps, new insights into people's motivations for choosing a route, and individual methods of orientation, can be explored. For the drawing of the maps, no graphical restrictions on how to represent elements were imposed; the only rule was not to copy it from a street map or image. In addition, they were asked to comment on what they had drawn, to record in-depth information on perception and important factors beyond the sketch. Figure 4.10 is an illustrative example showing the mental map drawn by the participant whose objectives *tracks* were shown earlier in Fig. 4.6 on page 66.

The paper for the mental map is prepared with a frame/box to further limit the space for drawing on, as our pre-study experiment has shown that this additional boundary line helps *inexperienced sketchers* to navigate on the white page. Participants tend to draw towards the edge, and then they realise that there is no more space left on which to draw the second half of the journey. The additional space outside the frame can accommodate some of this information, which is otherwise lost or drawn in a disconnected way.

A great deal of information is contained within the mental maps on how people perceive, use and ultimately how they create their space. As an abstract concept,

4.4 The Image of the City as Routine: Mental Maps

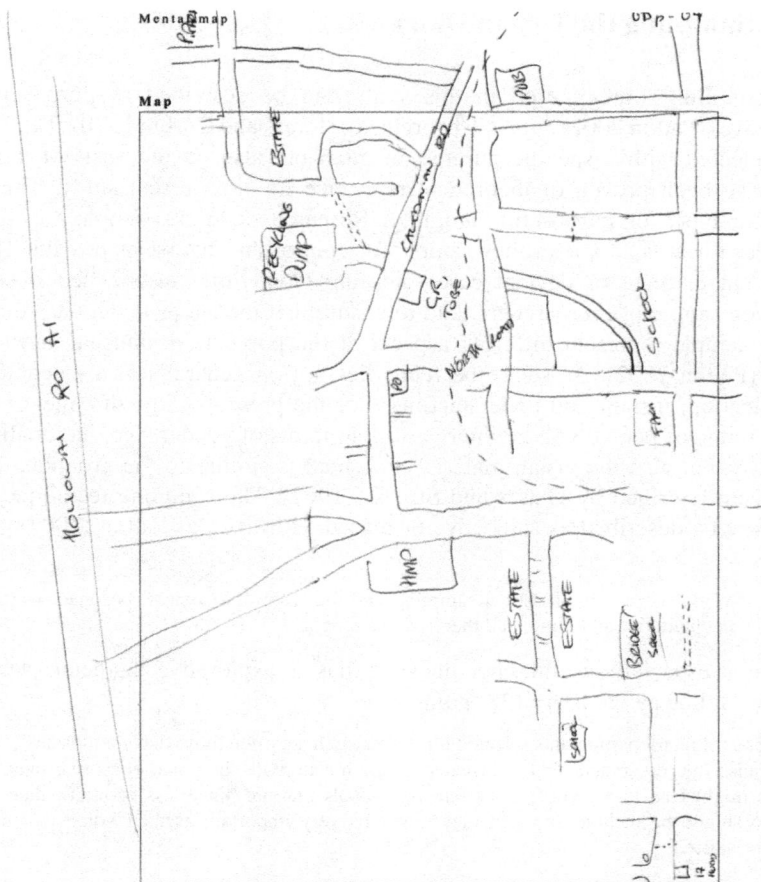

Fig. 4.10 A mental map of the routine commute from home to work and back home by one of the participants of the Urban Diary project

it could be compared to the technical creation of space in the virtual world as an orbit around the subject in time and space. The engine only renders a certain area or distance in a game scene and not the entire game world, city or house as a whole. In a similar manner, the temporary space people create in the real world could be described as a bubble. Space, as in social space or individual space, is not the same as Euclidean space, which is the way we normally think about space. If we describe space from personal perception and a temporal point of view, the concept of space is different from the space-in-a-box concept. The creation of space is something very personal that, through mental maps, can be accessed and recorded as a dynamic concept of temporal perception based on mood, concerns and circumstances.

4.5 Sampling the Urban Diary

The sampling strategy used in this study can be described as purposeful (or purposive) (Patton 2002) but also theoretical (Strauss and Corbin 1990). The sample was selected with a specific purpose in mind but also on the basis of concepts that have been proven of theoretical relevance for the present study. There are different types of purposeful sampling: Researchers might sample for specific attitudes towards a topic or they sample for opinions as diverse as possible (Patton 2002). In extreme or deviant case sampling (sometimes also called maximum variation sampling), researchers tend to ensure that the subjects who are included in the sample represent differing aspects of the population, unusual and special cases (Patton 2002). As such, the focus lies not on generalisation, but rather on specification, insight and understanding. For the present study, this means that a broad range of people will be interviewed, in terms of gender, age, nationality and ethnicity, but also the visual material presented is similar to the grounded theory approach described by Glaser and Strauss (1967). The sampling technique could therefore be described as subjective or biased. However, as Patton (2002, p. 230) argues:

> what would be *bias* in statistical sampling and therefore a weakness, becomes intended focus in qualitative sampling, and therefore a strength

This is especially useful when the study has an explorative character. Nevertheless, as Holloway (2010, p. 138) points out:

> These voluntary participants selected for the research are often those that are most articulate because the researchers find it easier to communicate with them and elicit rich data, but this might lead to a neglect of certain individuals that are powerless or inarticulate and who should be included; indeed they might be very important as their voices are often marginalised.

The Urban Diary sample is based on a chain sampling, identifying cases of interest through people who know other people with relevant cases (Bradshaw and Stratford 2005). In detail, the sample is a sort of typical case sample based mainly on participants who are pursuing a clear activity during the day, such as a job. This provides to some extent a given structure, only partially modifiable by the participant. The minimum criterion for the selection is to have a strong activity connection to the urban area in question. This more or less implies living or working in either London or Basel. There were no special requirements resulting from the use of technical equipment, for the GPS devices are very simple to use. They required the participants only to remember to always carry the unit with them and to charge them overnight. In the London case, participants were required to turn them on and off on leaving and arriving, whereas in the Basel study, the units recorded automatically. As an additional criterion, no children were included in the study and all participants were over the age of 18.

4.5.1 Sample Description

The sample size for the Urban Diary GPS tracking is $N = 48$ participants, where N is the total number of individuals or units in the study. The sample size of N between 40 and 50 is a compromise between an achievable high number of N and a manageable amount of detail in the time frame of the research with the limited resources of just one researcher. The sample size is, by nature, not statistically representative: However, the richness of data and the detail covered will qualitatively make up for this. This sample is composed of three spatially different groups. One group is based in the wider London area with a sample of $N = 23$, the second group is based in and around Basel with a sample of $N = 21$, and the third group is composed of individual participants from different locations with a sample of $N = 5$.

The limitations of the sample size are partly due to limitations in the availability of GPS units, but this nevertheless allows several different types of characteristics to be represented. The study contained overall 40 % female and 60 % male participants. Of the sample, 64 % are employed, of which 20 % work part time. One per cent of the participants are self-employed and 35 % are students. In terms of social structure, 46 % of the participants are single, 29 % are living in a relationship and 25 % are living in a relationship with children.

4.6 Individual Spatial Experience: The Interviews

The GPS devices record participants' movements automatically, and they are not required to keep a manual diary. For detailed trip information, participants are interviewed at a later stage of their data collection. The contact between researcher and participant during the 2-month period is relatively close. Meetings take place on a weekly or biweekly basis, ostensibly so that the researcher can download the collected data from the device. However, these regular meetings also allow for informal chats, with the personal routines and habits being the main topic of these discussions.

The aim of the formal interview is to collect information on how the participants perceive their activities and how they would describe these routines. From the GPS data, a schedule can be generated, but this might not reflect the intended plans of the individual. The interview was designed as semi-structured, with the main topics based around the personal schedule, transport and movement, experience of the city, orientation and memory. Participants were asked to undertake two additional tasks, which were, first, to write down a rough personal schedule on a daily, weekly and yearly basis and, second, the drawing of the mental map described earlier. The daily schedule helps with the interpretation of the data regarding patterns that are not synchronised with the individual's normal routine. It is also of use to learn more about an individual's organisation, both in time and space. The mental map, on the other hand, is directly related to space and visualises an individual perspective of the city. Both elements are regarded as important to the spatial narrative of everyday life in the city. In addition, this information is believed to be essential to interpret the GPS information.

From our experience, participants often present a perception of their spatial habits and will describe them at the beginning of the tracking as diverse and spread over a large area of the city. The first few times they see the collected data, it can be disappointing for them to see such a strong routine. Very often participants suddenly become aware of a number of routines they follow without having noticed them beforehand. Routine seems to be negatively perceived and participants often would describe themselves as active, flexible and spontaneous, implying a widely spread range of activities with a diverse movement pattern. Of course, one does not necessarily exclude the other, but the usual interpretation of a strong pattern tends to lead to this reaction. This phenomenon arguably has its origin in the modernist ideal conceptualisation of space and movement. It could be a late descendent of the illusion of the automatic and autonomous freedom that played an important part in modernist spatial concepts and encapsulated by positive feelings about the beauty of the machine and the associated freedom newly inherited by the middle class. As Bonnett (2000, p. 28) puts it:

> Thus *ordinariness* and *everydayness* are maintained as the provinces of the working class, ...

To describe their personal routine, participants often refer to someone else whom they think of as very flexible or very inflexible, to provide an example for comparison. It appears to be more convenient to define routine in terms of metaphor or other substitutes. It appears to be a personal subject where people prefer to make assumptions and live with stereotypes.

The participants were then asked to write down details about their schedule, focusing on important structuring events. Three scales were of interest: the day, the week and the year. It is not as simple a matter as to explain one's daily schedule. There are a lot of *ifs*, ands, *ors* together with *thens and woulds*. In short, routines are presented as a dynamic string of decisions with numerous dependencies. Nevertheless, there are strong elements of direction with an indication of structure within this pool of fluid decision making. Again, the major element is the working week versus the weekend, and then there are the clear western standards for a daily structure both on weekdays, illustrated in Fig. 4.11, and weekends, as illustrated in Fig. 4.12. The focus does represent the personal situation. There are big differences, though, between participants who have dependent children and those who have none.

Taking the two time frames together, these can be regarded as representing the participant's *mind map* of weekly activities. Regarding the information, one might expect large gaps between plans and activities; however, the two are largely similar. The *mental picture* of our routines is strong, and comparing this to participants'

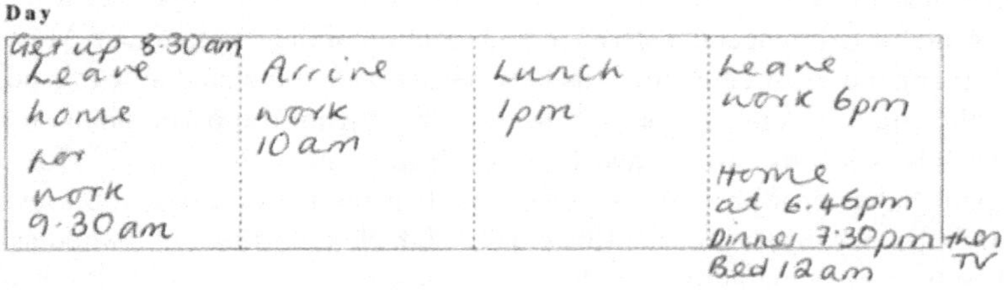

4.6 Individual Spatial Experience: The Interviews

Fig. 4.11 Participant's schedule for a day corresponding with the track record in Fig. 4.6

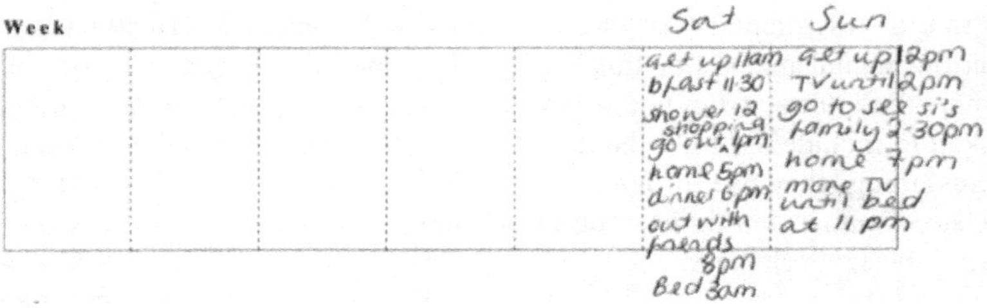

Fig. 4.12 Participant's schedule for a week corresponding with the track record in Fig. 4.6

perceptions of their spatial activities, this can be surprising as, in spatial terms, people often think their activities are much more flexible and they are travelling more than they actually are, as explained earlier.

Whilst working with these schedules, an unexpected finding emerged: The time spent interacting with the urban morphology, for example, by moving about in the city, is rather restricted. There are clearly defined time frames for each individual, of course, but generally time spent in the city is limited and certainly not random. From the examples in Figs. 4.11 and 4.12, activity involves spatial interaction on weekdays during the rush hour in the morning and the evening. Other than this, there is little activity. The weekend pattern is different, in that there is afternoon and evening activity, with Saturday being the more active day.

The information about yearly events did not generate much valid data, as for most of the participants, this category was too broad. It seems not to be a unit that people plan or even live in, even though in professional life, this is definitely an important time frame. In terms of personal activity, few have had planned activities other than the expected Christmas and Easter breaks. Birthdays and holidays were amongst the other named activities on a yearly scale. Regarding the city and its spatial morphology, longer time frames are of course interesting, but the connections probably have to be found elsewhere.

The topics of *space* and *movement* explore how participants use city space on a daily basis and how it is perceived in connection to their everyday routines. It is therefore of interest to see if and how individuals are able to connect the spaces they frequently visit in terms of their mental map. This is especially noteworthy in the context of London because, for example, travelling by tube might leave the traveller unable to connect the start and end location of the trip spatially. Movement, on the other hand, is directed towards how participants travel and how this becomes part of the routine. Again, it is interesting to hear from the participants how they see themselves in this respect and how much they think they travel. Most of the participants have clear preferences regarding their mode of transport. Some mainly travel by tube, because it is easier to navigate with clear destinations, whereas others would only very reluctantly use the tube, because it is narrow, underground or busy. Instead, they prefer the buses and describe them as flexible and close to their destination.

To explore the topics of *spatial experience* and *memory*, the participants were asked to draw the mental map of one journey in the sense described earlier in the section on mental maps. For this setting, the focus was specifically on the daily commute, the journey from home to work and back. To draw the map, participants are asked to include not only the direction in which they travel but also additional elements: things they use for navigation, orientation or simply as reminders. These can be street names, buildings or urban settings, and even views or atmospheres can play an important role here.

In one of the examples, as illustrated in Fig. 4.13a, only the top of the sheet was used. Participants are asked to comment on what they drew, and the transcript of this helps to interpret the drawing later, for example, regarding the sequence (in which the events were drawn, or in which the illustrated journey was undertaken, or both). Individuals' comments about their feelings in connection with a certain element or configuration within their trip can also be traced back. A frequent phrase, for example, is "This is not to scale", pointing out that there is an uncertainty about actual metric distances and global scales.

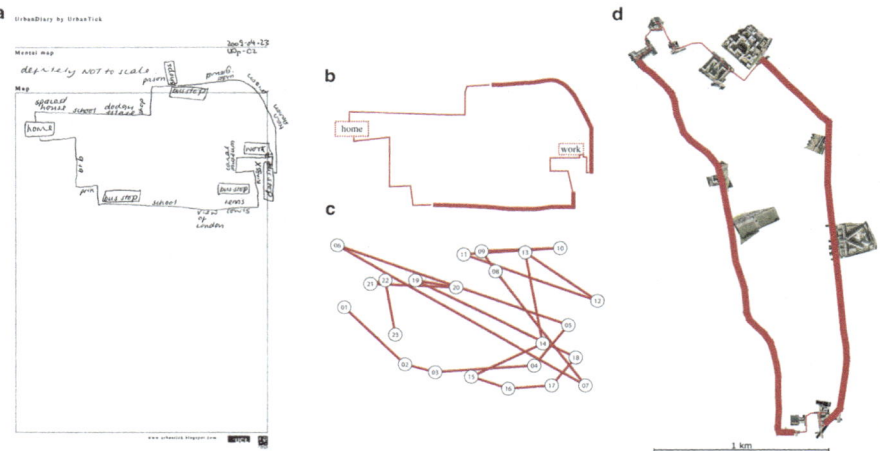

Fig. 4.13 UDp-02 mental map (**a**) compared to GPS record (**d**). A simplified mental map (**b**) shows the travelled route by bus (*thick line*) and walking (*thin line*). In (**c**) the drawing sequence of the map is shown

In the example pictured, Fig. 4.13a contains the mental map as drawn by the participant, where Fig. 4.13d is a reference map generated from the same participant's GPS record. In the middle, the two analysis diagrams look at the relationship of the map to the mode of transport used on the journey (vgl. Fig. 4.13b), and as the sequence of the map's creation (vgl. Fig. 4.13c), represented as a dot-to-dot doodle. Both are based on the participant's mental map. These examples will be discussed in detail in Chap. 8.

During the interview, aspects of daily activities came up that are of note, where the participants seem to feel that an explanation is required. Many participants feel the need to explain their activities and make an excuse for them. This seems to be related to the amount of movement, flexibility or distance. Generally, routine is viewed as negative, whereas flexibility and independence is deemed positive. This experiment has been insightful in this respect, as the recorded movement unveiled routines that seemed to be much stronger than the participants have so far realised.

This wealth of impressions and information has been collected with all participants in both locations. Of course, the data sets contain variations in quality of data and information provided, but overall the fieldwork has provided a very valuable source of detailed information that propels the project forward.

4.7 The Individual Spatial Shape

There are two main levels of interest: one is the level of the individual, in the context of personal routine and activity and the other is the collective level, looking at overall patterns and rhythms that point towards a spatial society and urban morphology. The first one is immediately present for the participants as individuals who experience the city, where the findings mainly reflect the context each perceives themselves to be in. The second, collective, level is something that can be constructed from the individual activities. The research was intentionally designed to capture individual data and not collective data, to gain insight into the routine of the individuals' base behaviour with a view to aiding the understanding of how the two levels interact.

Timewise there are big differences between work time and personal time. The pattern of this almost universal division determines all participants' actions. In addition, there is the pattern of the week, with its weekdays and weekends, structuring the time over a longer period. Working days are represented spatially as a back-and-forth movement between the home location and the workplace. The London morphological characteristic here loosely resembles a star shape (see Fig. 4.14); as discussed above, people live outside and travel linearly into the centre and back out again. To some extent, this might depend on the data sample, and the pattern observed here might not apply for all configurations. However, the routine in movement between a few fixed destinations shows up clearly. Over time each routine-defined location builds up a sub-local area. Depending on the activity and the time spent in this location, the area grows denser as the creating individual becomes more familiar with the location.

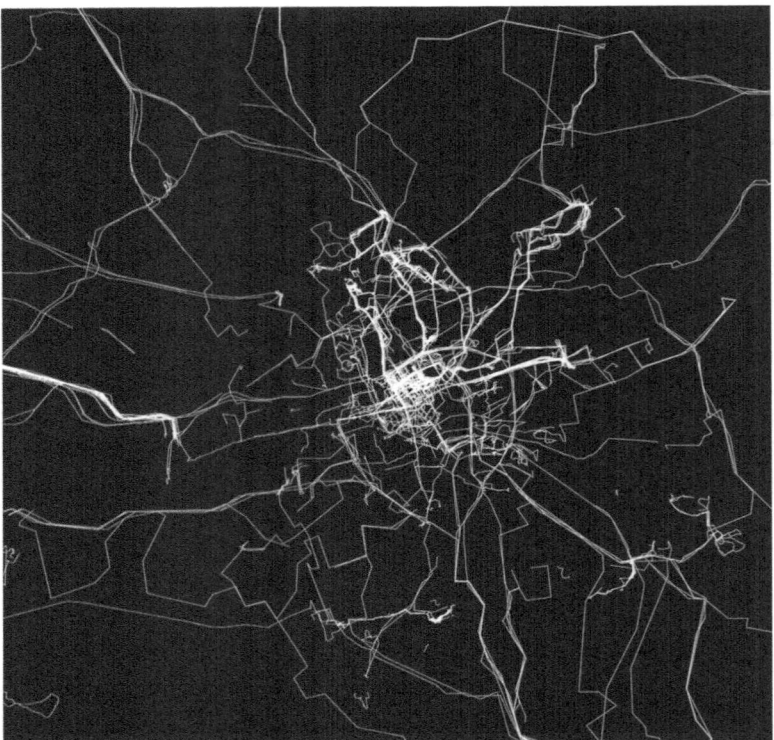

Fig. 4.14 A map showing the tracks as recorded by all participants in London. There is a series of strong linear connections from outside to the centre of the city, represented by a star shape

The pattern that emerges from the data very much supports what has been predicted in pre-studies and abstract diagrams in the 1970s. For example, Golledge (1978) in *Learning About Urban Environments* shows how individuals build up their personal city of experienced, e.g. *known* areas over time from scratch. With the Urban Diary data, the starting point is slightly different. The participants are not new to the city and already have a well-established spatial routine. Nevertheless, over the period of the 2 months of individual data recording, these patterns start to emerge in a similar way. In this sense, the data now available supports the hypothesis that individuals maintain a network of nodes of importance to their personal spatial habit. This will be explored further in Chap. 7.

The way the different destinations are connected spatially depends on the mode of transport and personal preferences. For some participants, the working week tracks are two islands on the map, as most of the travelling happens underground, whereas other participants travel to work by means of overground transport, and the map shows an intense, continuous collection of tracks. The weekend travel pattern, on the other hand, is mainly focused around the home location or tends to be directed

4.7 The Individual Spatial Shape

Fig. 4.15 What shape are you? Track records show distinct individual pattern of four participants over the same time period and represented at the same scale

outwards, away from the city. Very often this is directed by the location of friends and family but also the location of shops and amenities such as markets, parks or playgrounds play a major role as attractors.

By looking at a neighbourhood area where participants live, the local pattern becomes more obvious. The weekend pattern shows activities within the neighbourhood and local streets rather than the main routes. The localised activity of participants with dependent children is not only greater, but more intense. Local amenities and activities play a major role in their weekend planning. Their home location creates another set of destinations with clear directions.

It emerged from the recorded data that each participant has her/his unique trail pattern, illustrated in Fig. 4.15. Almost like an individual fingerprint, the shape created by daily movement is unique. The shape is determined by a number of factors such as the spatial relationship of destinations, the physical distances travelled, the amount of travel and the intensity of repetition. The first point, the relationship of destinations, makes for the overall shape and the last point, the intensity of repetition, makes for the character of the shape.

This study, to trace citizens' spatial habits, has provided the research with a great deal of valuable data, and a number of possibilities for mapping the data will be explored in the following chapters. More than two million records of locations, each with a time and space stamp, have been collected. The data quality has been good overall with a few outliers, where a participant was not quite consistent in terms of commitment over the recording period: This is to be expected with such a sample. With most participants, it was a matter of letting them integrate the recording, and especially the device handling, into their everyday routine. This worked well when it was emphasised in the first week. This initial week, as a buffer and a period of getting used to remembering to handle an additional device, is quite helpful and probably necessary in order to establish the routine for the duration of the study and achieve a valuable data set at the end. The contact times have also helped to maintain interest and commitment. It proved valuable to keep participants informed and share with them the data collected. The involvement of participants, through wearing a special device and *being in charge* of the collection, could in this respect be regarded as positive.

In terms of technology, GPS tracking has been successful and has proved to be a good method for collecting data about individual movement. The details recorded and the resulting maps work well on an urban, but also on a local, level demonstrating the versatility of individual spatial activity.

Preliminary findings so far have shown that people in the city seem to live according to rhythms that appear to be largely congruent, more so than might be expected. The sample of 20 participants is enough to demonstrate that there are emergent patterns at the level of collective activity.

The analysis of the project data so far has been very much personal and individual. Another step in Chaps. 6 and 7 will be to look at ways to combine the data, to move towards the analysis of the collective level. This will be closely connected to the body, both of the participants but also the city, looking at morphology as a product of rhythmic processes.

Regarding cycles and routines, the pattern as explored here in the urban context demonstrates how close the ties are between the urban fabric, the chosen route and the resulting personal trails and how this consistency builds up and remains over a longer period. This demonstrates the productive process and visualises how the city can be understood as an emerging result of the activity culture inhabiting it.

In the next chapter, we now switch back and examine movement patterns through various types of social media, at more aggregated levels where the behaviours of much larger numbers of individuals are mapped as New City Landscapes.

Bibliography

Axhausen KW, Zimmermann A, Schönfelder S, Rindsfüser G, Haupt T (2002) Observing the rhythms of daily life: A six-week travel diary. Transportation 29(2):95–124

Besten ONN-D (2008) Cars, dogs and mean people: environmental fears and dislikes of children in Berlin and Paris. In: Urban trends In Berlin and Amsterdam. Berliner Geographische Arbeiten. Geographisches Inst. der Humboldt-Univ, Berlin, pp 116–125

Bonnett A (2000) Buses. In: City A-Z. Routledge, London, pp 26–28

Borden I (ed) (2001) The unknown city: contesting architecture and social space: a strangely familiar project. MIT, Cambridge, MA

Bradshaw M, Stratford E (2005) Qualitative research design rigour. In: Qualitative research methods in human geography, 2nd edn. Oxford University Press, Melburn, pp 67–76

Carlstein T, Parkes D, Thrift NJ (eds) (1978) Timing space and spacing time. Edward Arnold, London

Chadha K, Osthimer S (2010) GPS technology white paper. SiRF Technology. Available at: http://www.hr-tews.de/GPS/sirf_en.htm. Accessed 17 July 2012

Chen C, Gong H, Lawson C, Bialostozky E (2010) Evaluating the feasibility of a passive travel survey collection in a complex urban environment: lessons learned from the new york city case study. Transp Res Pt A Policy Pract 44(10):830-840

de Certeau M (1984) The practice of everyday life. University of California Press, Berkeley

Gauntlett D (2008) Representing identities (Part 1: Method). [Online video], http://www.youtube.com/watch?v=LtS24lqluq0&feature=youtube_gdata.

Giddens A (1986) The constitution of society: outline of the theory of structuration. Polity Press, Cambridge

Glaser B, Strauss A (1967) The discovery of grounded theory: strategies for qualitative research. Aldine Transaction, Chicago

Glennie P, Thrift NJ (2009) Shaping the day: a history of timekeeping in England and Wales, 1300–1800. Oxford University Press, Oxford
Golledge RG (1978) Learning about urban environments. In: Carlstein T, Parkes D, Thrift NJ (eds) Timing space and spacing time. Making sense of time, vol 1. Edward Arnold, London
Golledge RG, Zhou J (2001) GPS-based tracking of daily activities. Technical report UCTC Grant # DTRS99-G-0009, University of California Transportation Center, University of California, Santa Barbara
Gould P, White R Mental maps. Pelican geography and environmental studies. Penguin, Harmondsworth (1974)
Hägerstrand T (1978) Survival and arena. In: Carlstein T, Parkes D, Thrift NJ (eds) Timing space and spacing time. Human activity and time geography, vol 2. Edward Arnold, London
Hamm B (1982) Einführung in die Siedlungssoziologie. C.H. Beck, München
Hanson S, Burnett P (1982) The analysis of travel as an example of complex human behavior in spatially-constrained situations: Definition and measurement issues. Transp Res Pt A Gen 16(2):87–102
Hillier B (1996) Space is the machine: a configurational theory of architecture. Cambridge University Press, New York
Hillier B, Hanson J (1984) The social logic of space. Cambridge University Press, Cambridge
Holloway I, Wheeler S (2010) Qualitative research in nursing and healthcare, 3rd edn. Wiley-Blackwell, Oxford
Kilroy B (1998) Resource applications of GPS technology. http://www.fs.fed.us/t-d/pubs/htmlpubs/htm98712324/index.htm
Kwan M-P (1999) Gender and individual access to urban opportunities: a study using space-time measures. Prof Geogr 51(2):211–227
Larsen J, Urry J, Axhausen KW (2006) Mobilities, networks, geographies. Transport and society. Ashgate, Aldershot
Lefebvre H (2004) Rhythmanalysis: space, time and everyday life. Continuum, London
Ley D (1972) The black inner city as frontier outpost: images and behavior of a Philadelphia neighborhood. Pennsylvania State University, Pennsylvania
Loew M (2001) Raumsoziologie. Suhrkamp, Frankfurt am Main
Lynch K (1960) The image of the city. MIT, Cambridge, MA
Massey D (2005) For space. SAGE, London
Matei S (2003) Mental maps: social and spatial research on emotional and affective implications of maps and space. http://www.mentalmaps.info/.
Miller HJ (2004) Activities in space and time. In: Hensher DA, Button KJ, Haynes KE, Stopher PR (eds) Handbook of transport geography and spatial systems, Handbooks in transport, vol 5. Elsevier, Oxford
Miller HJ (2005a) Necessary space-time conditions for human interaction. Environ Plann B: Plann Des 32(3):381–401
Miller HJ (2005b) A measurement theory for time geography. Geogr Anal 37(1):17–45
Miller CC (2006a) A beast in the field: The google maps mashup as GIS/2. Cartographica 41(3):187–199
Miller H (2006b) Social exclusion in space and time. In: AXHAUSEN KW moving through nets: the physical and social dimensions of travel. Selected papers from the 10th International conference of travel behaviour research H. Elsevier, New York, pp 353–380
Miller HJ, Ahmed N (2007) Time-space transformations of geographic space for exploring, analyzing and visualizing transportation systems. J Transp Geogr 15:2–17
Neuhaus F (2010) Cycles in urban environments: investigating temporal rhythms. Lambert Academic Publishing (LAP), Saarbrücken
Ortúzar JdD, Willumsen LG (1994) Modelling transport, 2nd edn. Wiley, Chichester
Patton MQ (2002) Qualitative research and evaluation methods. SAGE, Thousand Oaks, CA
Portugali J (1996) The construction of cognitive maps. GeoJournal library, vol 32. Kluwer Academic, Dordrecht/London
Portugali J (2004) The mediterranean as a cognitive map. Mediterr Hist Rev 19(2):16–24

Portugali J (2005) Cognitive maps are over 60. In: Cohn A, Mark D (eds) Spatial information theory. Lecture notes in computer science,vol 3693. Springer Berlin, Heidelberg, pp 251–264

Prasad R, Ruggieri M (2005) Applied satellite navigation using GPS, GALILEO, and augmentation systems. Artech House mobile communications series. Artech House, Boston

Ratti C, Williams S, Frenchman D, Pulselli RM (2006) Mobile landscapes: using location data from cell phones for urban analysis. Environ Plann B Plann Des 33(5): 727–748

Reades J, Calabrese F, Ratti C (2009) Eigenplaces: analysing cities using the space – time structure of the mobile phone network. Environ Plann B Plann Des 36:824–836

Sheller M, Urry J (eds) (2006) Mobile technologies of the city. The networked cities series. Routledge, Abingdon

Spencer J (2003) Global positioning system: a field guide for the social sciences. Blackwell, Malden

Steinitz C (1968) Meaning and the congruence of urban form and activity. J Am Plann Assoc 34(4):233–248

Strauss AL, Corbin JM (1990) Basics of qualitative research: grounded theory procedures and techniques. Sage, Thousand Oaks

Tolman EC (1948) Cognitive maps in rats and men. Psychol Rev 55(4):189–208

Tsui JB-Y (2005) Fundamentals of global positioning system receivers: a software approach. Wiley series in microwave and optical engineering, 2nd edn. Wiley, Hoboken

Tuan Y-F (1977) Space and place: the perspective of experience. University of Minnesota Press, Minneapolis

Yáñez MF, Mansilla P, Ortúzar JD (2010) The santiago panel: measuring the effects of implementing transantiago. Transportation 37(1):125–149

Chapter 5
New City Landscape

Abstract This chapter presents the second data collection, which consists of social networking data mined from the Twitter microblogging platform. This section provides a description of the methods and tools with a detailed discussion of the ethical considerations. With NCL, the focus is on the geolocated twitter data collected for 20 major cities from around the world. The data is used to investigate city activity patterns overall, based on location and time.

It reviews the rise and fall of Digital Social Networks from Myspace and Bebo to Virtual Worlds and Facebook, alongside the location-based services Foursquare, Gowalla and Twitter. Following this overview, the methods of data mining and crowd sourcing are discussed with, in the latter part, a focus on location-based data sets. This then leads to a discussion of the concrete technical aspects of the collection of Twitter data, discussing different strategies and possibilities.

This is followed by a short discussion of the complications with the Twitter data sample in general and specifically the location based data. The next sections are concerned with aspects of data analysis in time-space as well as networks. At the end, the aspects of privacy and ethics are discussed. This extends the general discussion of the ethics relevant to this publication, due to the urgency and grey areas in this field if applied to online social networking data.

This chapter presents the second data collection, which consists of social networking data mined from the Twitter microblogging platform. This section provides a description of the methods and tools with a detailed discussion of the ethical considerations. With NCL, the focus is on the geolocated twitter data collected for 20 major cities from around the world. The data is used to investigate city activity patterns overall, based on location and time.

Section 5.2 reviews the rise and fall of Digital Social Networks from Myspace and Bebo to Virtual Worlds and Facebook, alongside the location-based services Foursquare, Gowalla and Twitter. Following this overview, the methods of data mining and crowd sourcing are discussed with, in the latter part, a focus on location-based data sets. This then leads to a discussion of the concrete technical aspects of the collection of Twitter data, discussing different strategies and possibilities.

This is followed by a short discussion of the complications with the Twitter data sample in general and specifically the location based data. The next sections are concerned with aspects of data analysis in time-space as well as networks. At the end, the aspects of privacy and ethics are discussed. This extends the general discussion of the ethics relevant to this publication, due to the urgency and grey areas in this field if applied to online social networking data.

5.1 Introduction

Urban settlement is on the increase. Since 2007 (UN News Centre 2009) half of the world's population now live in urban settlements. Millions of people are living together in dense agglomerations requiring a high level of organisation and discipline. This complex arrangement of functionality, performance and individual freedom has in the past led to various attempts to conceptualise the model of organisation of the city as a single artefact. As discussed earlier, concepts were described as mechanical, in line with the modernist idea of a functional city leading to a description of the city as *Urban Machine*, discussed in Chap. 2; also as organic (Batty 2005), relating the city organism to forms of arrangement and organisation in nature, or describing the city as the physical result of social organisation (Hillier and Hanson 1984) influenced by individual actions expressed in the form of local culture.

Chapter 4 outlined the study settings, investigating city spaces from a subjective perspective, following individuals through these spaces. In a complementary perspective presented here in Chap. 5, the focus is on large networks of individuals, tracing activity on a collective level as it unfolds across an urban area. It is no longer the individual's steps that are tracked as a form of linear progress, but the full complement of activity sparks across the urban area. As such, this is moving the perspective from a linear to an aerial view. As the individual is tracked linearly in time, the collective is tracked as a shape with an area. This description serves as an illustrative explanation of the differences between individuals. It is not, however, intended to serve as a theory or paradigm interpretation of what is being looked at. Previously, it was emphasised that even though tracks technically appear as lines, such a description greatly neglects the essential aspects of individual movement and creation of space whilst on the move. Similarly, as will be shown later on in this chapter, traces of collective activity in the city can be seen as patterns of a unique urban configuration, characterising each urban area individually.

The term Digital Social Networks (DSN) describes a range of Web-based networking platforms that allow the personalisation of online platforms and the building of personal connections between different users, including the sharing of digital content. In essence, it is a digitised circle of friends and contacts. To trace collective activity across the city, location-based digital social network data is used. In the dataset used for this analysis, each data point has a location tag and a time stamp associated with it, and it is interpreted as an activity in time and space. The messages are virtual, e.g. digital, but sent from a real-world location. The actual content may or may not be related to the local area, but the fact that the individual sending it is at the location creates a reference point for local activity

5.1 Introduction

in a wider sense. Activities on Digital Social Networks are, beside the locational context, embedded via connections to other users. This aspect provides a glimpse of collective connection in a spatial context. It is not the individual message, but the overall activity in the city as a whole that is of interest. When does the city wake up, what time is the peak activity and what time do people go to bed? This is the rhythm of the city as observed from a perspective of Digital Social Networks.

In regard to social networks, Smith (Smith, 1986 cited in Bruggeman 2008) explains:

> Humans are more cooperative with non-kin and exchange more information than all other animals, with large ramifications for their society, which eventually became a globe-spanning network. Their pro-social activities enables them to specialise in few activities, and render them dependent on others for their remaining needs and desires to be fulfilled.

With the rise of network science and social network theory in the late 1980s and the early 1990s (Barabási 2002), combined with the growth of the internet, investigation of networks in social systems has become a specific discipline. Within a social network, individual people are represented as nodes of a graph, and edges between the nodes represent relationships between them. For example, as illustrated in Fig. 5.1, Mrs Sanders knows Mr Neuberg and this results in two nodes, Mrs Sanders and Mr Neuberg, who share an edge, a connection, between them. However, Mr Jett is an independent node since neither Mrs Sanders nor Mr Neuberg link with him. This figure expresses a simple social network, indicating participants and connections between them.

Social configurations and connections, as we know from our everyday experience, can be complicated and entangled. The social networking graph as a graphical simplification of reality is based on two main elements, nodes and edges. It provides a simple method for visualising large and very complex configurations of connection. As Jeremy Boissevain (1979, p. 392) states:

> Network analysis asks questions about who is linked to whom, the content of the linkages, the pattern they form, the relation between the pattern and behaviour, and the relation between the pattern and other social factors.

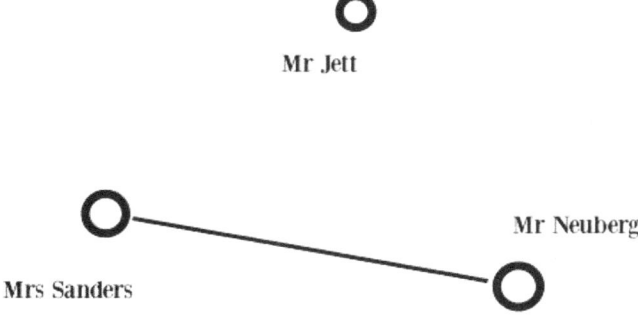

Fig. 5.1 Simple network graph showing the relationships between three individuals. Each node (*ring*) represents an individual, and each edge (*line*) represents a connection (relationship) between the individuals

The social connection, it could be argued, is what creates us as individuals, both in terms of distinctiveness and the creation of self. Social networks are important for learning and collective support. On the other hand, social networks can also cause great harm and destruction both socially and environmentally (Tinbergen 1968). In short, social networks are constantly at work pulling and pushing the strings of relationships.

Of specific interest to this study is how the social network relates to the spatial condition. Space is not an isolated phenomena: It is directly shaped by the culture and the society which creates and uses it on a daily basis (Lefebvre 2004). As discussed in Chaps. 2 and 4, Urban Diary, Hillier and Hanson (1984) in *The Social Logic of Space* focuses on this close relationship between the social configuration and the spatial morphology. This book introduces a concept through which the social connection is embedded in the spatial configuration at a societal level. Using a number of examples, they show how social structures can be inferred from spatial structures. By examining social networks, we have the potential to understand the spatial configuration and/or the meaning and identities of spaces. The question is no longer, to come back to the previous network example, merely why and how Mrs Sanders and Mr Neuberg know each other, but what this tells us about the space they live in. How does the space relate to the connection or vice versa, and are the conditions static or fluid as to how exactly the two things influence one another? Space plays an important role in the individual's meeting and keeping in touch.

For example, Appleyard et al. (1981) shows this in his study on social relationships across the road in relation to traffic conditions (see Fig. 5.2 for illustration). In three different street settings in San Francisco, Appleyard studied the number of social connections between neighbours across the road. The settings differed only in the number of cars passing by on the road. The first road showed little traffic, with about 150 cars per day and an average of 8.4 social connections from one side of the street to the other. The second road had a traffic volume of 2,150 cars per day and had 5.4 social connections linking both sides. In the third setting, the road was crammed with 21,340 cars per day and the personal connections across it dropped to a very low 1.5.

The study illustrates one aspect of the close relationship between the social network, activity pattern and spatio-structural conditions. It is expected that the social networking data used here will allow the investigation of exactly these aspects about the spatial configuration and the social conditions in detail.

It is clear that with today's technology and media, the meaning of spatial distance and configuration have changed. This alters the game, since instant and live meetings with people located in completely different places anywhere on the planet are possible. Furthermore, social connections have changed. These can be established, maintained and dissolved without any real-world contact, e.g. in virtual worlds, chat rooms or Digital Social Networks. The importance of physical space is still present and active nevertheless. It is simplified to the fact that the individual is present in space and therefore interacting and acting. This can be both virtual and real for the individual, even if remotely present.

5.1 Introduction

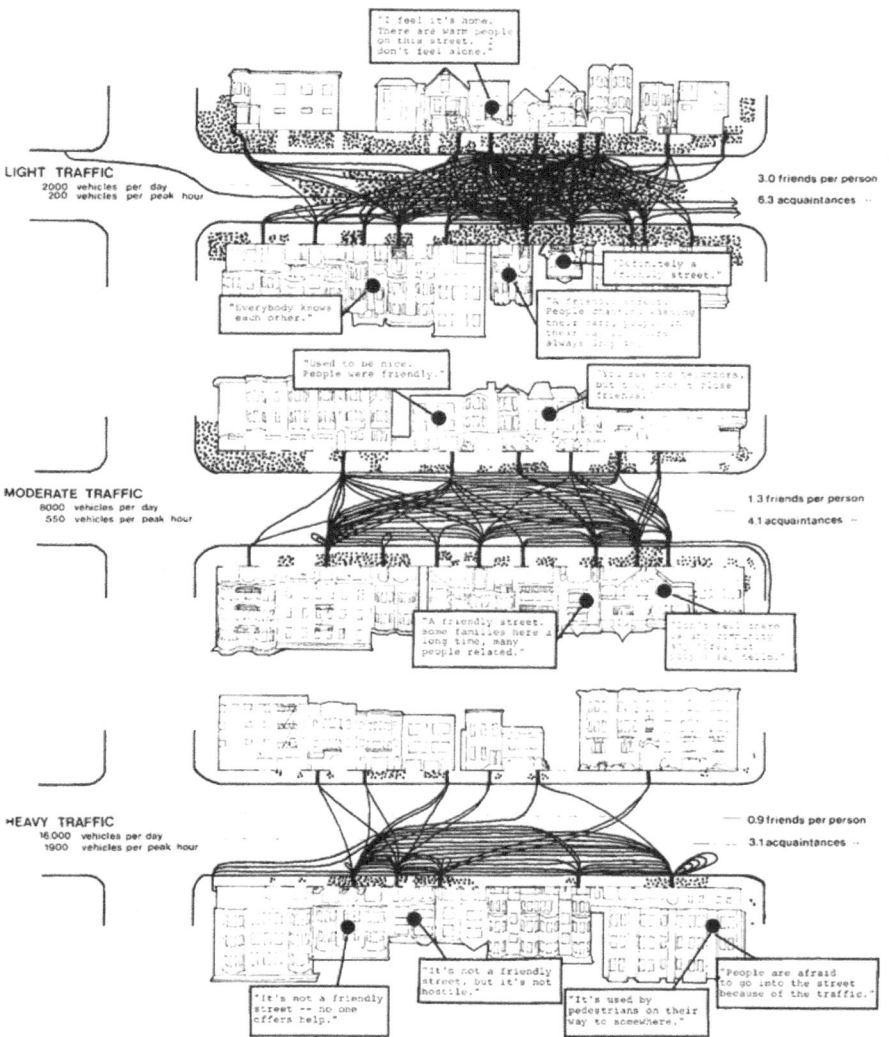

Fig. 5.2 Appleyard's study into social connections across the road, undertaken in San Francisco, USA, in 1982. Three similar roads with very different volumes of traffic. Here shown, *top* to *bottom*, from *low* to *high* volume of traffic. *Lines* indicate where people said to have friends in the road. The number of lines indicates the density of friendships against traffic volume on the road (Illustration taken from Appleyard et al. 1981)

In relation to the overall hypothesis, the study is focused on location-based aspects of the Digital Social Network data, allowing a glimpse of its rhythms and cycles. Of interest are the routines and the temporality of social connections and activity from dawn in the morning to dusk in the evening as well as throughout the night. How does the city tick socially in space? We look at how social networks can be used as a data source for urban analysis, specifically in relation to the investigation into rhythms and temporalities in urban environments. The idea is to generate a temporal-spatial image of the city via the social network, using spatial location at places around the urban centre, thus making the transition from individual to collective activity. In the following section, the collected data is detailed with an introduction to social networking and Digital Social Networks, with millions of people sharing an online presence and connecting with virtual and virtual-real-world friends.

In the following section, the current Digital Social Network scape, with a brief look at the context and the history, will be introduced. It will discuss the rise and fall of different networks over the past 10 years and how this activity relates to everyday social experiences, pointing out the importance of shared experiences of spatial details: These are a relevant source of data for spatial analysis. Subsequently the aspects of data mining and metadata are introduced together with the location data of digital social networking. This is followed by a discussion of recent examples of the mapping and visualisation of such data. This includes the discussion of data mining aspects together with recent examples of analysis and visualisation across different Digital Social Networks. The aim is to cover a range of different current source platforms, specifically Foursquare, Facebook, Flickr and Twitter, in order to provide an overview of the different aspects, possibilities and challenges.

Subsequently the focus will shift to introduce the specific work carried out using Twitter data and the creation of what we will term a New City Landscape (NCL). The technical challenges, the data sample and the mapping of the data are discussed in separate sections. This allows for a detailed introduction to the methods and technologies, providing a good picture of the possibilities. In addition, a section is dedicated to the discussion of the ethical and privacy implications and considerations in relation to the work with and handling of digital social networking data. Since the case differs from the Urban Diary project, it is important to highlight the differences and outline the changes of approach. The specific aim of this chapter is to provide an in-depth discussion of the specific steps and considerations in order to enable a detailed discussion and development of the method. The chapter concludes with a summary and outlook section to lead into the presentation of the findings in the following Chaps. 6 and 7.

5.2 Digital Social Networks: A Recent Phenomena

The popularity of Digital Social Networks is a comparatively recent phenomenon with a history of under 20 years, closely connected to technological developments. It has been especially influenced by the growth and change of the Internet as the manifestation of the networked society, as noted by Wellman (2001) and Livingstone and

5.2 Digital Social Networks: A Recent Phenomena

Brake (2010) and more recently by Bryant (2011). In 1997, sixdegrees.org[1] started as a networking platform, one of the first Digital Social Networks on the Web using a profile page (see, e.g. Boyd and Ellison (2008) on the history of Social Networks). Earlier projects were mainly based on email addresses such as classMates.com[2] started in 1995. 2003 brought the introduction of two new services MySpace[3] focusing on personal presentation and LinkedIn as the professional counterpart focusing on business connections. The two were followed in 2004 by Facebook and Bebo. Facebook, 8 years on, is currently the dominating Digital Social Network globally. New networks are invented almost on a daily basis, trying to attract the crucial threshold of active and returning users. The latest large addition to the online networking ecosystem is Google+[4] in mid-2011. It shows how attractive the Digital Social Network market has become to large companies such as Google.

A Digital Social Network service nowadays is preliminary based on a personal page acting as a platform to present oneself, with text, details, images and links. The network allows the user, after signing up, to connect to other users and maintain a pool of contacts, *friends*, with whom the user can share items and content specific to different levels of privacy. The personal page also shows activity of friends and followers, automatically updating. Specific tools to communicate, e.g. message and comment, allow for online social interaction between users. With the construction of the online gateway around the personal page, it could be argued that these platforms are individual centred as presented by Papacharissi (2011). This means they put the individual at the centre of its interests and activities.

It is estimated that the global average of people with internet access who use social networking sites on a daily basis is 46 % (Asrianti 2011). It can be assumed that this number will rise as more and more people have access to a computer and the Internet. In addition, social networking becomes more popular as mobile phones give users instant access to these platforms anywhere, anytime. In this context, new terminologies for the activities were introduced. For example, *real-time information sharing* allows users to send and receive messages instantly, called *live feeds*, but also to *check in* to places that they are frequenting at this particular moment and the sharing of this information as *location-based information* with accurate GPS

---

[1] See for details http://www.sixdegrees.org/, the documentation on Wikipedia at http://en.wikipedia.org/wiki/SixDegrees.com or a blog post on The Dallas Morning News http://www.dougbedell.com/sixdegrees1.html.

[2] See their online page on http://www.classmates.com/ and a story on CBSnews online http://www.cbsnews.com/stories/2003/05/05/60II/main552363.shtml.

[3] See it online at http://www.myspace.com/.

[4] The Google+ platform was opened for public beta, based on invitation only, on June 28, 2011. After a 2-month test phase on September 20 2011 the service was officially opened to the general public (Cellan-Jones 2011). It can be found at https://plus.google.com/ with an introduction available at http://www.google.com/intl/en/+/demo/.

coordinates. In this context, more recently *geotagging*, the adding of location-specific information, such as metadata of videos, photographs and messages, plays an important role.

The rise of social networking sites has caused concerns regarding individual wellbeing and safety. Topics such as user privacy, online violence, social isolation and child safety are becoming more prominent. Although most sites have features that allow users to adjust their privacy settings, people are fearful of making too much information available to large cooperations or indeed to strangers (Gross and Acquisti 2005). The current service model Digital Social Network platforms run on is centralised, meaning that one service provider has complete control of all data traffic. In response to this, decentralised projects have been started. One of the more prominent efforts is the Diaspora[5] Some providers, such as MySpace and more recently Google+, encourage users to use their real name in order to be able to link with other users. This move has been criticised for putting identities at risk.

The amount of time people spend on social networking sites and the Internet in general can be socially isolating. Internet addiction can lead to loneliness and depression but also child safety issues, especially in relation to online bullying (or cyber bullying). Violence and sexual abuse are contextual problems in relation to the rise of Digital Social Networks (CBS News 2009). These aspects cannot be discussed in detail here; however, the wider issue of ethics and privacy, specifically in the context of specific location information in connection with the accessing and use of this data, is discussed later on in Sect. 5.9. The matters of data mining being used to collect the user and/or activity-related information will be discussed in detail in the following Sect. 5.3.

Despite such a large, and growing, catalogue of concerns, social networking sites are highly popular (Utz et al. 2011; Kaplan and Haenlein 2010). At the time of writing, Facebook has approximately 750 million users, Twitter approximately 100 million, and Google+ more than 20 million.[6] They create virtual communities that connect people living spatially distant from each other. This allows people to engage in activities and discussion from their home that they otherwise would not have had access to. Specifically, social networking sites allow users to find and add people with similar interests but also to post and receive information instantly. In addition, social networking sites give users the feeling of membership and belonging, as people are able to establish relationships (Blanchard and Markus 2002).

---

[5]Diaspora was started by a group of five college students in 2010. The development was widely supported by the New York Times (Dwyder 2010) featuring the project prominently during the extended privacy discussion around issues with Facebook's privacy changes in late 2009 and 2010 as, for example, reported by Anon (2009), Kincaid (2009), and Gilbertson (2009). In 2011 Diaspora has released a private beta, which is expected to go live in early 2012. The difference from existing platforms is that Diaspora is based on a distributed network with each user hosting his own page, providing complete control. No data will be held centrally. It is a sort of peer-to-peer network around social connections.

[6]See article on BBC online at http://www.bbc.co.uk/news/technology-14985494 and the Guardian online at http://www.guardian.co.uk/technology/2011/jul/21/google-plus-20-million-users.

5.2.1 Sharing and Shared Experiences

Central to the social network is the aspect of sharing and the creation of *shared experience*. This practice of sharing and collaboration in the widest sense is what give the platforms the name social networking. The individual data put online at the portal is being shared with a selected group of *friends*, and they are able to comment, *like*, resend or respond to and extend the content in some other form (Carroll and Romano 2011). This practice creates an entangled and interaction-like environment, where content is very fluid. It is in constant change every instant, as it is being newly linked to other content or individuals.

A number of sharing platforms link in closely with the rise of online social networking. The two groups share a number of mutual features and can be distinguished by their main focus. Where sharing platforms focus on presenting material to visitors, the social networking platforms focus on connecting with other users. Representatives of the sharing platforms are, for example, Flickr, established in 2004, and Picasa, established by Google in 2001, both being photo sharing platforms. YouTube[7] and Vimeo[8] are similar platforms focusing on the sharing of videos.

A third group of online platforms focusing on shared creation is the Wiki-type platforms. These tie in with both social networking and sharing. These platforms contain shared knowledge Web resources, where a large group of contributors can all contribute to the page content. The largest project is Wikipedia,[9] a lexicon created by the contributions of its users. In this group, the sharing of online content is more formal and structured, but it has also implemented a democratic voting structure to create a community-based decision-making process (Boldi et al. 2011; Forte and Bruckman 2008).

The focus is the connection and the linkages individuals build between themselves when forming groups and networks. This practice has recently led to two things, one of which Kelsey (2010) describes as *Personal Digital Archaeology*, concerned with the digital self that is created and reinvented on a regular basis. The other aspect is the impact these Digital Social Networks have on social sciences because of their existence as a data source (Wellman 2001). What is available to social scientists is networking data, describing in tremendous detail how individuals act and link socially (Webmoor and Neuhaus 2012). Millions of individual records could potentially be accessed, analysed and used to test previously theoretical social models of interaction and linking (Kaplan and Haenlein 2010).

---

[7] YouTube was established in 2005 and later bought by Google. It is accessible at http://youtube.com.

[8] Vimeo was established in 2004. It is available at http://vimeo.com.

[9] Wikipedia was founded by Jimmy Wales and Larry Sanger in 2001. Today Wikipedia is developing in 282 languages and has developed a number of subprojects such as Wikimedia, a repository for shared media items, and Wikimaps, a platform for shared maps. It is accessible online at http://wikipedia.org or the English version of Wikipedia at http://en.wikipedia.org.

In traditional social science research, this kind of data would have been recorded using interview techniques and questionnaires. A lot of work was required in the data collection phase involving a number of researchers to gather a data set containing a practical maximum of a few hundred people (Webmoor and Neuhaus 2012). With these new information sources, data on millions of people is available from a desktop computer. This is expected to revolutionise the social sciences in the coming years and the digital shared experiences will become more important as both a concept for users as well as researchers (see, e.g. Woolgar (2006) and Halfpenny and Procter (2010)). This of course has many implications for existing research procedures and guidelines, most dramatically those related to ethics and privacy. It will require new tools and practices that will need to be installed and learned.

Shared experiences are usually focused on specific content to establish the shared aspects. This can take on different forms, very much depending on the content as well as the medium or level of privacy. The shared content can, for example, be around an image where a comment written in response to it is created by a second party. Depending on the level of privacy, linking and tagging parts of the image, for example, are possible in Wikipedia. Images on Wikipedia can be tagged with comments and remarks, as well as links. On Flickr, areas in an image can be tagged with comments, marks and links.[10] On Facebook, tags are tightly connected to the level of privacy.[11] Users can tag themselves too, if they have the permission of other users. The image will automatically be associated with the tagged person and also appear in their personal image section. The tagged person's friends will then in turn also be able to see the image, even if they do not know the original owner of the images (depending on the privacy setting, but this is the basic idea). This is not only an outsider's comment on the image, but it references part of the image content. With this standard, an area of an image can be marked to link to a different page or a face in a photograph can be tagged with the name of the person, linking to their profile page.

The same is possible with videos on YouTube, where video responses can be posted, in addition to comments, likes or dislikes. Furthermore also text-based elements such as notes statements or status updates and similar shared elements can be created. Links on Facebook, for example, can also be commented upon and *liked*. A further element for shared experiences is spatial location. This is a relatively newly added feature, that was developed since 2006–2007, but it was only made available to the masses, e.g. integrated in popular Digital Social Networks, from 2010[12] onwards (Singel 2010).

---

[10]See also documentation on *People in Photo* option on Flickr at http://www.flickr.com/help/people/.

[11]Details on Facebook tags can be found on the help page http://www.facebook.com/help/?page=121363771279781.

[12]The feature was introduced on the Facebook blog in August; see http://www.facebook.com/blog.php?post=418175202130.

5.3 Metadata and Data Mining

The amount of data related to individuals that is held and searchable on the internet is dramatically increasing (IGS 2011). A recent project by the Sociable Media Group at MIT's Media Lab has created a project called Personas[13] to demonstrate name-related search capacities. The project asks for your name and based on this provision searches the internet for associated information, presenting the result in a neat colour bar chart augmenting the range of topics the system was able to find and associate with the search term. See results for the search *urbanTick* shown in Fig. 5.3. As the project developer Aaron Zinman put it:

> In short, Personas shows you how the Internet sees you.

The project demonstrates how information is stored, is held and is accessible throughout the Web on any subject, name or topic. This information might or might not be related to the initial search terms. Similar to this Web search, the search for data on more specific sections or platforms can deliver more specific results. For example, the database of worldwide live air traffic, Plane Finder[14] can return the flight movement in a certain area or time sequence. This process of online

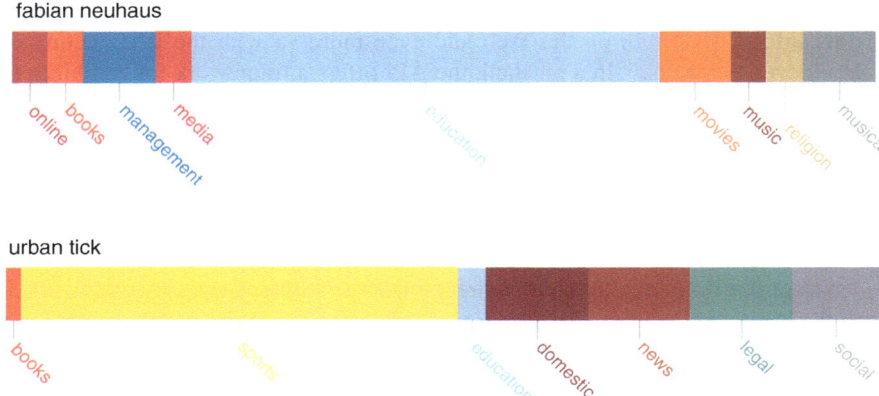

Fig. 5.3 The *Persona* Web application lets the user enter a name for which a crawler searches the World Wide Web for matches analysing the context the name is used in. Based on these broad categories of content analysis, a characterisation of the name is provided and the final result visualised as a coloured bar chart (Image generated using the web page http://personas.media.mit.edu/personasWeb.html)

---

[13] See the website at http://personas.media.mit.edu/personasWeb.html, accessed 2011-06-20 and find project details at http://personas.media.mit.edu/.

[14] The platform is accessible at http://planefinder.net/, was set up by Pinkfroot and covers commercial air traffic from across the world based on ADS-B data transmitted by aircraft and received by ground base stations.

search result, generally resulting in a specific data set that can be used further, has been termed *data mining* or also, in connection with the internet, *online data mining*. See, for example, Han et al. (2011) for a detailed discussion of data mining including techniques and methodology. The term describes the activity of digging for information as raw material for further processing.

The challenge is to make sense of the result and treat the results as a created dataset as opposed to readily available and curated. The difference is that the mining conditions are creating a very specific context which can be, in some cases, unique, meaning the dataset cannot be recreated even if the same parameters are being used. This is especially true for temporal and dynamic data.

In our case, the subject is more specific than in the context of general Web information, but the temporality and the social context present the challenges. The availability of this data has attracted researchers, specifically in the social sciences, interested in analysing and visualising the information. However, the data is not readily available just to download. Instead the data has to be collected, requested or constructed by whoever requires it. Mining can be done manually. However, it is most often automated though a script or specific software to crawl the information sources and extract the relevant data, often via the use of an Application Programming Interface (API).

Digital Social Networks provide a wealth of information in respect of individual details. The providers have themselves pushed the importance and distribution of this information both on the Web and even more so with the introduction of public developer APIs.[15] In a nutshell, the API offers a framework for machines to communicate and exchange information. Via the API requests and results can be pushed back and forth. Originally this was developed and introduced to allow third-party developers to access the central database and allow their software clients to exchange information with the central service, as, for example, to allow the integration of smartphones as a Digital Social Network interface to read and create content.

This integration has led to the combination of technologies such as photo and video as instant sources to be shared directly from a mobile device, integrating them with the Digital Social Network platform. A photo can now be posted to the platform and be shared instantly with friends and followers. The same is possible for video footage or written statements and comments. Another interesting aspect of the use of mobile platforms is the location capacity of smartphones with integrated GPS. This location can be automatically shared either as the information itself or as integrated information attached to the media of text, photo or video as part of the metadata.

Web mining is based on the information available, findable and accessible across the Internet. With the data in a Digital Social Network being rich in personal and highly individual details, the method posed more than just technical challenges. Principally it raised ethical questions regarding the gathering, management and

---

[15]See for details of the individual API documentation for Facebook on http://developers.facebook.com/docs/, for Twitter on https://dev.Twitter.com/docs, or for Flickr on http://www.flickr.com/services/api/.

5.3 Metadata and Data Mining   95

use of the data but also regarding the research setting in general, as, for example, discussed in Webmoor and Neuhaus (2012). These questions are discussed in detail in a subsequent Sect. 5.9.

5.3.1 The Location Tag

Location features lead to a new breed of Digital Social Network being created with the introduction of location information. So far only a high-level location has been included in the profile. This feature would describe, for example, the country or the city individuals set as their home location. With the introduction of GPS-enabled mobile phones and smartphones, Digital Social Networks have quickly adapted to mobile networking and learned to make use of location-based information to develop new application features. Google launched their Google Latitude[16] service at a large scale in 2009. This breed of services has developed from a pure location-sharing aspect towards a more playful interactive social networking experience (see Fig. 5.4 for a map of recorded user locations).

A group of platforms has specialised in location and is building networks around the concept of connecting people based on their location as well as their social ties. Platforms such as Gowalla, launched in 2007, or Brightkite, launched in 2007, or Foursquare, started in 2009 are completely based on the user's location, linking to local services and activities as well as *friends* nearby.

These location-based Digital Social Networks are mainly based on mobile platforms. Most services offer a free application for users to download to their mobile device through which individuals can access the service. This app, as a service client, can directly access the smartphone internal GPS chip or aGPS (assisted GPS) data. The client shares the location with a central database and receives real-time updates of what is happening nearby, whilst sending updates of activity or related information to the server for sharing with other users in the area. In addition to the mobile access, most services have a complementary Web platform, to provide an overview and updates and allow access to non-smartphone users.

The popularity of these platforms[17] has quickly attracted a large user group, which in turn has led to large investments in the companies providing the services. The success is a combination of technology, the availability of smartphones and the

---

[16]Google Latitude was launched on February 5, 2009, by Google as a successor to their earlier SMS-based service Dodgeball. See the website of Google Latitude at https://www.google.co.uk/latitude/.

[17]Foursquare The company raised $1.35 million (Frommer 2009) in its Series A and $20 million (Ante 2010) in its Series B round, and the user base rose from 60,000 in October 2009 (Wortham 2009) (7 months after launch) to 5 million in December 2010. See details on the Facebook blog About Facebook on http://aboutfoursquare.com/foursquare-hits-five-million-users/ and 8 million in March 2011. See Foursquare tweet from 2011-03-30 on http://aboutfoursquare.com/foursquare-hits-8-million-users/.

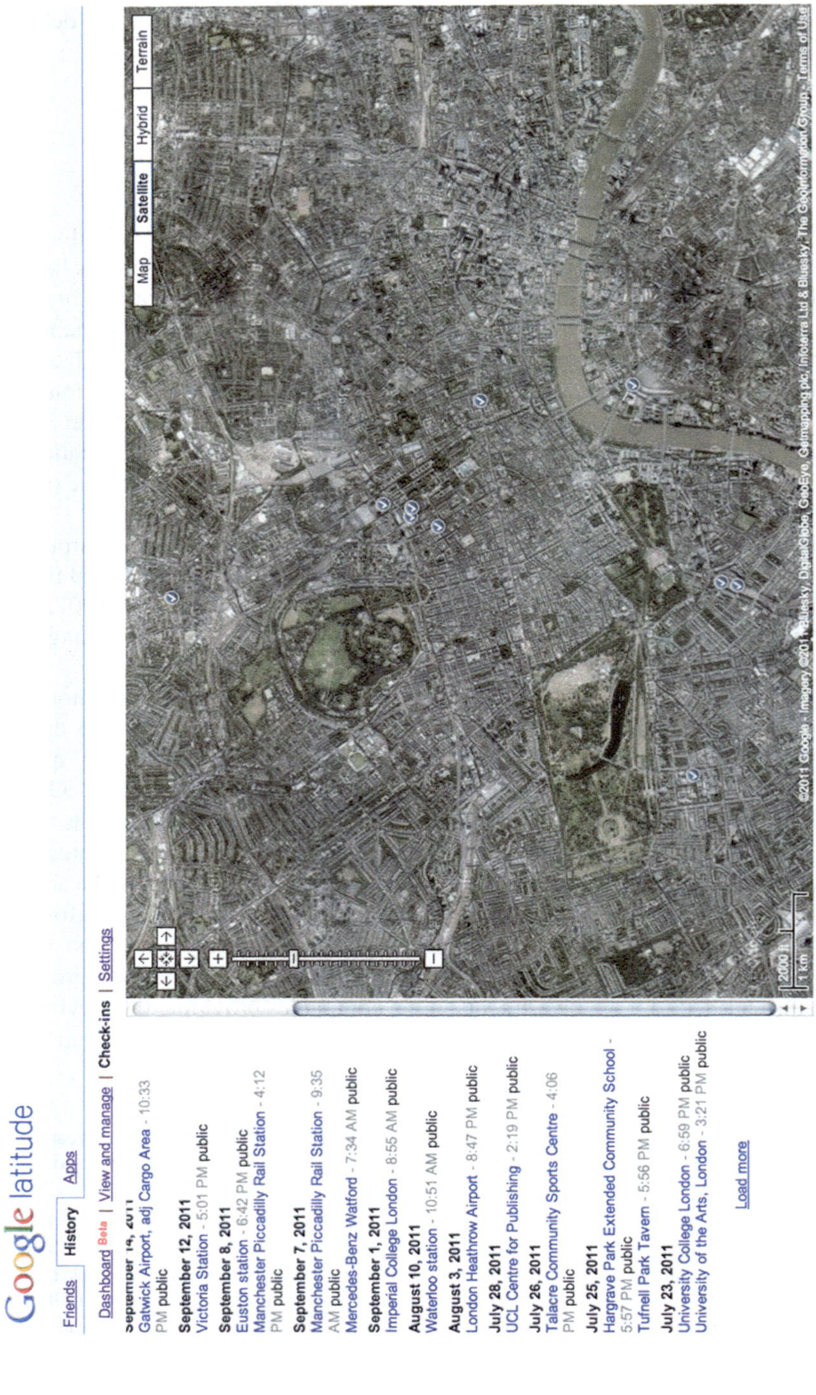

Fig. 5.4 The Google Latitude Web interface shows a summary of the most recent locations a user has checked in with a *green tick mark*. The service also attempts to analyse the recorded data providing a breakdown of activities organised by categories, e.g. time at home, time at work and time out. However, this is guesswork and, based on frequency, not actual information

users' curiosity. There are a number of reasons why one would want to join in and share the location with friends, for example, to meet up if one happens to be close by, or the user can show off as to where one is and so on, but none of these aspects are quite convincing. It is arguably more of a combination of different factors. The concept of self-identity and sharing of experience constituting the personal image, as Goffman (1959) describes it, is definitely an important factor. Adding one's location to the requirements of the individual presentation on the stage is an interesting aspect. This also requires a social classification and recognition of space and place as part of the repertoire. This may not be new, but it has become more specific.

The integration of location as a detail of Digital Social Network platforms has made the user-generated data of interest for any sort of spatial research, but specifically for urban studies. The fact that the data describes not only who does what in connection to whom but also where they are in time and space is a dramatic change in the landscape of available data sources. For the present research, this source became interesting as an extension to the detailed data collected with the GPS as part of the Urban Diary tracking. With the data mining of Digital Social Network platform, a much larger number of individuals can be observed in space and time.

The research presented here focuses on the Twitter platform due to access to a simple API, the popularity of the platform and a strong focus on mobile integration. The platform has managed to increase its popularity, the user base as well as the usage, constantly since its introduction. Similarly the integration of location data has extended quite substantially since its launch. At the start of this research project in 2010, the integration of the location information as part of the Twitter metadata was only a few months into the full service. Most platforms now support the location option, and users are getting used to a routine of sharing their current location; this is also inspired by the emerging culture and cult around location-sharing across different platforms and media.

Social networking is now not only Mr Neuberg knows Mrs Sandberg,[18] but where Mr Neuberg is located in space as he communicates with Mrs Sandberg. The same is true obviously for the location of Mr Jett, and space can act as an additional factor assessing the relationships and proximities of the three of them as well as the social and spatial context. See Fig. 5.5 on page 98.

The data is also time tagged. This allows the data to be analysed in time and space, in comparison to other user data.

5.3.2 *Examples Using Digital Social Networks Data*

The social networking industry has seen a massive push towards location data in many areas over the past 3 years. This ranges from software and application to products, mapping and visualisation but also to technology and platforms.

---

[18]See Fig. 5.1 on page 85.

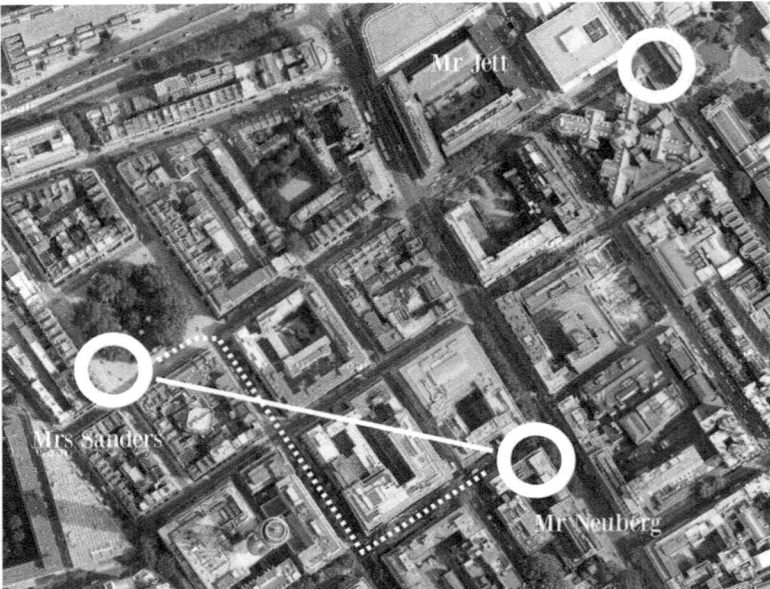

Fig. 5.5 Simple network graph showing the relationships between three individuals visualised on a map with the current location of each individual. Each node (*ring*) represents an individual and each edge (*line*) represents a connection (relationship) between the individuals

The development of devices and the usage of this feature go hand in hand, fuelling one another. An important aspect is the push towards free and open data. Only if the data is accessible can the mapping and visualising communicate its complexity, in turn shaping the desire of a large user group to share their information and location details with fellow users. The push towards open data has mainly been part of the public debate with calls to open up government databases but also calls for access to data held by private companies such as the providers of Facebook, Twitter or Google Latitude. Most of these companies now provide developer access to social networking data on various levels.

Some of the data is accessible via an API. This service is intended for third-party services and application, accessing the main data stream to provide an extended service of the core functionality. For example a software app such as TweetDeck[19] or HootSuite[20] provides client-based user access to different social networks from the same app utilising API access to different services.

---

[19]TweetDeck was developed by Iain Dodsworth, starting in 2008. The software runs across platforms on the computer and on the mobile device and lets user manage a range of different social networking accounts such as Foursquare, Facebook and Twitter.

[20]HootSuite was launched in 2008.

5.3 Metadata and Data Mining

Most of these analysis and visualisation projects using Digital Social Networks data as a data source are developed in a loose research context with very different interests and backgrounds, and sometimes with an undefined context. One of the early book publications to pick up on this trend was *Data Flow* Klanten (2008) published by Gestalten, featuring a handful of projects concerned with Digital Social Networks data.

Whilst focusing specifically on location-based visuals of Digital Social Network data, it is interesting to note how the different Digital Social Networks are used globally. There are dramatic differences between Internet access across the globe, and this is reflected in the use of Digital Social Networks. A global view is provided by the Global Web Index, mapping the activities across the world by country. Such an overview provides, for example, *Digital Social Networks the global view as of 2011* put together by Global Web Index[21] for Trendstream. It shows the global distribution of the use of Digital Social Networks including the popular platforms.[22]

One of the source platforms which implemented the location information early on, as well as API access, is the photo sharing platform Flickr. It is a popular platform allowing users to tag their photographs with a location. Flickr allows access via the API to the stored data, and researchers have started to look at locations of photographs.

Crandall et al. (2009) undertook research at Cornell University into automated image recognition and the detection of landmarks. The project[23] looked into image recognition at a large scale, using images from online photo sharing platforms such as Flickr and Picasa and trying to automatically detect the location where the photograph was taken. As a by-product of the research project *Mapping the World's Photos*, a place map was created (see Fig. 5.6 for illustration). In this particular context, around 35 million images were used, collected from Flickr via the public API. The main hypothesis of the project is *that geospatial information provides an important source of structure that can be directly integrated with visual and textual-tag content for organising global-scale photo collections.*

By using the time stamp and the geolocation, where available, the movement of the photographer can be traced. Similar to a rough GPS track, the different locations where a photo is taken can be mapped as a sequence in space and time. Crandall et al. (2009) plotted this information, and the result was a series of urban tourist movement maps. In their paper, they published two city maps, one of Manhattan, New York and one for the San Francisco Bay area. For London, the team published a map showing the main features of the UK capital, based on point data of locations

---

[21]Global Web Index (see it online at http://globalwebindex.net/) is a project by Trendstream, can be found online at http://www.trendstream.net/ a marketing company providing research data. The Global Web Index is one of the biggest Digital Social Network studies updated three times a year providing very detailed data.

[22]For more details, refer to http://globalwebindex.net/thinking/new-globalwebindex-infographic/.

[23]See details on urbanTick at http://urbantick.blogspot.com/2009/06/movement-mapping-using-flickr.html or on the project web page at http://www.cs.cornell.edu/~crandall/photomap/.

Fig. 5.6 By plotting the locations of geotagged images from the photo sharing platforms Picasa and Flickr, a rough map of London becomes visible. Especially the Thames with the bridges can be clearly seen (Image by David Crandall taken from Crandall et al. 2009)

where photographs shared on Flickr are taken. The resulting map is recognisable as London with the shape of the Thames, the bridges and parks highlighted in the data.

The plotting of locations from photo sharing platforms is a first step towards developing further analysis of the data. Eric Fischer has visualised this concept using location-tagged online shared photographs, to show how urban areas are being photographed. He developed a set of 100 world cities, i.e. a geotagged world atlas, drawing data from the photo sharing website, using the geotag field to map the images. The visualisations map out, by the click of cameras, how users capture the city and redraw its morphology with each picture (see a map of Vancouver in Fig. 5.7 original image on Flickr).[24] By accessing the API's of both Flickr and Picasa, Fischer was able to process thousands of images per location. He calls the set *The Geotaggers' World Atlas*.[25]

Fischer traces the movement of individual photographers by plotting lines between the points of photographs taken by the same user. In addition Fischer attempts to classify the mode of transport by speed, derived from the time stamps. Using the time stamp of the image, an estimated travel time is calculated and

---

[24] Accessible at http://www.flickr.com/photos/walkingsf/4622369372/sizes/l/in/set-72157623971287575/.

[25] To be found online on Fischer's Flickr page at http://www.flickr.com/photos/walkingsf/sets/72157623971287575/.

5.3 Metadata and Data Mining

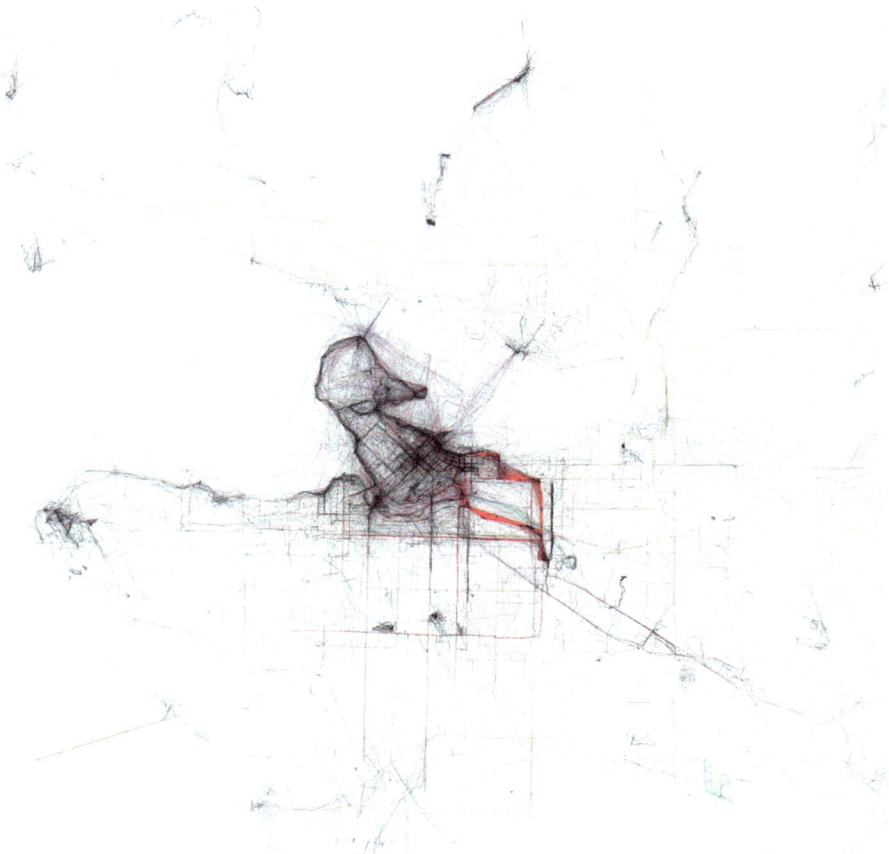

Fig. 5.7 A map by of Vancouver with the locations of geotagged images shared on Picasa and Flickr. The different colours indicate the speed of travelling between taking the photo. Essentially, this gives a rough idea of the mode of transport used (Image by Eric Fischer)

assigned to a specific mode of transport. Each mode of transport is indicated in a different colour. The different colours read as follows: black is less than 7 mph (11 km/h), red is less than 19 mph (30 km/h), blue is less than 43 mph (69 km/h)—car, and green is faster. The entire collection is rendered on top of an Open Street Map (OSM) background layer. This OSM service provides a flexible *open source* mapping context with its Wiki style data contribution setup. Everyone with a GPS device can record geographic features and contribute the spatial recording to the online map. The digital map can then be used under the *Creative Commons Attribution-ShareAlike 2.0 license* for other projects.

Processing the Flickr data, Fischer merges the individual details of each photograph with the metadata for each photographer and blends it in with the overall activity in the whole area. The result can be described as a mashup (Yu et al. 2008;

Fig. 5.8 Moscow with the locations of geotagged images shared on Picasa and Flickr. The two different colours, *blue* and *red*, indicate whether the image owner is a tourist or a local. This is determined by the profile location of the photo sharing platform user (Image by Eric Fischer)

Liu et al. 2007; Wood et al. 2007), a combination of different data sources on different layers showing in one frame. It shows details on various scales, creating an overall image despite the data not being aggregated. The vast number of details is responsible for blending the individual information.

In a second visualisation, Fischer created a set of maps of the world cities entitled *Locals and Tourists*[26]; see Fig. 5.8 original image on Flickr,[27] using the same sources. This time the focus was on who took the picture and their relationship

---

[26]See the set on Flickr http://www.flickr.com/photos/walkingsf/sets/72157624209158632/ for more details and all the images produced.

[27]Accessible at http://www.flickr.com/photos/walkingsf/4671497033/sizes/l/in/set-72157624209158632/.

5.3 Metadata and Data Mining

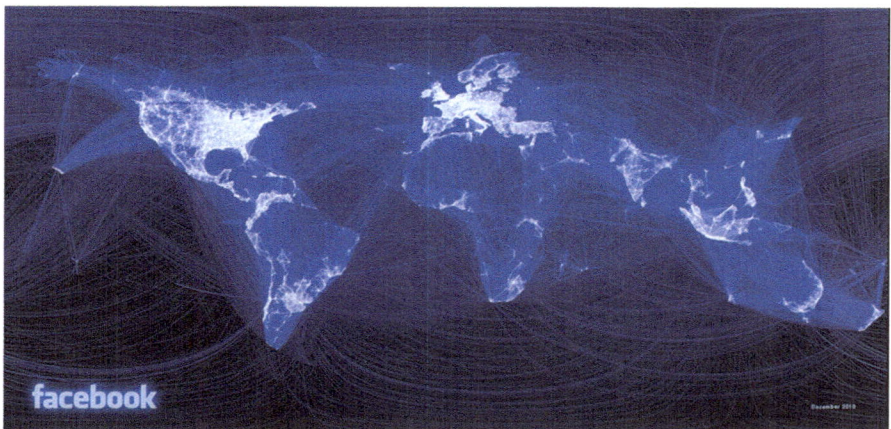

Fig. 5.9 Showing the worldwide social connections between cities. The visualisation is based on a sample of 10 million pairs of friends (Image by Paul Butler)

to the place. Fischer used three categories, local, tourist and unknown. The resulting image highlights the areas of which only tourists take snaps in red and in blue, areas where predominantly locals take pictures.[28]

One example of spatial network visualisation using Facebook data has been developed at Facebook by Paul Butler[29] using a sample of 10 million pairs of friends plotting their location on a map. It resulted in a flow map visualising the connections between the cities, whilst outlining the shapes of the continents; see Fig. 5.9.

For the rendering, a colour ramp from black to blue to white was used to indicate the intensity of connections between different cities. The resulting map shows strong connections in white, outlining a local network. The blue lines are of less intensity and span further across counties and continents. Essentially what it confirms is that location still plays a role in Digital Social Networks.

Similar in nature to the Facebook map of the word is the project *Twitterlandschaft*[30] (Twitter Landscape); see Fig. 5.10 for illustration. The project aim is to show the proximity of cities based on the connection established on Twitter. Thus, the project focuses on Twitter users and not the messages. The visualisation is based on the followers of active users and the geographic location of each user. Based on this, the proximity of cities is calculated, and connections are established. The world map is biased by the distribution of the technology and the accessibility of the service. Nevertheless, the application provides an interesting insight into how

---

[28] See the details of the project at http://urbantick.blogspot.com/2010/06/geotaggers-world-atlas.html.

[29] See detail of the project online at http://www.facebook.com/note.php?note_id=469716398919.

[30] See the project details online at http://incom.org/projekt/1666. The project was developed at the Technical University of Potsdam by Gero Fallisch and Mischa Neubauer, supervised by Till Nagel.

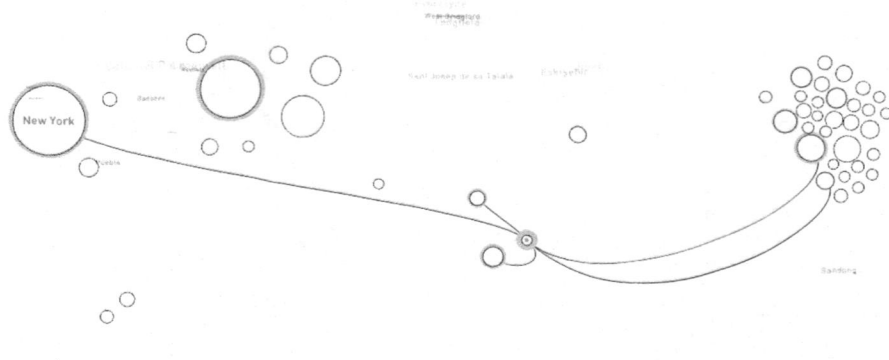

Fig. 5.10 *Twitterlandschaft* (Twitter landscape) showing the connection and proximity between actively tweeting cities (Image by Gero Fallisch and Mischa Neubauer)

close actively tweeting cities actually are and the importance of geographic distance versus celebrity status and number of followers.

Jer Thorp in 2009 at BLPRNT collected tweets based on the key phrase *Good Morning*.[31] Visualising this on a spinning globe illustrates how different parts of the world wake up and send out good morning tweets. The outcome is predictable but clarifies the point that time of day, content and location are connected and make of useful tags for mining of Digital Social Networks. For an illustration, see Fig. 5.11.

The term captures the rhythmicity of everyday live and, visualised on the rotating globe, illustrates the repetitive activity of starting a new day every day. Even though it could appear very superficial on a global level, the individual context in which the message was sent is very likely to be specific. This balance between the scales and contexts is an interesting characteristic of these new Digital Social Network datasets.

Another example is the *Just Landed*[32] visualisation also by BLPRNT (see Fig. 5.12). This time Thorp used the search term *just landed* and matched the message with the home location of the user's profile to determine the origin and destination of each connection using the Twitter API for tweets and MetaCarta[33] for lat/long information. The visualisation, developed using the software package Processing, is an intuitive visualisation of global travelling based on air traffic.

---

[31] See the original post on BLPRNT at http://blog.blprnt.com/blog/blprnt/goodmorning.

[32] See the original post on BLPRNT at http://blog.blprnt.com/blog/blprnt/just-landed-processing-Twitter-metacarta-hidden-data.

[33] MetaCarta was a company providing geographic solutions.
  The consumer facing application was acquired by Nokia in early 2010. Details on http://www.metacarta.com/index.htm.

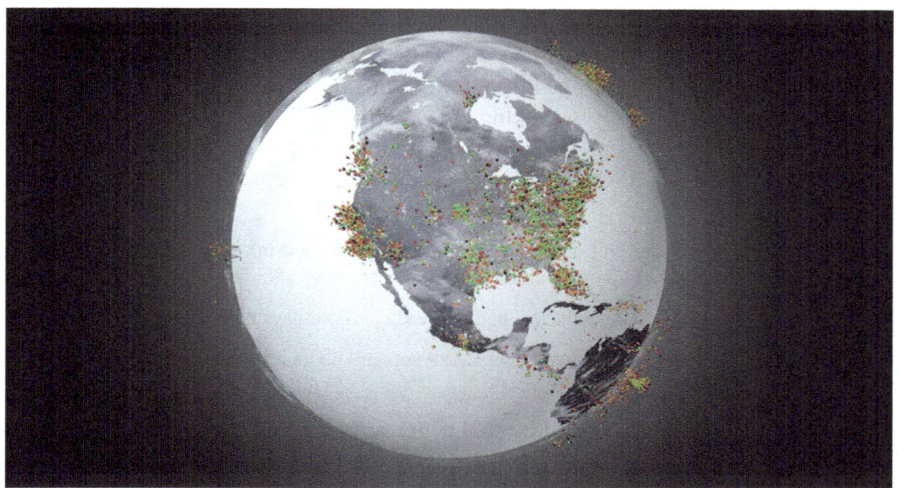

Fig. 5.11 Tweets containing the key phrase *Good Morning* visualised on a rotating globe showing how the mornings roll over the world every day (Image by Jer Thorp)

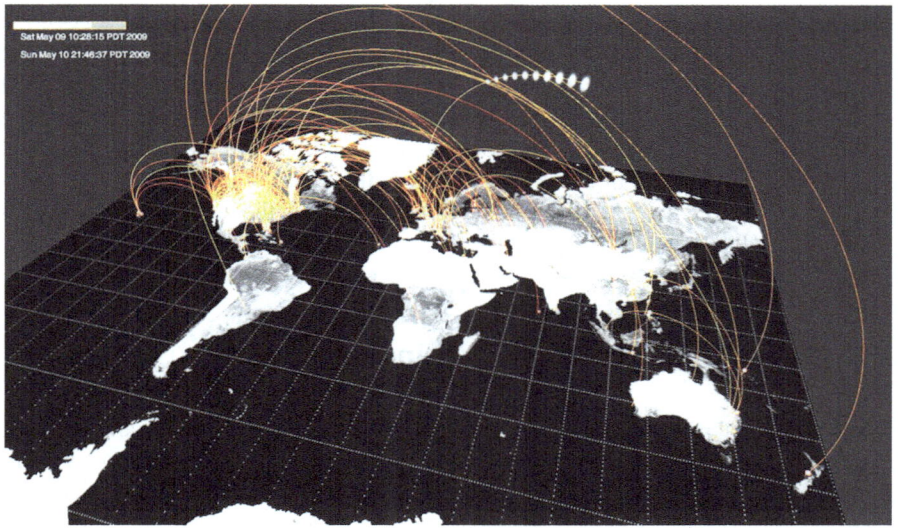

Fig. 5.12 Worldwide travel connections plotted from Twitter data using the search term *Just landed*.... The visualisation connects the destination with the profile location to show the distance travelled (Image by Jer Thorp)

Statistics[34] from 2009 show 40 % of usage in the states and 60 % in the rest of the world. Japan has a share of 36 % and the rest of the 60 % is divided almost equally by the western European countries. In this sense, there is a strong bias in this visualisation if one is looking for overall world movement. Nevertheless, this animation is an informative usage of this informal *I let the world know* tool. In a short snippet BLPRNT[35] illustrates the sort of conversations this data is sourced from:

> Queen_Btch: just landed in London heading to the pub for a drink then im of to bed... so tired who knew hooking up on an airplane would be so tiring =S
>
> jvirgin: Just landed in Maui and I feel better already... Four days here then off to vegas
>
> jchecrothers: Just landed in Dakar, Senegal... Another 9 hours n I'll be in South Africa two entire days after I left... Doodles

There are a whole series of additional wider services making use of Digital Social Network and location information. They operate much in the sense of mashups, bringing different data sets together: See, for example, Miller (2006), Wilde (2006), and Merrill (2006). Also, it is used to populate maps with time series data and increasingly with real-time data. Some Twitter-based services are, for example, NearbyTweets, TwitterMap, TrendMap, or Twittervision. They all provide a global overview, some focusing on trending discussion topics such as TrendsMap, and others, like Nearbytweets, plot tweets as they occur by location.

All these examples illustrate just how experimentally insightful the new Digital Social Networks are as a datasource for mining location-based activities and extending them into the realm of social networks. These visualisations begin to paint a picture of the complexity and the connectedness of places through both the social networks and the interaction and activities of individuals.

Using this as the background, in the following section, the details of specific uses of Twitter data as the New City Landscape concept are introduced and discussed.

5.4 From Twitter to New City Landscapes

Increasingly people use digital or online networks to communicate and interact. The urban environment is changed through the remote nature of these new communication technologies which are available on the go. Everywhere is here and here is everywhere. Users can engage in interaction en route, they leave messages, distribute news and respond to conversations. Even though the activity might not be physically linked to the current location, the activity takes place in space. As such, interactive hot spots change and fluctuate throughout the city as individuals follow the narrative path of their everyday routines.

---

[34] Twitter did not communicate its user numbers in detail back in the early days.

[35] Conversation taken from BLPRNT original blog post at http://goo.gl/4W5Ns.

Where is the city active and does it physically change over time? Urban areas are not the static artefacts that they are often described as in texts and theories. Urban areas are hot spots of human activity, which, to some extent, manifest themselves as built structures but are largely temporal and ephemeral. They are not *artefacts* existing in a constant state, but past aggregates, telling tales of memories and rumours. In an attempt to listen to these stories and narrative as they unfold through the streets, alleyways, in courts, buses, on roof terraces, etc., the social networking platform Twitter was employed. The aim is to reconstruct the city's activity hot spots as a time-frozen *New City Landscape*, drawing out the ever-changing locations of people's presence and the power of spatial creation through narratives and activity.

Twitter, here, is used and discussed as a data source to monitor location-based activity. As shown recently in a number of publications related to the activities on Twitter, for example, on the Iranian election in 2009 by the Web Ecology Project (Leavitt 2009) or the national mood by the Mood Map project (Mislove et al. 2010), Twitter is a versatile source of information. At the same time, it is a novel way of gaining access to large-scale, detailed and indeed complex data sets. It provides a unique picture of activity, connection, location and content. It is furthermore interesting as a data source for social networking and interaction between users as, for example, discussed and outlined by Huberman and Wu (2009) in 'Social Networks that Matter: Twitter Under the Microscope'.

Recently, researchers have become interested in location information on a range of scales from city, national to international level, for example, as discussed in *You Are Where You Tweet: A Content-Based Approach to Geo-Locating Twitter Users* (Cheng et al. 2010) for the detection of location-based on the tweet content or the *Semantic Twitter: Analyzing Tweets for Real-Time Event Notification* (Okazaki and Matsuo 2011) for natural hazard detection. However, the location information on a detailed scale using the actual GPS latitude/longitude is a relatively new addition to the Twitter service, introduced in early 2010. With New City Landscape, we focus solely on the location information dataset coming from the Twitter feed.

Users of this Twitter service can send any kind of information in the form of a 140-character message, known as a *tweet*, and it can contain both links and keywords. In this sense, it is a message left on a public message board. The service allows users to maintain a pool of followers with whom one shares the tweets. On the service platform, users receive a constant *live* feed of messages sent by other tweeters whom they follow. This means that, in practice, not everyone will actually receive the message the user sends. In general only people that follow or subscribe to the user's feed will have these messages included in their personal live stream. However, it is possible to collect a much larger sample of sent tweets via the open API (application programming interface) provided by Twitter. Via the API, the public stream of twitter messages is accessible. Limits are imposed by Twitter, and these are mainly related to infrastructure in terms of volume and numbers. This open API was initially provided for third-party developers to provide applications that make use of the Twitter feed, to extend the user experience and offer user flexibility. For example, it is possible to follow the personal feed on a mobile device or integrate it into a blog.

As part of the meta-information sent with each tweet, additional tokens of information are sent. This contains the name and Twitter identity of the sender, date and time but also, for example, language and platform used. It can also include the physical location, via the GPS, of the smartphone (if the user has enabled this feature). This information is part of the selection criteria used for the NCL sample. For NCL we are only working with actual geotagged tweets sent from within a certain radius around an urban centre.

From the collected data, a new landscape based on density of tweets is generated. The features of this landscape of digital activity correspond directly with the physical location of their origin, but at the same time represent with hills the locations from where a lot of messages are sent. The flanks and valleys stand for areas with lesser activity and vast plains and deserts of no tweets stretch across the townscapes. These New City Landscape maps do not represent physical features, but users' temporal interaction with physical features. The digital realm has become as much part of the urban environment as the physical features, and with these tweetography maps, it is made visible for the first time. The maps allow us to make a direct comparison between real-world activities, physical locations and digital messages. In a globalised world, this local reference develops an increased importance as a sense of place, a source of identity and memory. The digital social media data allows us to investigate many groups' social location interaction, combining the global scale with its local source.

5.5 Technical: Twitter API, Criteria and Storage

The collection of Twitter data is based on the technology developed for the Tweet-O-Meter (T-O-M). This tool was developed as part of a JISC Grant at the Centre for Advanced Spatial Analysis by Steven Gray. T-O-M was developed as a visualisation of the tweet volume in urban areas. Via the Twitter API, the T-O-M web page visualises via dials the number of Twitter messages sent per hour, calculated on a minute-by-minute basis. The search for the calculation is based on a 30 km radius around the town centre. Initially T-O-M started with the comparison of London, New York, Paris and Munich but was later updated and currently also shows San Francisco, Barcelona, Oslo, Tokyo, Toronto, Rome, Sydney, Moscow, Hong Kong, New Delhi, Shanghai and Sao Paolo. See Fig. 5.13 for illustration.

Similar to the T-O-M service, the data for the NCL maps is collected using the Twitter Search API. Twitter offers two different services through the API. One is the Streaming API and the other one is the Search API. For NCL we are using the Search API because of the built-in spatial search function it offers. With this spatial search function, the messages for a specific urban area are requested from Twitter. For the NCL maps, we have defined the spatial parameter consistently as the area within a 30 km radius around an urban centre. The search query requests from Twitter all messages which fit this criterion. Twitter offers three levels of API

Fig. 5.13 Tweet-O-Meter with a set of 16 dials showing real-time number of tweets per hour in 16 urban areas from around the world defined as a 30 km radius around the city centre. Results are calculated over a 1 min sample and continuously update at an average of 3–4 s. The web page also provides basic stats over the last hour (Website accessible at http://casa.ucl.ac.uk/tom/)

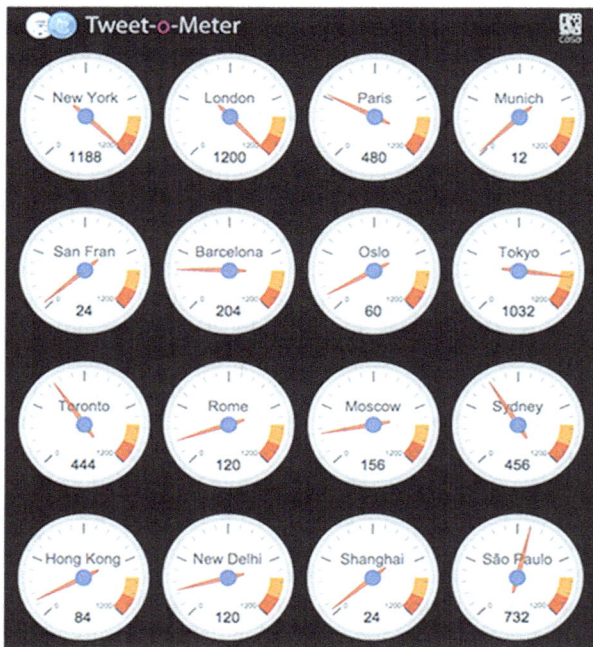

access.[36] They differ in data volume, with the *Firehose*[37] the large-one providing nearly full data access, the *Gardenhose* providing a medium access level of roughly a random sample of all public statuses with around 10 % and the *Spritzer* providing a random sample of public statuses at around 1 %.

It is the *Spritzer* we have free access to, and this is what the data collection is based upon. However, there is room for the creative sourcing of feeds. For example, parallel collections can be running, or parallel collection can be feeding into the same search. Due to IP limitations imposed by Twitter and infrastructure limitations, we are only able to run four parallel searches and collect queries at the same time. Depending on the search location, the resulting amount of data can be large, putting pressure on the infrastructure. In order not to miss out on messages, the response times of the system cannot be compromised, and therefore

---

[36] See Twitter API documentation or details on access level description https://dev.Twitter.com/docs/streaming-api/methods.

[37] For access to the full Twitter, steam companies are paying Twitter a lot of money. Ante (2009) reports that Twitter had finalised a deal with Google over $15 million and with Microsoft over $10 million for access to the Firehose service, allowing the tweets to become searchable through the two companies' search engines Google Search and Bing, respectively. Since then, Twitter has set up Gnip http://gnip.com/ as the official selling point for API-based data streams. Here also a *Halfdose* with about 50 % of the public messages and a *Decahos* offering about 10 % of all public messages.

we limited the data collection per location to 1 week, 7 days, of consecutive logging of messages sent using the Twitter service. One week provides good comparison data over a number of days but also shows the different patterns between weekdays and weekends. Furthermore, because of the IP and infrastructure limitations, we continuously have to make way for new collections. The focus was on getting a broad idea of Twitter activity in different cities from around the world. In a second stage, detailed and longer-term investigations will be done. The collected material needs to be post-processed because the location information quality is not the same for all messages. Some messages are reverse-geocoded from profile information, which generates generic place information using Google's location information. To do this, Twitter uses the profile location of the individual user and reverse-geocodes it via Google. For example, the profile location *London* will get interpreted by the service with a Latitude 51.5001518 and a Longitude −0.1262355, which is just outside the Houses of Parliament in Westminster, London, UK, on a traffic island. This is the generic spatial location for London as defined by Google. As a result, all the reverse-geocoded messages pile up in this location.

For a detailed mapping on an urban scale, this resolution of information is insufficient. As such, we are utilising the actual GPS location as latitude/longitude provided by the device in a manner similar to the Urban Diary but using data mining methods to collect traces and locations. The user can turn on location sharing and, if the device supports this option, automatically obtain the most accurate location information, as a geotag is included in the actual message meta-information. A secondary local filtering process of the raw data acquired from Twitter takes care of this. The resulting dataset contains only messages tagged with real GPS or latitude and longitude coordinates. With this information, accuracy is within the normal range of GPS accuracy of some 5–15 m. The data set is point based: However, taking the time stamp logged with each message into account gives information about flow data. The very basic characteristics are user, content, time and space information, who said what, when and where, as a multidimensional matrix.

5.6 Data: Numbers, Sample, Implications and a Comparison

The amount of data collected varies dramatically between the different locations. There are the Twitter-loving cities such as New York and London with more than 600,000 location-based messages sent over the course of 1 week. On the other hand, there are a lot of places, especially non-western countries with far less activity, where the number of geotagged tweets can be down to a few hundred. Additionally, the total location-based tweets and the actual GPS-tagged messages diverge. Furthermore, it is not the case that more messages sent results in more GPS-tagged messages. It can well be that an active place turns out very few latitude/longitude stamped tweets. This is the case, for example, in Sydney, Jakarta and Sao Paulo, where the percentage of geo-tweets is below 1 % of all location-based messages. However, in general the ratio of geotagged messages is about one

5.6 Data: Numbers, Sample, Implications and a Comparison

in ten (Smith and Rainie 2010). There might be a number of reasons for this. The messages may not be sent from a GPS-enabled device. This in turn might be based on the high cost of a smartphone and the fact that high data usage charges apply in some countries. Whereas data allowance in the UK is included in the monthly fee, providers in other countries charge per kilobyte or other unit. Such arrangements will lead customers to worry more about how they use the service, resulting in lower active online users.

Twitter is a relatively new service, having been around some 4 years. It started in 2006 with the idea of a group SMS service. The number of users is continuously growing dramatically. In early 2010 (Website-Monitoring 2010) Twitter registered 106 million registered users, with a growth rate of 300,000 a day. Per day, the site gets more than 3 billion requests. One year later, early 2011 (Chang 2011), Twitter counted nearly 200 million accounts, sending about 110 million messages a day. This fact puts some constraints on the comparability of the data samples. Also, the short-term usage of the service is loosely connected to large media events and numbers frequently fluctuate considerably over the day (Fig. 5.14).

In a recent study, by Pew Research, Smith and Rainie (2010) looked at the Twitter user group in the USA in more detail. The key findings or characteristics were: more women than men use the service, African-American and Latino internet users are more than twice as likely to use Twitter as are white internet users, and Twitter is about twice as likely to be used in urban areas as in rural areas. These characteristics are based on data collected in October and November 2010. This study also found that about 10 % of the tweets are geolocated. Based on the percentage of geo-tweets collected as part of NCL, New York and Amsterdam are by far the most accurate data set with around 50 % of the collected messages being geotagged. In other words, New York has a lot of mobile device-based Twitter activity, compared

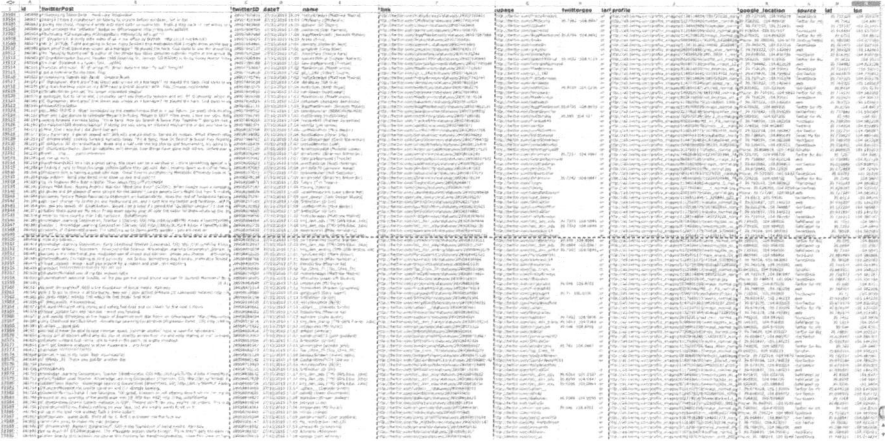

Fig. 5.14 Screenshot taken from one of the data tables showing the amount of metadata detail that is collected per individual tweet and the number of lines, here lines 120,000–120,040, showing the quantity of data collected

to London, Paris or Beijing, with only between 20 and 30 % of lat/long based messages. Since the mappable messages are a subset of the location-based filtering process, it is difficult to characterise the sample of users. For a start, it is difficult to characterise the average Twitter user, and the subset of users using location-based services is almost impossible to characterise or reference. Whether the fact that women lead in the use of Digital Social Networks still applies to this location-based data set is not clear; however, in some of the cities, the spatial pattern shows that lower-income areas are very much present on the map. This concurs with the Pew research finding that lower-income households use Twitter more than medium-income households. What we can say is how active the users are. As an average each tweeter sends approximately 11 geo-referenced messages over the period of 1 week. We can visualise the movement pattern over the same period. This will be discussed in more detail in the section on temporal aspects.

To get the location information included with the message, the user has to enable the service in the account settings and furthermore send the message from a location-aware device. This additional information can be understood as personal, and this leads us to a wider discussion on ethics and aspects of privacy in Sect. 5.9.

A total of over 100 cities have been monitored. To look at the details, it is helpful to break the data down to a manageable number of specific locations. From the data collected, a sample of 21 cities has been selected for the analysis. The selected cities, alphabetically, are: Amsterdam, Bangkok, Barcelona, Bogota, Buenos Aires, Calgary, Den Haag, Denver, Dubai, Hong Kong, Lagos, London, Mexico City, Moscow, Mumbai, New York, Paris, San Francisco, Seoul, Singapore and Tokyo. The selection is based on location as well as quality of the data and population size: All cities have populations over 1,000,000 inhabitants. The data quality varies considerably between the different locations, and the selected sample contains only locations that are comparable in terms of the numbers, both overall messages and geolocated messages. Figure 5.15 shows a world map with the selected locations.

Each location was monitored for a period of at least 1 week, including the weekend. This temporal unit provides a framework for comparison in the sense of a structural pattern, since the week is a common time interval guide for activities. The data is recorded at UTC world time but for the purposes of analysis is transferred to local time, representing the local specialities. Nevertheless, it is interesting to look at the comparison in UTC since this shows the time shift between the locations and how the peak times for activity shift around the globe. They are, in effect, passed on from city to city as the day progresses in sync with the rotation of the world.

The monitored area in each city is defined as within a 30 km radius around the city centre (Fig. 5.16). This condition is the same for all cities. The central location is chosen based on structure, distance and contextual conditions. This means that there might be a shift if the geographic conditions have strongly shaped the city and the search area would be empty to a high degree. For example, if there is an uninhabited mountain range on one side, or the city has grown linear along a coast, the centre is moved to favour inhabited areas. This has been done, for example, for Barcelona, Lagos or Mumbai.

5.6 Data: Numbers, Sample, Implications and a Comparison

Fig. 5.15 Map showing the 24 sample cities used for the NCL analysis from around the world

Fig. 5.16 The image shows the 30 km radius of the collection area, in this case around the urban centre of Istanbul, as a *red circle*

Thirty kilometres is an arbitrary distance chosen to capture the urban structure as precisely as possible. Observations (Neuhaus et al. 2006, p. 134) have shown that 1 h walk will allow an individual to reach an area within about 5 km.[38] Beyond this one can expect to find a different identity and morphology of the urban fabric. It is expected that at such a distance, no one single centre will fill the whole area, but at the distance of between 30 and 60 km, the collected data will include a number of subcentres relevant for local activities.

The data set contains a total of 3,356,520 geolocated messages sent from all sample locations together. These messages were sent by a total number of 212,972 unique users. In comparison, the overall number of messages sent via Twitter per day is about 200 million.[39] It can be expected[40] that about 5 % of them are georeferenced (Fig. 5.17): This would give a total of 10 million geolocated messages sent globally per day.

The comparison between the number of messages and the number of unique users shows, across all cities in the sample, a relationship of about 10 geolocated messages

---

[38] The historic centre of Paris covers an area of 10 km diameter from one side of the wall to the other. This is an equivalent of 1 h walk from the centre to reach the gate at the wall to leave the city. In today's London, 1 h walk will bring you from Tufnell Park in the North to the British Museum in the heart of London, covering about four different identities as demonstrated by Neuhaus et al. (2006).

[39] See post on the Twitter blog from 30 June 2011 at http://blog.Twitter.com/2011/06/200-million-tweets-per-day.html or also the TechCrunch post the same day on http://techcrunch.com/2011/06/30/Twitter-3200-million-tweets/.

[40] Estimates for geolocated messages vary Neuhaus (2011) a lot and very much depend on the location and the local tweet culture.

5.7 Kernel Density, Contours and Colouring

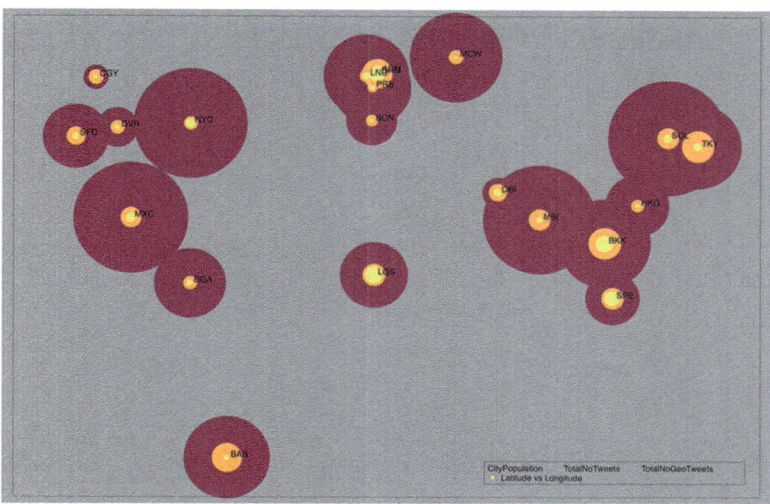

Fig. 5.17 The illustration shows the data collected by location on an abstract world map. The *dot* size represents the city's overall population in *purple*, the overall tweets recorded in *orange* and the geolocated tweets in the sample in *yellow*

per users per week. See Fig. 5.18 for an illustration of the graph data. All cities share similar relationships between unique users and geolocated tweets. There are on average between 8 and 12 messages per user. There is however an overall trend between total number of tweets and geolocated tweets: Groups of cities move closer together. The size of the circle indicates the total number of tweets recorded. The bottom graph shows the total number of tweets collected versus the total number of geographically located messages. Here the cities show a lot more differences. More messages do not mean more geolocated messages. The dot size indicates the population size of the urban area.

This is the average, but activity varies, with a lot of users sending only very few messages and a few *power users* sending higher numbers of tweets.

On the other hand, there seems to be no data-driven correlation between overall tweets and geolocated tweets. The percentage of geolocated messages in each city shows a unique pattern for each location and seems neither to be influenced by overall number of users nor messages.

5.7 Kernel Density, Contours and Colouring

The point cloud of Twitter messages is drawn from the database and mapped using a Mercator projection. This universal projection allows for recognition and readability of urban areas located around the globe. For the mapping, the individual points are being aggregated as a density surface. A *density* is an amount of *something* per

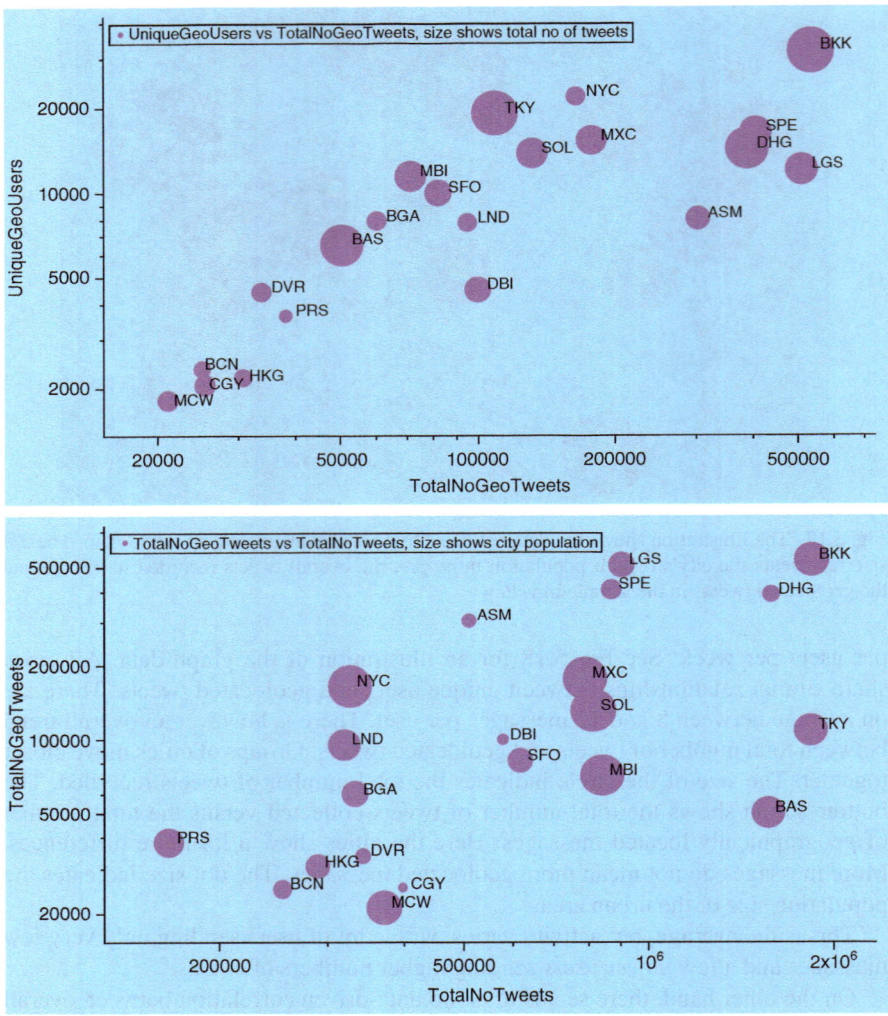

Fig. 5.18 The two graphs show the detailed comparison between the 21 locations. In the *top graph*, the number of geographically located Tweets (x) and geographically located users (y) is compared

unit area (Quantdec 2003). For an interpolated density, the *kernel density* is used to achieve a smooth continuous surface. This will calculate a magnitude per unit area from point or polyline features using a kernel function to fit a smoothly tapered surface to each point or polyline. Using a kernel density, the points are transformed into a raster map based on a density value. The Kernel Density Estimate (KDE) for the transformation is 0.02 km. This gives a resolution ensuring that medium- to small-range differences in the urban fabric are picked up and represented on the map. There are a total of nine threshold steps used to visualise the different densities.

5.7 Kernel Density, Contours and Colouring

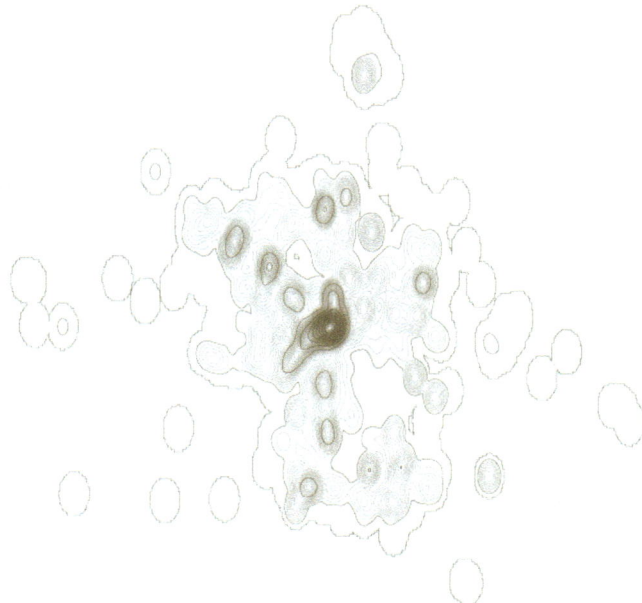

Fig. 5.19 Mapping out the contour lines for one of the urban areas as generated from the individual message densities. This illustration is mapping out Calgary in Canada, showing activity on the Twitter platform as a density landscape for the period of 1 week in May 2011

In a next step, the density surface is transferred into contour lines corresponding to the intervals (see Fig. 5.19 for illustration), resulting in a contour map of the surface. Here, however, each of the previous steps is interpreted in 12 steps. This ensures a detailed modelling of the landscape and enables a good qualitative reading of the three-dimensionally interpreted landscape, similar to topography maps as discussed by Kraak and Ormeling (2010, p. 105). This map communicates the active locations via a 3D landscape with high and low points, where high means a lot of tweets and low means very few tweets sent. Here the mountains rise over active locations and cliffs drop down in to calm valleys, flowing out to tweet deserts (see Fig. 5.20 for an illustration of the final map).

The design of the maps is borrowed from classic hypsometric landscape maps, using a sate colour scheme from green to brown to light brown with a light blue base. The colour scheme is modelled to imply the rising of mountains and dropping of cliffs, ranging from a blue green, to dark green to brown and finally white at the very top of the hills and peaks. The different colour steps have been borrowed from a scheme developed by Rudolf Leuzinger (1826–1896) for the *Carte physique et géographique de la France* published in 1880 (Jenny and Raeber 2008). The colouring corresponds to the nine intervals of the kerned density calculation, and each step up represents a reduction of 10 % in the message group (Neuhaus 2011).

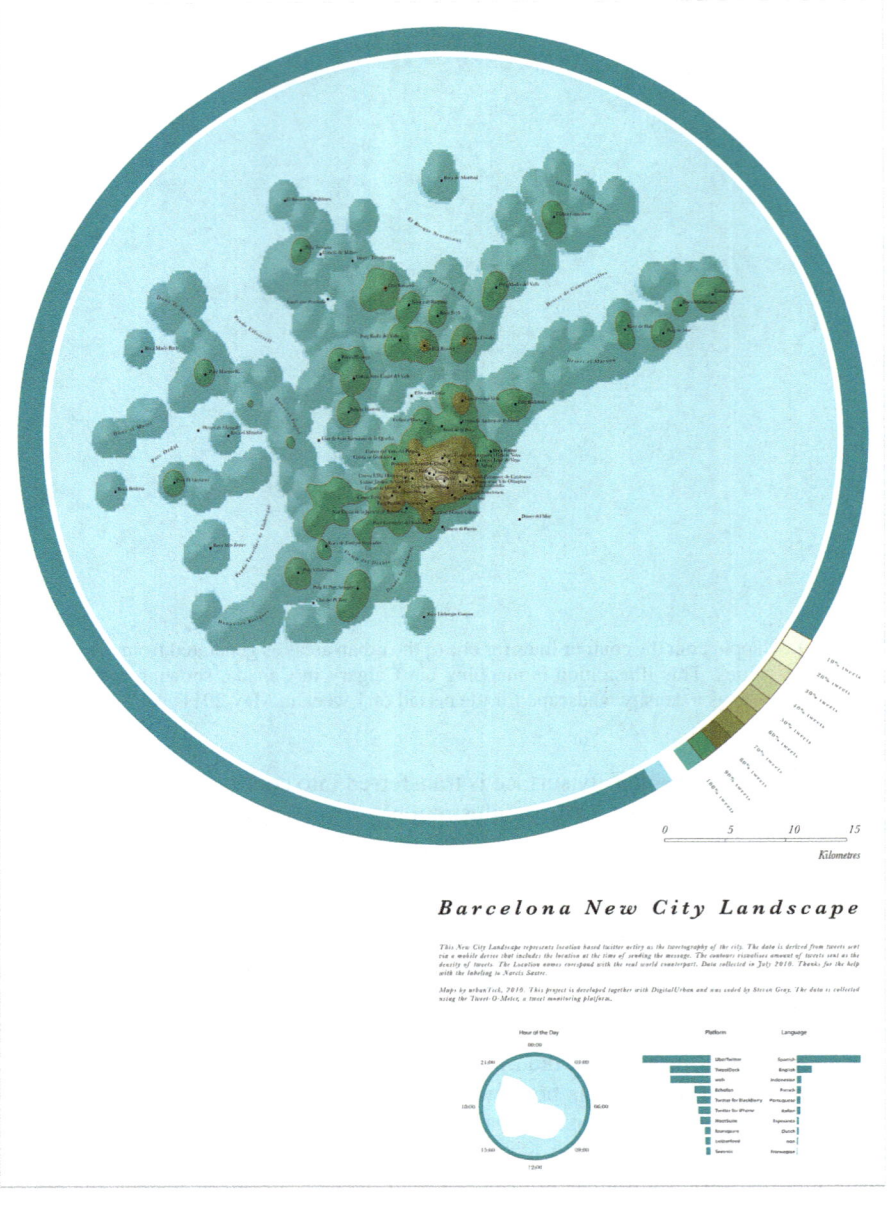

Fig. 5.20 Barcelona New City Landscape Map (See http://urbantick.blogspot.co.uk/2010/11/new-city-landscape-barcelona.html for an online version)

Throughout, the emerging landscape features have been renamed to reflect these conditions. The new names are fabricated using the real-world names in combination with a landscape description of the virtual surface overlaid. This could be *mountain* or *eak* for high points, *slope* or *valley* for descending features or *desert* and *meadow* for average and consistent areas. Inactive areas are termed, for example, *quarry* or *ditch*. Together with the familiar real-world element, the locations become tangible and memorable points of orientation and maybe identification.

This secondary link via the description of place through the name establishes the connection between the virtual location and the real-world place. This works for both the familiar observer as well as the novice, new to the place. As such it puts a stronger emphasis on how we talk about place and how it is picked up in everyday language. There are very often many more associations with a name than there is space to represent on the map. In this sense, the labelling plays a central role for the understanding of the maps (Neuhaus 2011).

5.8 Networks

Taking the aspects of *Network* from the rather simple example containing only three individuals (nodes), earlier illustrated in Fig. 5.1 on page 85, to a more complex level allows us to demonstrate the power of network graphs. Network graphs are capable of handling large numbers of elements broken down to simple two-type arrangements. There is either a connection or there is none. One of the origins of this research area is not as such a theory but a play and a film *Six Degrees of Separation* by Guare (1992). These publications popularised the small-world idea of social psychologists Stanley (1967) and influenced the concept of social connections as presented, for example, in Knoke and Yang (2008). There was also the *Kevin Bacon Game*, where the connections between movie stars appearing in the same film were discussed amongst film enthusiasts in terms of degrees of separation.

In the past 50 years, the use of the term *social network* has increased dramatically, especially in the social sciences (see Knoke and Yang 2008, p. 1). The network concept was simultaneously applied and developed in a number of fields including social sciences but also mathematics and physics (Watts and Strogatz 1998; Barabási 2002; Watts 2003).

In the social sciences, the methods and theories were mainly developed by Jacob Moreno and Kurt Lewin after studying the *Gestalt Theories* (where Gestalt means form), a subfield of psychology that looks specifically at the interplay between perception and the larger structure of the human mind (Prell 2012). The *sociogram* (an early graph representation), presented by Moreno in 1933 (Wasserman and Faust 1994) and the *matrices* (a table matrix), mainly promoted by Forsyth and Katz as more *objective* compared to the sociograms in 1946, were developed to represent social relations. Moreno's work on networks after founding the journal *Sociometry* in the 1940s was then summarised in his 1953 book, co-authored by Jennings: *Who Shall Survive*.

Even though developed in different fields, the achievements of the different researchers from other disciplines are often contested or not recognised depending on the perspective. For example, Barabási (2002, p. 52) claims that the paper by Watts and Strogatz (1998) presented the first challenge to the assumption that real networks are fundamentally random. Social scientists contest this on the ground that the social sciences never really assumed any randomness in the configuration of social networks (Scott 2011).

In *Social Network Analysis* (Knoke and Yang 2008), three different underlying assumptions are identified for the analysis of social networks.

> First, structural relations are often more important for understanding observed behaviours than are such attributes as age, gender, values, and ideology. ...Second, social networks affect perceptions, beliefs, and actions through a variety of structural mechanisms that are socially constructed by relations amongst entities. ...A third underlying assumption of network analysis is that structural relations should be viewed as dynamic processes. This principle recognises that networks are not static structures, but are continually changing through interactions amongst their constituent people, groups, or organisations. (Knoke and Yang 2008, p. 25).

Knoke and Yang (2008) discusses, in *Network Analysis*, different concepts for the definition of a network sample. They propose the topics of *realist and normalist strategy*, *positional strategies* and *relational strategies* to define the boundaries of a network very practically from an identified dataset. In addition to this, by using the Twitter data, the present text proposes to add an additional category of *spatial strategies* to identify the social network's data set boundaries.

In the development of theory, and especially the analysis of networks, advances in technology and software have played an important role. Today, powerful software packages like Gephy, Cytograph, Pajec, UCINET or Siena are available at little or no cost and can handle a range of data and formats whilst offering an increasing palette of built-in analysis options. Essentially, off-the-shelf social network data management, analysis and visualisation is available. This enables researchers from very different fields to use those methods in their specific tasks, applying social network analysis methods.

Nowadays, with the availability of large data sources such as the Digital Social Network platforms, these practices are becoming more and more interesting. The integration of the spatial context is the most relevant development: Beyond the who and why, the where is becoming an important attribute in the mix (see for example illustration Fig. 5.5 on page 98). Here it is used as the defining element of the dataset by creating networks from data collected at a specific location with clear geographical boundaries.

The social networks from the Twitter data used in this study are based on the interaction between the different users. They are not based on the followers and following option. Instead a link between users is established if they interact with a retweet (RT:) or an at (@) tweet. This creates a network reflecting the activity and the flow of information in the respective dataset (see Fig. 5.21 for an illustration of a Twitter-based network).

Fig. 5.21 A network graph showing the interactions on the Twitter platform as connections between users located in the city of Doha, capital of Qatar. The representation is nonspatial, based solely on the intensity of connections between users of the Twitter platform. However, the sample is defined spatially as that it only contains tweets sent from within a 30 km radius around the city centre of Doha. Important nodes are represented as larger, as are more active links

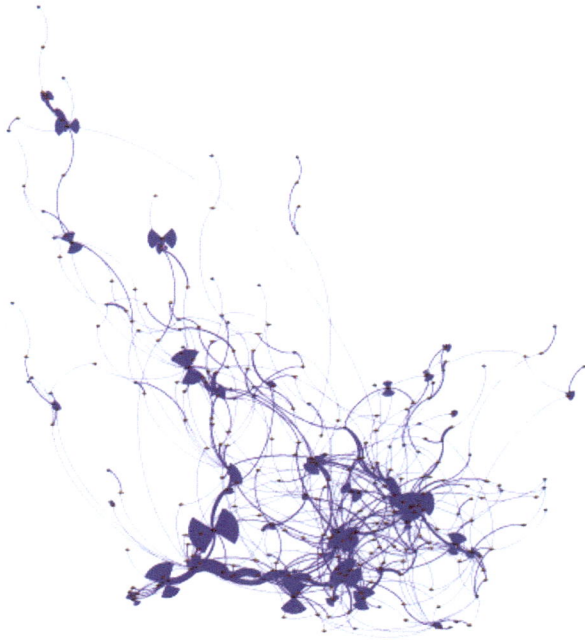

5.9 Ethics: Privacy, Ownership, Agreement and Protection

Issues and concerns surrounding privacy and ethics have been raised continuously, around the data mining projects being developed in recent years to visualise the rising quantity of freely available data on the internet. The discussion about best practice and privacy protection in this emerging context is only just beginning and has only started to formulate guidelines and regulations. For the time being, the provision of privacy for individuals being represented mainly relies on the individual data operator. The questions arise over how far the users of online services agree to *their data* being used for further research or analysis. In this case, the data used, i.e. geolocation data, is being unknowingly generated whilst online. The data collection runs remotely through the internet without any direct consent from the *user*. The user does not know his or her tweets are being used for anything other than to be read by his or her followers. However, they have agreed to the Twitter terms and conditions, stating that every message is sent publicly and not private unless specifically indicated.

With each NCL map, we are working with around 150,000 Twitter messages sent by about 45,000 individual Twitter users over the period of 1 week. The data is collected through the public Twitter API, provided as an additional service by Twitter. Using the API, Twitter packages the outgoing data stream of tweets for third-party developers of Twitter applications. The issue in the case of Twitter, and likely with other similar services, lies in the perception of private and public. With Twitter the user can set up a personal profile and start sending 140-character messages. These messages are generally undirected statements that are sent out to the world using the Twitter platform (see illustration Fig. 5.22 for a screenshot of the Twitter interface as of 2011). To get other people's messages delivered onto the personal Twitter account page, one has to start *following* other users. This needs to happen in order for other users to see one's messages, and they then can start *following*. Each user can manage the list of followers manually. However, whilst this setting creates a sense of closed community, and could and probably does lead one to believe the information or data sent using this platform can only be read and accessed by the circle of followers (e.g. friends), this is not the case. Every Twitter message sent, unless deliberately sent as a private message, is public.

For example, late 2010 saw the first prosecution of a Twitter user (Booth 2010), for tweeting a joke to his friend:

> Robin Hood airport is closed. You've got a week and a bit to get your shit together, otherwise I'm blowing the airport sky high!!

The Twitter user was planning to fly out to meet his friend, but the airport was closed due to heavy snow. How this message got him into trouble is not quite clear. The news article only states that a member of airport staff had by chance found the message using his home computer.

This example shows that the tweets can be picked up by anybody via different channels, but most likely through the Twitter API. Twitter also has a privacy page[41] where they attempt to explain the company's privacy guidelines and considerations. It states:

> We collect and use your information to provide our Services and improve them over time (Twitter 2012).

Twitter clearly states that the concept of the service is to publicly distribute messages. It further states that the default setting is set to public with the option to make it more private. This is not true, however, for the location information. In this case, the user has to activate this feature if one chooses to include this information. In this sense, every user whose location information is mapped on the NCL maps has chosen to share this information with the world. Nevertheless, there is an option to opt out of this and delete the location information of all messages sent in the past:

> You may delete all location information from your past tweets. This may take up to 30 minutes (Twitter 2012).

Twitter makes it, not perfectly but, clear what the implications are for using the service:

> What you say on Twitter may be viewed all around the world instantly (Twitter 2012).

---

[41] The Twitter Privacy Page can be accessed online at http://twitter.com/privacy. Previous versions of the Twitter Privacy Policy can be accessed through http://twitter.com/privacy/previous.

Fig. 5.22 Screenshot taken from Twitter web page, showing the continuous feed and closed interaction space presented to the individual user. The *left column* contains the feed of messages coming in continuously. It can be noted that all 13 messages shown have been sent within the last minute. This gives an idea of how quickly the content can change on the platform from an individual user perspective

Ethical considerations and privacy protection are an important element of the work ethos applied with the NCL project, and we are aware of the implications. One of the most important is that we take responsibility for the data set in the sense that we are responsible, now that we have the data, for protecting the privacy of the personal data contained in the data set. This is especially important because essentially we are creating an isolated, *frozen* data set and the information and content is to some extent removed from its original context. Because of the ephemeral nature of the Twitter service, messages flow constantly and new content buries old information: The data is lost after a relatively short period. Public logs, like the Twitter search engine, normally only date back a few weeks. In this sense, the NCL data provides a glimpse into the past, but this time discrepancy as well as contextual isolation creates a distortion, which has ethical as well as privacy implications.

With this research, we are not at all interested in spying on individuals' activity, but in focusing on the space and time aspects in the wider context of groups in an aggregated sense. New City Landscape has a number of mechanisms in place to protect the individual identities of the message authors. For one thing, the raw number of messages and individuals already generate a sort of blurring filter and the individual message drowns in the sea of information. Regarding the location-based data, the GPS accuracy is not reliable enough to actually pinpoint a specific location. As mentioned earlier, and as also discussed in the previous Chap. 4 in Sect. 4.3 on GPS technology, there is a 5–15 m uncertainty to be expected within the GPS information, even though it comes as an accurate set of latitude and longitude coordinates. The signal could have been reflected or obstructed by elements of the immediate or wider surrounding. As an additional measure, the NCL data is not published in its raw data format, e.g. a metadata list. Furthermore, all visualisations are based on processed data in mainly aggregated form, from which the individual message can no longer be identified. This is shown in Fig. 5.20, for example, with the NCL map density surface. These are crude measures, applied at the consumer end, but help to protect the privacy of the individuals generating the data used in this study.

5.10 The Collective Data Pool

The emerging question is why map this data and what does it mean? As discussed, one of the important elements is the creation of a link between the virtual and the real world. A number of different aspects of such links have been described, the most direct one being the labelling of locations. But this also includes characteristics such as infrastructures or natural features, or the time path the individual users produce. This is partly suggested in the title, a New City Landscape. Of course there is little new in the sense of addition, but there is a lot new in the sense of difference. A different perspective, a different shading, a different orientation and a different understanding of the spatial perceptions are what the research is focusing on. Take, for example, the sample and the fact that the Twitter service is predominantly used by certain income groups, and other groups are almost absent. This adds to the image created and offers—as, for example, in the case of London—a different perspective. The lower-income areas are not just inactive, for suddenly they play a role in a wider context and an image of the city. Furthermore, the urban morphology visualised on the basis of an activity morphology in reference to the physical features is a novelty but also merely a confirmation of individual daily experience. The different places do show a range of patterns, both spatially and in their temporal linking of the ephemeral virtual flows to the grounded cultural values of the place.

There are some ethical challenges involved in this type of research. This is in relation to the technical situation as well as the management and communication of data and visualisation. The discussion of these aspects is still in its infancy and needs a dramatic effort in order to establish and maintain value, reliability and trust for this emerging social science. In any case, these technologies and the resulting New Landscapes are part of a new image of the city currently emerging from a massive surge of data, generated as a by-product of regular digital activity. This is only the beginning of a new piece in the puzzle of urban identity.

There are numerous aspects that have not been discussed or even yet explored. It is vital to the emerging discussion to also include the opportunities as well as complications, because of the timeliness of the topics and the inexperience with a new field. In this sense, a few points of further investigation will be shared here. One of the largest aspects is probably the exploration of the social networks contained in these data sets. This is not only a spatial limitation (Neuhaus 2010) to these activities but also a social or in this case social network limitation. We do not know every street in the city and similarly we do not know everybody in the city. Even in the neighbourhood, we do not know everyone or maybe we have not met some people in the same block of flats. There can be many reasons, but such data sets might start opening avenues for explorations of this kind with the benefit of combining spatial and social features at a large scale. They might create the possibility of looking at patterns beyond the boundaries of established political boundaries of neighbourhoods and approaching the definition of units from the bottom up via the individual, allowing for communities to be defined in terms of spatial-social interaction.

As highlighted numerous times in the sections so far, aspects of time, temporal patterns and temporality play an important role, not only in the conventional sense of clock time but in additional meanings such as being, identity and flow. These combine memories, desire and ideas in the individual's activities as a whole, which enables a richer understanding of the urban context. Via tools such as the Time Rose diagram and multilevel analysis of pattern, these investigations have to be extended.

Two more points worth mentioning in this context are without doubt the aspects of spatial classification in relation to characteristics and features, as well as the establishment of fixed points to relate and validate the data against established datasets. However, with this, we can see a whole range of applications for these kinds of data sets, ranging from planning and design to security and sustainability. Currently the main interest is in the interface between virtual and real worlds in a theoretical but also very practical sense. Where do users tweet, when and what? Another track can be the modelling of networks, social networks in this case, in relation to location. This could lead to possible applications in the modelling of geographical spreading behaviour through visualising the network. In a very practical sense, the data can visualise infrastructure pressure, for example, in comparing the exigencies of major events with everyday performance. An alternative use might be security planning, with the potential to recognise patterns leading to trending locations versus trending topics. On a larger scale, all these topics contribute to a better performing urban area in sustainability terms.

Of notable interest is the potential link between the virtual mental map and the real-world mental map of the city. With the increased usage of digital tools and representations of spatial context, this potential for a new image of the city is rising. The early understanding and reading thereof can be key to upcoming planning questions relating to the shift in the usage of the city in the near future.

In the next two chapters, we will attempt to combine both levels of investigation. On the one hand, this is the Urban Diary discussed in Chap. 4 consisting of the tracking of individuals' everyday habits, and on the other hand, the New City Landscape discussed in this chapter. These next two chapters, Chap. 6 on *Time* and Chap. 7 on *Space*, will refer back to the methods and setups presented here but will extend aspects that are only hinted at here, examining the temporality of the data in detail. The comparative setting of the two studies will bring up similarities as well as points of difference, leading into the discussion of temporality and the routine in defining a spatial *habitus* in Chap. 8 on *Temporality*.

Bibliography

Anon (2009) Facebook faces criticism on privacy change. BBC. http://news.bbc.co.uk/1/hi/8405334.stm

Ante SE (2009) Content-search deals make Twitter profitable. BusinessWeek. http://www.businessweek.com/technology/content/dec2009/tc20091220_549879.htm

Ante SE (2010) Foursquare locates new funds to expand. Wall Str J. http://online.wsj.com/article/SB10001424052748704846004575333222375027784.html

Appleyard D, Gerson MS, Lintell M (1981) Livable streets. University of California Press, Berkeley

Asrianti T (2011) Cheap smartphones change RI internet behavior: survey. The Jacarta Post. http://www.thejakartapost.com/news/2011/05/31/cheap-smartphones-change-ri-internet-behavior-survey.html

Barabási AL (2002) Linked: the new science of networks. Perseus Publishing, Cambridge, MA

Batty M (2005) Cities and complexity: understanding cities with cellular automata, agent-based models, and fractals. MIT, Cambridge, MA

Blanchard AL, Markus ML (2002) Sense of virtual community-maintaining the experience of belonging. In: Proceedings of the 35th annual Hawaii international conference on system sciences (2002) HICSS, Big Island, pp 3566–3575

Boissevain J (1979) Network analysis: a reappraisal. Curr Anthropol 20(2):392–394

Boldi P, Bonchi F, Castillo C, Vigna S (2011) Viscous democracy for social networks. Commun. ACM 54(6):129–137. ISSN:0001-0782

Booth R (2010) Twitter joke trial man's bomb threat was 'hyperbolic banter'. http://www.guardian.co.uk/uk/2010/sep/24/twitter-joke-trial-bomb-threat

Boyd DM, Ellison NB (2008) Social network sites: definition, history, and scholarship. J Comput-Med Commun 13(1):210–230

Bruggeman J (2008) Social networks: an introduction. Routledge, London

Bryant M (2011) Frequent mobile social networking grows 67 % in a year in key European markets. The Next Web. http://thenextweb.com/socialmedia/2011/11/21/frequent-mobile-social-networking-grows-67-in-a-year-in-key-european-markets/

Carroll E, Romano J (2011) Your digital afterlife: when Facebook, Flickr and Twitter are your estate, what's your legacy? New Riders, Berkeley

CBS News (2009) Social networking: an internet addiction? CBS News. http://www.cbsnews.com/stories/2008/06/24/earlyshow/main4205009.shtml

Cellan-Jones R (2011) Google+ opens service to everyone. BBC. http://www.bbc.co.uk/news/technology-14985494

Chang O (2011) Twitter hits nearly 200M accounts, 110M tweets per day, focuses on global expansion. http://blogs.forbes.com/oliverchiang/2011/01/19/twitter-hits-nearly-200m-users-110m-tweets-per-day-focuses-on-global-expansion/

Cheng Z, Caverlee J, Lee K (2010) You are where you tweet: a content-based approach to geo-locating Twitter users. In: Proceedings of the 19th ACM international conference on information and knowledge management, Toronto, pp 759–768

Crandall DJ, Backstrom L, Huttenlocher D, Kleinberg J (2009) Mapping the world's photos. In: Proceedings of the 18th international conference on World Wide Web, Madrid, pp 761–770

Dwyder J (2010) Four nerds and a cry to arms against facebook. New York Times. http://www.nytimes.com/2010/05/12/nyregion/12about.html?ref=business

Forte A, Bruckman A (2008) Scaling consensus: increasing decentralization in Wikipedia governance. In: Proceedings of the 41st annual Hawaii international conference on system sciences, Big Island. IEEE, pp 157–157

Frommer D (2009) Foursquare raises $1.35 million, led by union square ventures. Business insider. http://www.businessinsider.com/foursquare-raises-13-million-from-union-square-ventures-2009-9

Gilbertson S (2009) Facebook privacy changes hint at a brave new, Twitter-Like, world. Wired. http://www.wired.com/epicenter/2009/03/facebook-privac/

Goffman E (1959) Presentation of self in everyday life. Doubleday Anchor Books, Garden City

Gross R, Acquisti A (2005) Information revelation and privacy in online social networks. In: Proceedings of the 2005 ACM workshop on privacy in the electronic society, WPES'05, Alexandria. ACM, New York, pp 71–80

Guare J (1992) Six degree of separation. DVD Dramatists Play Service

Halfpenny P, Procter R (2010) The e-social science research agenda. Philos Trans R Soc A Math Phys Eng Sci 368 (1925): 3761–3778

Han J, Kamber M, Pei J (2011) Data mining: concepts & techniques. The Morgan Kaufmann series in data management systems, 3rd edn. Morgan Kaufmann, San Francisco/London

Hillier B, Hanson J (1984) The social logic of space. Cambridge University Press, Cambridge

Huberman DM, Romero BA, Wu F (2009) Social networks that matter: Twitter under the microscope. First Monday 14(1):8

IGS (2011) Internet growth statistics. Online, Internet world stats. http://www.internetworldstats.com/emarketing.htm

Jenny B, Raeber S (2008) Rudolf leuzinger. http://www.reliefshading.com/cartographers/rleuzinger.html

Kaplan AM, Haenlein M (2010) Users of the world, unite! the challenges and opportunities of social media. Bus Horiz 53(1):59–68

Kelsey T (2010) Social networking spaces: from Facebook to Twitter and everything in between. Apress, New York

Kincaid J (2009) The facebook privacy fiasco begins. Techcrunch. http://techcrunch.com/2009/12/09/facebook-privacy/
Klanten R (ed) (2008) Data flow: visualising information in graphic design. Gestalten, Berlin
Knoke D, Yang S (2008) Social network analysis. Number 07-154 in Sage university papers series, 2nd edn. SAGE, Los Angeles
Kraak MJ, Ormeling F (2010) Cartography: visualization of geospatial data, 3rd edn. Prentice Hall, Harlow
Leavitt A (2009) The iranian election on Twitter: the first eighteen days. Web ecology project. http://www.webecologyproject.org/2009/06/iran-election-on-twitter/
Lefebvre H (2004) Rhythmanalysis: space, time and everyday life. Continuum, London
Liu X, Hui Y, Sun W, Liang H (2007) Towards service composition based on mashup. In: IEEE congress on services 2007, Salt Lake City. IEEE, pp 332–339
Livingstone S, Brake DR (2010) On the rapid rise of social networking sites: new findings and policy implications. Child Soc 24(1):75–83
Merrill D (2006) Mashups: the new breed of web app. IBM Web Archit Tech Libr 19(2):1–13
Miller CC (2006) A beast in the field: the Google maps mashup as GIS/2. Cartographica 41(3):187–199
Mislove A, Lehmann S, Ahn Y-Y, Onnela J-P, Rosenquist JN (2010) Pulse of the nation: U.S. mood throughout the day inferred from Twitter. http://www.ccs.neu.edu/home/amislove/twittermood/
Neuhaus F (2010) Cycles in urban environments: investigating temporal rhythms. LAP Lambert Academic Publishing, Saarbrücken
Neuhaus F (2011) New city landscape – mapping urban Twitter usage. Technoetic Arts 9(1):31–48
Neuhaus F, Mittal A, Haepp J (2006) ArKwAy – the floating city. Technical report, The Bartlett School of Architecture, UCL, London
Okazaki M, Matsuo Y (2011) Semantic twitter: analyzing tweets for real-time event notification. In: Breslin J et al (eds) Recent trends and developments in social software. Lecture notes in computer science. Springer, Berlin, pp 63–74
Papacharissi Z (2011) A networked self: identity, community and culture on social network sites. Routledge, New York
Prell C (2012) Social network analysis: history, theory and methodology. SAGE, Los Angeles/London
Scott J (2011) Social physics and social networks. In: The SAGE handbook of social network analysis. SAGE, London, pp 55–66
Singel R (2010) Facebook launches 'Check-In' service to connect people in real space. Wired. http://www.wired.com/epicenter/2010/08/watch-facebooks-location-sharing-announcement-live/
Smith A, Rainie L (2010) 8 % of online Americans use Twitter. Technical report, Pew Research Centre, Washington, DC
Stanley M (1967) The small world problem. Psychol Today 1(1):61–67
Tinbergen N (1968) On war and peace in animals and man. Science 160(3835):1411
Twitter (2012) Twitter privacy policy. Policy May 17, 2012, Twitter. http://twitter.com/privacy
UN News Centre (2009) Half of global population will live in cities by end of this year predicts UN. http://www.un.org/apps/news/story.asp?NewsID=25762
Utz S, Tanis M, Vermeulen I (2011) It is all about being popular: the effects of need for popularity on social network site use. Cyberpsychol Behav Social Netw 12(2):10
Wasserman S, Faust K (1994) Social network analysis: methods and applications. Number 8 in Structural analysis in the social sciences. Cambridge University Press, Cambridge
Watts DJ (2003) Six degrees: the science of a connected age. Heinemann, London
Watts DJ, Strogatz SH (1998) Collective dynamics of 'small-world' networks. Nature 393(6684):440–442
Webmoor T, Neuhaus F (2012) Agile ethics for massified research and visualization. Inf Commun Soc 15(Special issue, edited by A. Carusi):43–65
Website-Monitoring (2010) Twitter facts & figures (history & statistics). http://www.website-monitoring.com/blog/2010/05/04/twitter-facts-and-figures-history-statistics/

Wellman B (2001) Physical place and cyberplace: the rise of personalized networking. Int J Urban Reg Res 25(2):227–252

Wilde E (2006) Knowledge organization mashups. TIK Rep 245, 8(245):1–16

Wood J, Dykes J, Slingsby A, Clarke K (2007) Interactive visual exploration of a large spatio-temporal dataset: reflections on a geovisualization mashup. IEEE Trans Vis Comput Graph 13(6):1176–1183

Woolgar S (2006) Social shaping perspectives on e-Science and e-Social science: the case for research support. Unpublished paper, National Center for e-Social Science. Retrieved November, 13

Wortham J (2009) Face-to-Face socializing starts with a mobile post. New York Times. ISSN:0362-4331. http://www.nytimes.com/2009/10/19/technology/internet/19foursquare.html

Yu J, Benatallah B, Casati F, Daniel F (2008) Understanding mashup development. IEEE Internet Comput 12(5):44–52

Chapter 6
Structuring Time

Abstract This chapter focuses on the aspects of the temporal by bringing together both data sets from the fieldwork under the aspect of clock time but also experienced time. Both the individual and the collective as well as cultural aspects of time and time conception are discussed. Here the processing and visualisation of the data with a focus on the impact on morphology with respect to sets of flows are included. The different data sets, interviews, mental maps and tracking will be discussed in relation to the individual level. The NCL work focuses on the collective level. As a second stage, the individual and collective part are related and a synthesis is generated. The main focus will be on the relationship between activity and temporal patterns.

Overall this chapter extends the time aspects of the literature review and critical discussion of the results in the context of the presented theories. A specific focus here is on the time geography concepts and anthropology discussing the value and applicability of habitus, constraints and path, to name a few.

This chapter focuses on the aspects of the temporal by bringing together both data sets from the fieldwork under the aspect of *clock time* but also *experienced time*. Both the individual and the collective as well as cultural aspects of time and time conception are discussed. Here the processing and visualisation of the data with a focus on the impact on morphology with respect to sets of flows are included. The different data sets, interviews, mental maps and tracking will be discussed in relation to the individual level. The NCL work focuses on the collective level. As a second stage, the individual and collective part are related and a synthesis is generated. The main focus will be on the relationship between activity and temporal patterns.

Overall this chapter extends the time aspects of the literature review and critical discussion of the results in the context of the presented theories. A specific focus here is on the time geography concepts and anthropology discussing the value and applicability of *habitus*, *constraints* and *path*, to name a few.

6.1 Clock Time

This chapter examines in detail the aspects of time and the visualisation of time in the fieldwork data and results. Time is one of the two aspects looked at as part of the routine and rhythm of activities in everyday life, which form a habitus. The second aspect, space, will be discussed in the next Chap. 7.

To distinguish different aspects of time, specific terms are introduced to extend the details of time in an urban context as discussed earlier in Sect. 2.2 Time Space. The overarching term used to describe time in the context of measured units grouped together is termed as clock time. This is a term used also, for example, by Glennie and Thrift (2009) in *Shaping the Day: A History of Timekeeping in England and Wales, 1300–1800*, Zerubavel (1989) in *The Seven Day Circle* and Richards (2000) in *Mapping Time: The Calendar and its History*, describing specifically the time as counted by the clock. Similarly, Elias (1992, p. 116) uses the term *physical time* for time represented by discrete quantities.

Time is also discussed as a social phenomenon or *social time* by Elias (1992), for example. Elias explained his definition of time-space as:

> Timing thus is based on people's capacity for connecting with each other two or more different sequences of continuous changes, one of which serves as a timing standard for the other (or others). (Elias 1992, p. 72)

> ... the word *time* (emphasis in original), one might say, is a symbol of a relationship that a human group, that is, a group of beings biologically endowed with the capacity for memory and synthesis, establishes between two or more continua of changes, one of which is used by it as a frame of reference or standard of measurement for the other (or others). (Elias 1992, p. 46)

Key to the concept of time-space are two experiences: that of continuity/discontinuity and that of recurrence (Tabboni 2001).

We will move on from clock time being a social construct to time as a social element, to the section on time as it is experienced. Under the topic of rhythmic time, we will be looking at the repetition pattern of time as discussed by Lefebvre (2004) in *Rhythmanalysis: Space, Time and Everyday Life*.

The presentation will always take both studies' data, the Urban Diary and the New City Landscape, into account: aspects of the Urban Diary and the New City Landscape data will be presented together. As introduced in Chaps. 4 and 5, they represent different aspects, scales and foci of the urban environment and therefore present different viewpoints. The aim of this research is to bring those aspects together and thus overcome the common isolation and limitation of perspective in the research setting.

This is a challenging setting, which will be approached step by step. Each aspect is discussed for both studies and then summarised, before the next aspect of time is introduced. The chapter starts with aspects of clock time as the definition we are all most familiar with. This section presents the data as collected, unit by unit along a directed time axis. Individual beats of time are set against the GPS records of participants as well as the overall collection of messages in the urban areas used.

This clock time section is followed by a section on time-space and time as it is represented in society as part of social interaction. This involves interaction between individuals and groups but also interaction with the urban morphology whilst travelling in the city. In other words, there are two parts to time-space, the individual and the social aspect. In this context, travelling is seen as a physical activity in sync with a mental activity. This is based on the fact that one needs constant decision making and navigation in order to be able to move. Moving is discussed, for example, by Portugali (2004) as a cognitive act of spatial embodying, thus merging action and perception into a process of travel.

The two parts are in line with the discussion of time in sociology and philosophy, where time is often distinguished between time-space and individual time. As Tabboni (2001) points out in her paper *The Idea of Time Space in Norbert Elias*, Emile Durkheim worked extensively on this topic. Also Henri Hubert and Marcel Mauss, after Durkheim, picked up on this setting of time as both social and individual.

The discussion around time, the individual and the social is of key importance in the art movements of the last century, especially in modernist approaches. From Picasso to Hockney, time and space play a major role. The understandable dilemma created between clock time and time-space is illustrated in these artworks, for example, by Hockney's photo collages.

6.2 Urban Diary Times

Time is often interpreted as the purely mechanical ticking of a clock and the counting of seconds, minutes and hours to make up a day. In this section, the argument is made that clock time, as referring to time measured by a clock device, is only a specific area of time or time measurement. It is however an important one that has entered everyday life to the point where it is taken for granted and universal (Hughes and Trautmann 1995; Tabboni 2001; Glennie and Thrift 2009; Brand 2000).

In today's life, clock time has become so immersed and integrated; it is generally regarded as natural as *breathing* (Glennie and Thrift 2009). Often it is perceived as an additional sense, a time sense, of natural fact. However, this is debated and often opposed by scholars. Nowadays, aspects of time are everywhere, everything is on time, timepieces rule every single step we take in the city, and we actively surround ourselves with devices to measure time wherever we are. For example, most household kitchens will include some five different fixed installed units, each telling the time of day, including mobile gadgets. Of course, there are many more timepieces to be found in everyday life, with phones and computers, cars and train stations all relying on hyper-accurate timepieces synchronised by atomic clocks, without which many aspects of everyday infrastructure would not function. Examples are to be found in navigation, especially with GPS; in most communication technologies, for example, mobile phone communication; and also in financial and data transactions.

We have grown used to the timing devices to such an extent that they dominate most areas of our lives as the key reference point. These range from economic aspects (market and trade), technical aspects (electronics, mechanics and transport) and devices to social interaction and connections. The implied linearity of endless stacking beats moving in one direction has reshaped our perception of both distance and duration (Andersen and Grush 2009).

This understanding of time as clock time is, however, recent. The short history of clock time is discussed in detail by Glennie and Thrift (2009) in *Shaping the Day: A History of Timekeeping in England and Wales, 1300–1800*. The authors point out that clock time is a social convention rather than a universal given, as it is casually perceived to be in everyday life. The units of 60 s, 60 min and 24 h, 7 days and 12 months could easily be different. Nevertheless, these divisions make up a clever system, counting quite accurately[1] over longer periods of time.

It is a standard that was adopted internationally and has now been globally implemented. Since 1967, the International System of Measurements bases its unit of time, the second, on the properties of caesium atoms. SI defines the second as 9,192,631,770 cycles of that radiation which corresponds to the transition between two electron spin energy levels of the ground state of the 133Cs atom. In principle this property is based on the discovery by Galileo that a pendulum swings regularly regardless of the distance it covers. There were however timepieces long before the discovery of the pendulum's properties. Sundials and water clocks are as old as 6,000 years[2] and amongst the earliest timepieces that did not depend on the observation of celestial bodies. These timepieces were less reliable and informed than the timekeeping or horology we know today. Since Galileo's discovery around 1,600, the concept of the clock developed and gained increasing significance in everyday life, particularly in social life, as a regulator of practices. This manifests today in time discipline, enforced in institutions such as schools or workplaces but also in leisure activities, entertainment and shop opening hours.

The linear concept of time, however, is older than the practical concept of clock time especially in social and religious organisations where temporal frameworks have been established in the early days. For rituals and religious practices, linear time has played an important role for a much longer time (Zerubavel 1985; Reichholf 2008). It has defined the cultural conception of time for our society for over a few thousand years. The linear understanding of time can be traced to Christianity, with the ideas of the *Beginning*, God's creation of Adam and Eve and

---

[1] See ISO/TC154 (2004) for accuracy of time measurement.

[2] Known in Babylon and in Egypt around the sixteenth century BC, but also in China and India around 400 BC (Cowan 1958).

the *End* as the *Second Coming of Christ* or the *Last Judgement*. A one-way, linear procession dominates the scene and leaves a lasting impression in the collective memory.

Nevertheless, time is not always regarded as linear. In other cultures and religions, the conception of time is structured differently. It is, for example, cyclical in Buddhism and in Egyptian culture (Richards 2000). In Greek culture two terms for time are used, namely, the formal objective *Chronos* and a subjective time *Kairos*, representing an opportunity as gap or timeout (Rämö 1999; Miller 1992; Smith 1986, 1969). In the culture of the Hopi Indians, natives in the south west Americas, the concept of time does not cover the same three tenses, present, past and future, as discussed in Chap. 2 *Time Space*. Instead the Hopi use only two times (Whorf 1950), an objective time and a subjective time.[3] Those are manifested (object) and manifesting (subject). Figure 2.9 illustrates the concept. Objective elements are all the things accessible to the sense, being physical.

What we call future would be in the realm of the subject, in desires or in the mental state (Lynch 1972, p. 131). The Hopi culture also relates time to space in the sense that distance is part of what is manifest, as a duration. As a result simultaneity does not exist. Something far away cannot have the same time since the *distance* lies in between (Tuan 1977, p. 120). Spatially of much interest is how the two concepts meet in the distance. With such a concept far away can merge with the world of dreams.

In the following two subsections, the patterns and rhythms of activity will be presented under this clock time aspect. This means that the activity is visualised in reference to this linear system of exact unit by unit stacking. Linearity, however, is not represented as such in all visualisations. The aspects of habit and cycle will also influence the development of representation, but the progressive linearity of continuous counting is dominating throughout. Therefore the analysis of the data only works in this rather closed context; it cannot unfold any direct meaning outside the influence area of clock time.

The graph visualisation chosen in the following discussion to visualise activity over time focuses on the quantitative aspect of the data. The concept is to look at the schedule of information contained in the records. The aim here is to visualise personal routines in a spatial context over time. The graphs visualise the amount of activity over a specific time frame, analogous to the scale in spatial maps. The time frames or scales chosen are 1 day, 1 week and 1 month. Using these reference frames helps to establish an appropriate framework for the presentation of the data. Furthermore, the participants are all understood to use these time frames, which makes the translation and comparability of the data collected a straightforward process. More specific units, individual to each participant, relate to religion, culture, job or specific responsibility.

---

[3]The terms used here might appear confusing; however anthropological research used time terms in this fashion in the early twentieth century.

In the graphs shown in this Sect. 6.2, the x-axis represents time, whereas the y-axis refers to the amount of activity drawn directly from the GPS log; this is measured by the number of log points the GPS device has stored for the time period in question. The graphs do not provide information about time spent in one location, because they focus solely on travel time between destinations.

6.2.1 Activity Over Time

The data collected for the Urban Diary project in London with 20 participants serves as the pool for a first exploration of the time aspect of the GPS tracking data. The group of participants from London can be considered a subsample of a comparable size to other subsamples, such as the sample of participants recorded in Basel.

The graph illustrated in Fig. 6.1 details the recorded activity of 12 participants over a period of 1 month. Activity is in this case visualised as movement in the city, represented by the count of recorded GPS track points.

The graph shows the weekly pattern in 1 month both by day and by participant, with 4 peaks for each of the 4 weekends included. Overall, weekends tend to record

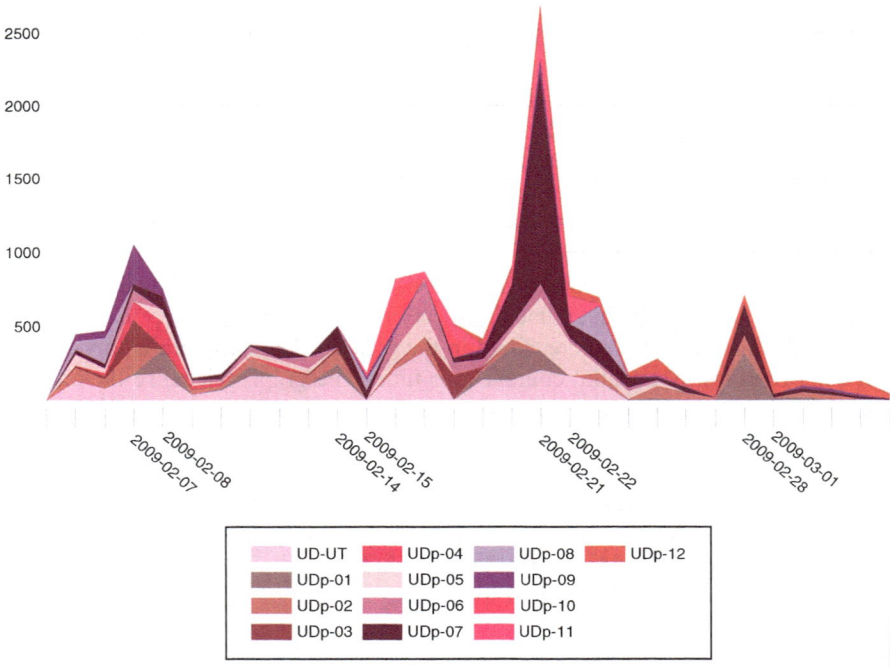

Fig. 6.1 Diary graph plotting GPS records as the number of recorded location points over the period of 1 month, showing 12 participants of the Urban Diary London study

6.2 Urban Diary Times

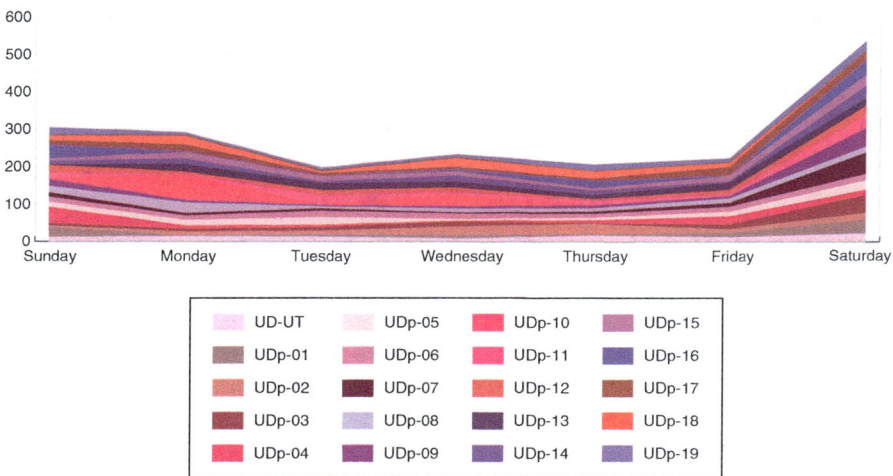

Fig. 6.2 Diary graph plotting GPS records as the number of recorded location points over a week showing 20 participants of the Urban Diary London study

more activity than weekdays. Broadly, the peaks in this example match. However, one peak is slightly moved into week 3 in the graph. This is explained by the fact that this was the UK school midterm week, a holiday break for families with schoolchildren. Participants who have children or work in a school as teachers spend more time travelling during the normal weekdays. This is down to the fact that they were less bound to specific activities dictating their daily schedule during normal working hours.

If we now look at the same activity data but on the graph in Fig. 6.2, showing activity over the 7-day week, the pattern emerges as to how active each of the days from Monday to Sunday actually is. Clearly, activities seem to accumulate on Saturdays. This shows up in particular in the week's graph. Saturday has more than double the amount of points than other days of the week. This is not only due to the one participant who does intense sports activity on Saturday, which is also visible in the graph showing the activity over 1 month in Fig. 6.1. Across the range, all participants tend to record significantly more activity on Saturdays. Other than that, the weekdays are fairly even in terms of activity, with a tendency to lower activity midweek.

Compared to the regularity of the week, as illustrated above in Fig. 6.2, the 24 h graph, in Fig. 6.3, shows an activity pattern with a number of distinct ups and downs. The graph shows two main periods, starting at midnight on the left. The first period shows little activity, followed by a second period later, from 9 a.m., of significantly increased activity. There is, as might be expected, an overall pattern. During the night there is little activity, with most of the activity concentrated during the day.

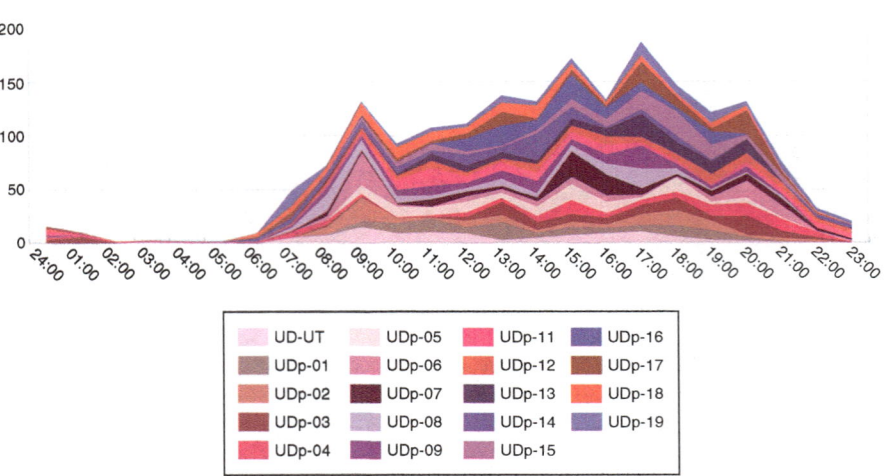

Fig. 6.3 Graph plotting GPS records as the number of recorded location points for 20 participants of the London study. All records are superimposed per participant into 24 h – 1 day

Over the early morning hours, 2–5 a.m., an expected flat portion represents little activity whilst participants are still asleep. The next day starts with a first peak during the morning rush hour around 9 a.m. After 7 a.m., participants start leaving the house, but the recorded activity takes off from 8 a.m., peaking around 9 a.m. and reducing thereafter to a first low point at around 10 a.m. Participants have reached their destination and spend time in one location, often the workplace.

By looking closely at the participants involved in these first two peaks, one can see that actually there are two groups, one generating the first *rush hour peak* and the second group creating a second, similar peak about 1 h later. This can be described as an early group and a late group. Those two are visibly distinct during the day. For the London sample there seems to be always an early and a late option for each task, and participants fall into one group or the other.

The activity during the morning, between rush hour and lunch, is recorded by a group of participants with small children, not yet of school age, who undertake morning activities. Their daily structure is adapted to the needs of young children and toddlers with morning activities, snack time and midday nap. As such these patterns, with four or five batches of activity, are distinct from the work rhythm, where usually only two batches take place: one in the morning and one in the afternoon.

The key orientation point is midday with the lunch break. In the recorded data, this is then also the second peak of the day with a high point around 1 p.m. This peak in activity, where the participants move again between places to get lunch, meeting up with friends and colleagues or going out for a walk, is an important cornerstone of the daily structure. It is referred to frequently both for personal use and in the professional context. After lunch, around 2 p.m., there is the low point of the day, with the least activity during this 24 h period apart from the early morning hours.

After the lunch break, there is a notable afternoon/evening peak. This is the result of a combination of weekend and weekday activities, such as work and outdoor sports mentioned above. Included in this peak are a first evening rush hour high point between 5 and 6 p.m. and a smaller second peak around 8 p.m., possibly pointing to visits made to the pub after work. Generally this resembles the expected daily routine pattern of a western city. More surprising is the accuracy the pattern shows, rather than any unexpected results. Although the sample is not representative, it was not expected that such regularity would be found.

6.2.2 Distance Over Time

The data recorded by the GPS units is attributed to every data point with an exact time stamp. Not only is this important for the actual GPS position, but it is also the listing characteristic ordering the data points as a string of events. The GPS tracker is programmed to record data points according to predefined settings. In this study the logging interval for track points was set to 7 s. This was chosen as a suitable trade-off between the storage capacity, the battery life and the accuracy of the recording.[4] To process the temporal data in the following, the time is set in relation to the spatial dimension. The unit is no longer focusing on the amount of activity, as in the previous examples in Sect. 6.2.1, but focusing on the distance from home at any given time. In detail, the following visualisations show the average distance from home at a certain time of day, providing a processed time-space measure for movement. The analysis in the following is taking into account the entire Urban Diary tracking sample, as described in Chap. 4 earlier.

Distance has long been a subject of much importance in geographical analysis, especially in connection to movement and migration. The origin of the *distance concept* is attributable in parts to Ravenstein (1885) who presented *The Laws of Migration* to the Royal Statistical Society in 1885. However, there is mention of two other researchers looking into the same question, one of which is Longstaff (1893) and the other is the French statistician (Loua 1885). The discussion around the importance of distance continued, and in the 1940s (Stouffer 1940) picked up on the topic of the relationship between mobility and distance. However, contrary to scholars emphasising the importance of cost – as in time and energy governing mobility – Stouffer argues that there is no direct relation between distance and mobility. Instead he proposes something called *intervening opportunities*. With this, he is saying that the distance depends on the number of opportunities one is given closer to the destination (Noulas et al. 2012).

---

[4]The frequency of location recordings affects the three aspects of spatial accuracy, battery life and storage capacity. Whilst a high level of accuracy is desirable, this will fill up the local storage as well as drain the battery quicker. The chosen trade-off of 7 s allows for a battery life of around 24 h and the internal memory to hold data for about 3 weeks worth of recordings.

This discussion on distance for both camps is mainly important in relation to modelling human behaviour. In this context however, it points out the importance of distance – although we are looking at actual travelled distance, not estimating the potential distance. By relating distance to time, we bring the time geography concept of *constraints* into the picture. In this way, the analysis extends the discussion in so far as *habitus* and *agency* are included as part of the concept of mobility.

The home location is taken as a base for each day starts anew. In this context, the distance from the base at a given time represents movement and as such draws out the pattern over time. Distances between home and work, or home and shops and so on, are used to visualise the reoccurring patterns in a temporal context. As temporal units, the hour and the day segments are being used in the context of a 24 h day.

The main urban areas visualised in this subsection are London and Basel, but the sample also includes tracks from Moscow, Plymouth and Zurich that serve as benchmarks. In this context part of the analysis will look into how the different urban areas compare and how the characteristics show similarities or differences.

6.2.3 Particular Distance

In great detail, this relationship between time and distance shows, if looked at for one participant at a time, the habitus as a rhythmic up and down or in and out. By plotting a graph over time and distance using all GPS track points, the progress and the inscribed activities are clearly distinguishable. Whilst processing all GPS track points, they are overlaid on the same 24 h frame. Even though this is a time-referencing visualisation, the spatial parameter is inscribed and mobility becomes activity.

The graph used for these distance plots shows time on the x-axis, starting at midnight and showing 24 h. On the y-axis, each GPS location is plotted by its distance in km from the home location: thus home is the base line.

The general emerging pattern is a bell-shaped curve, where around midday the greatest distance from home is reached and thereafter the distance decreases until the participant arrives back home. In the morning of a weekday, the commute as an increasing sequence of dots shows the progress that is being made whilst travelling between home and work. Once at work the distance levels out and remains the same. There might be a trip to a meeting at a different location or a trip to the sandwich bar around the corner over lunch, which can change the distance. Similar to the morning commute, the commute back home is indicated by a decreasing distance, with a dent where the participant stops to shop for dinner and finally returns back home.

Examining this in detail has shown that the story of the participants' everyday movement can be read in great detail from these graphs. Three examples will be looked at and examined in detail to demonstrate how the ghost of individual activities can be traced in the recordings through these visualisations. These cases represent on one hand the scales of time aspects in the data sets as well as space on

6.2 Urban Diary Times 141

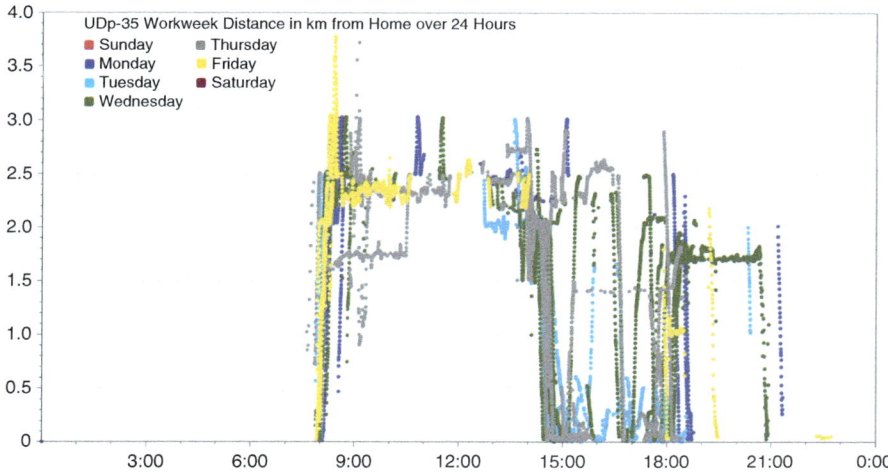

Fig. 6.4 Graph showing average distance in km from home over 24 h for participant UDp-35 on weekdays

the other. A range from the individual scale, to a group scale, to a collective scale across the urban area can subsequently be discussed. The examples are chosen out of the total sample to discuss the context of the records in detail, in order to examine the entire sample and make sense of the results. It proves important to learn more about the details, and learn about the details of activities from individual cases, to later on understand the results across the sample.

The first example – UDp-35 – is a participant living in Basel. Her personal situation can be summarised as a working mum with two children. She lives in the city of Basel and works part time. For the commute between home and work, she usually uses the bicycle or public transport. Only seldom does she commute by car. Figure 6.4 illustrates the emerging activity pattern over the working day.

Her out of home day starts around 8 a.m. when she leaves for work and returns home around 9 p.m. However, there is a complex pattern tied to the different weekdays defined by part-time work, dropping off and picking up of children at the nursery, meetings outside the office, outdoor activities with the children and educational activities in the evenings.

On workdays, three distinct locations beside the home location can be observed in the graph, which are frequently revisited and serve as the main activity hubs. The most dominant one is the workplace. A more or less horizontal line emerges between 8.30 a.m. and 2.30 p.m. at a distance of 2.4 km from home. The second location is about 2 km from home and shows as a dent in the morning commute around eight and then over and after lunch during the commute back home. This is where the participant drops off and picks up the children at the nursery. This does not show on all weekdays because of arrangements between the participant and her partner as to who drops off and picks up the children. Nevertheless, it is clearly outlined as a destination in this graph on Tuesday, Wednesday and Thursday.[5] These are the days the participant is working until after lunch and then picks up the children from nursery to take them back home. On Monday and Friday she works all day, and this is visible in the illustration through the horizontal line which continues after lunch and remains at a constant distance of 2.4 km from home. The participant still brings the children to the nursery in the morning before work, and this is shown in the graph with a dent in the commute at a distance of 2 km. The third specific location is the location for work meetings she goes to frequently, sometimes several times a day. This is related to her job as an architect where she meets with clients, builders and partners related to specific projects.

Afternoon activities in the second part, on the days with the children when she works part time, vary. According to the explanations by the participants, these depend on weather, mood and sometimes friends and family. However, the local area is popular. A trip to the local park in the street, or to meet up with other children in the same street, is indicated in the graph with activity close to the home location. There are also a number of trips to meet up other locations *in town* during these afternoon periods. These are, however, to be classified as one-offs in the recording period over 2 months, since these are not repetitive.

The weekend activities for this participant, shown in Fig. 6.5, reveal a different pattern compared to the weekdays, shown in Fig. 6.4. The activities are hardly repetitive, certainly not within the same time frame. A few locations are revisited, but not at the same time of the day. Overall the days start later on weekends, with

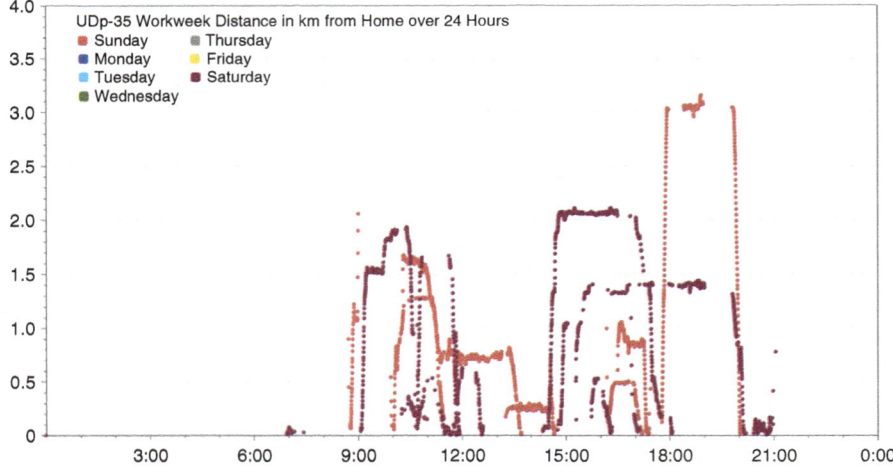

Fig. 6.5 Graph showing average distance in km from home over 24 h for participant UDp-35 on weekends

---

[5] See colour key for each day of the week in Fig. 6.4.

6.2 Urban Diary Times 143

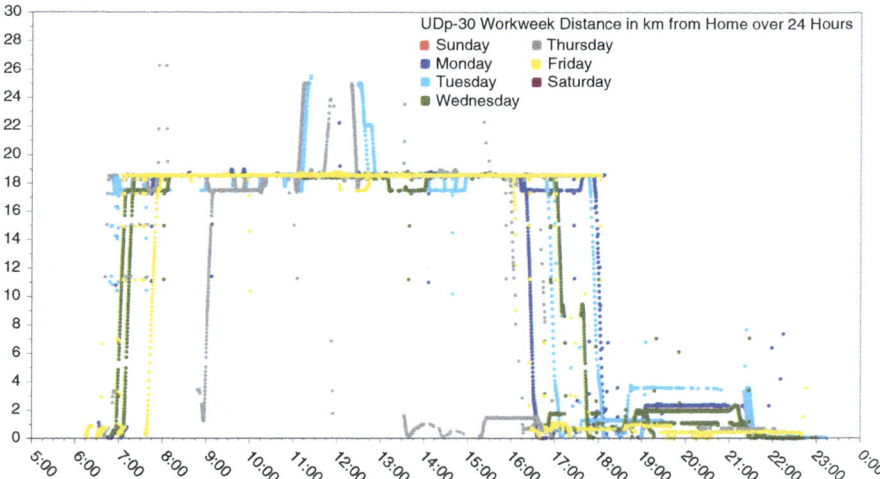

Fig. 6.6 Graph showing average distance in km from home over 24 h for participant UDp-30 on weekdays

trips starting at 9 a.m. or 10 a.m. Lunchtime is still important and on weekends in this example a family event at home. Because of this, the day is also on weekends structured in two parts, the morning and the afternoon, both with their individual activities. Because of the family commitments, days tend not to end much later on weekends either. Activities wrap up at home by 9 p.m. on Saturdays and Sundays.

Overall, the activity radius for this particular participant is well within the city boundaries and does not exceed 4 km at any time during the recording period. There is one exception for a weekend away with relatives, for which the participant travelled beyond the 30 km radius around the home location. As explained above, the more than once frequented locations too are close to half a dozen in total.

The second example – UDp-30 – is again a participant living in the city of Basel. She is single and works full time. The corresponding *home distance* graph is shown in Fig. 6.6. Her workplace is in the suburbs of Basel; therefore her commute is long compared to other participants in Basel. She usually travels by train, sometimes by car. Her home is close to the train station in Basel, but the short walk to the train station is still visible on the graph Fig. 6.6 as a small dent in the early morning hours after she leaves the house. This is followed by the train journey, indicated by a rapid increase of distance over time. The day starts shortly before 7 a.m., and she arrives at work shortly after 7 a.m. Commuting time is 30–40 min in total. The work location shows in the graph as a constant, at a distance of 18.1 km from home between 7.30 a.m. and roughly 5.30 p.m. Additional locations for this participant also show places for meetings out of the office. Lunchtime booked with sports activities occurs on Tuesdays and Thursdays between 11.30 a.m. and 1 p.m. This shows as a long trip over lunch. The time she leaves to commute home varies between 3.30 and 5.30 p.m.

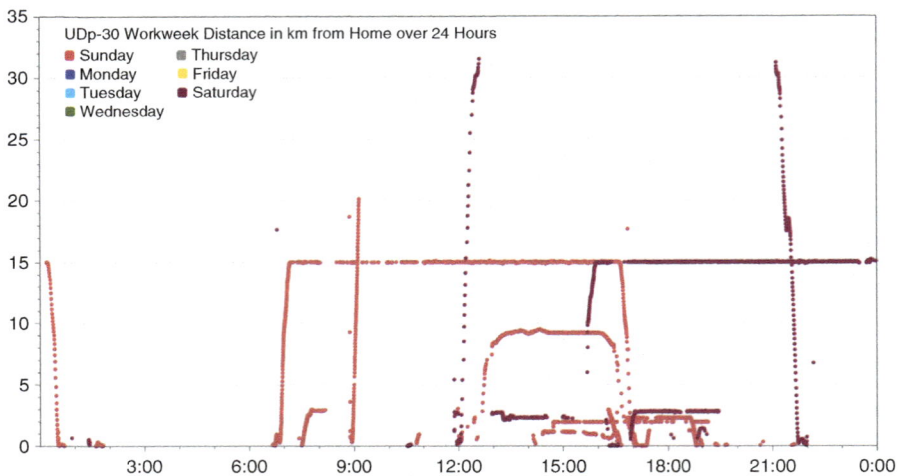

Fig. 6.7 Graph showing average distance in km from home over 24 h for participant UDp-30 on weekends

This depends on the workload, where working hours are extended if required – which, as can be seen from the graph, happens frequently. This participant is flexible in this respect and adjusts her evening programme accordingly.

The evenings after work are also packed with activities, but these are located in the neighbourhood of the home location. The activities include training with a sports team and meeting up with friends in the local bar on a regular basis, hence the repetitive pattern here too.

Weekends, as shown in Fig. 6.7, are less structured compared to the working day in Fig. 6.6. However, there are elements, like sports events with the team, that take place regularly on Saturday or Sunday. These show a similar pattern in terms of the timings. Often this means an early start to the day. However, spatially or in terms of distance, they vary because the location is not always the same.

For this participant, there are a number of trips that go beyond the city boundary, more than 30 km, over the weekend to visit friends in another city or visit family and parents who live out of town. In general however, activities take place in the afternoon on weekends. There is also a fair bit of local activity in the neighbourhood.

As a third example, participant – UDp-06 – lives with his partner in London. He works full time in central London. Most of his commute is underground on the tube. Therefore there is no GPS information on this section of the trip. However, there are details on the journey between home and the tube stop and the exit tube stop and work. The graph is illustrated in Fig. 6.8. During the week the day starts after 9 a.m. with a commute of a good 45 min in total. Work starts around 10 a.m. There is activity recorded over lunch between 1 and 2 p.m. where this participant goes out of the office to grab a sandwich and a coffee.

6.2 Urban Diary Times

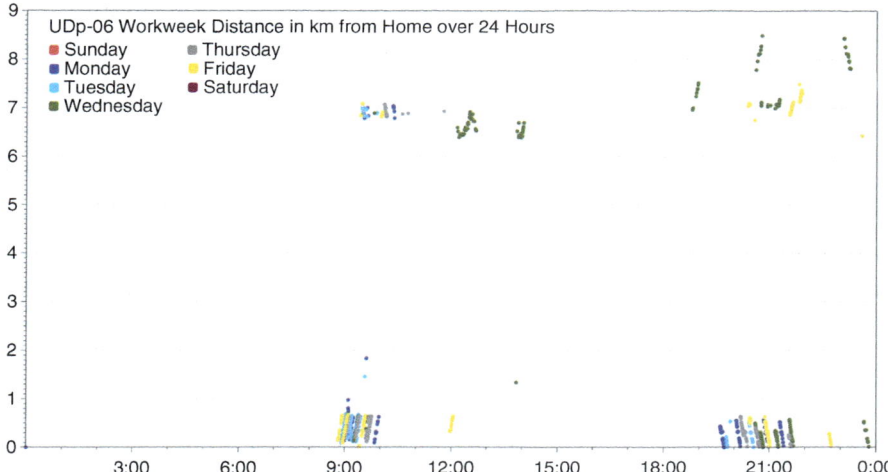

Fig. 6.8 Graph showing average distance in km from home over 24 h for participant UDp-06 during working days

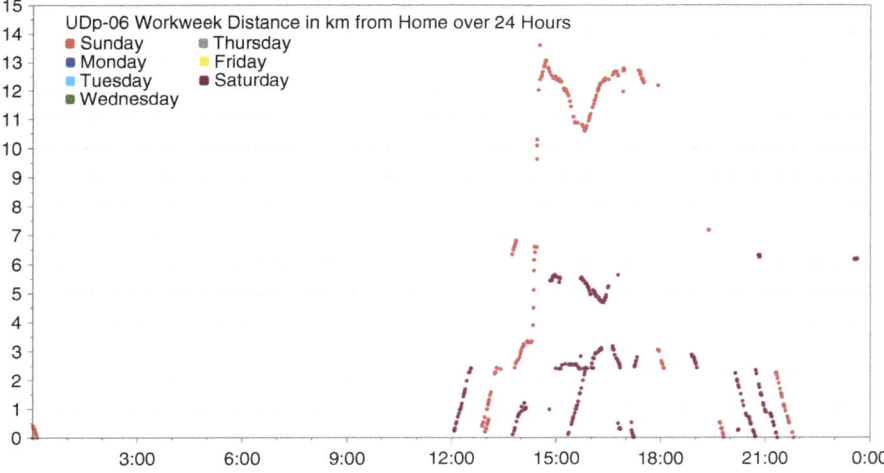

Fig. 6.9 Graph showing average distance in km from home over 24 h for participant UDp-06 during weekend days

The commute home is late and the participant arrives back home between 8 and 10 p.m. However, this includes the after-work trip to the pub, usually near the workplace. Given the late arrival back home in the evenings, few activities are recorded after this. There are a few one-off trips to evening seminars or the cinema.

Weekends for this participant, as shown in Fig. 6.9, are strictly afternoon and evening events. The day outside the house starts only after 12 noon. Locations are both in local, around the home, and in central London. Interestingly, this participant returns home around 9 p.m. during the weekend also. This seems to have become a routine too. There are, however, some late nights on weekends, where the participant is out until 1 a.m.

The three examples presented demonstrate the temporal patterns as they emerge from the data collected by individuals. Most striking of all is the fact that there are only so few activity locations per person over the course of the recording period. Although this period is 2 months, little variety and no obvious changes to the pattern are recorded. Clearly the routine is strong and guides the activities clearly throughout the day, the week and the month. The strong difference between working days and the weekend is to be expected and is a strong pattern in itself. Similarly, the randomness of the weekend afternoon activities is a pattern in itself. There are, to sum up, routines in routines forming pattern inside pattern, depending on the scale and sequence looked at. Fractals, as discussed as spatial patterns, for example, by Batty and Longley (1994), seem also to play a role in time analysis.

Time does show up in these visualisations as a valuable aspect of the data. Distance does not provide the whole picture and neither does time. However, if shown in relation to one another, the two paint a meaningful picture and the story of a journey or a day emerges. It is important to note that what is discussed here are patterns based on movement and spatial activity on the scale of the city, not activity in general. In other words the GPS trace represents movement in the urban context, in the street and between locations.

In some cases, as seen in the examples above, the distances from home to important locations are similar: measuring distance alone would not pick them up as different. In combination with time, however, differences can be seen, and clearly activities can be distinguished according to the times they fit into. To some extent it can be argued that, in this visualisation, when time is brought into the picture, it no longer makes sense to speak of location but activity, as the place has more attributes and additional dimensions.

This method of summarising travel activity as a function of distance from home over time paints an interesting picture of the daily routines and maps out the temporal pattern discussed earlier as habits. The next section looks at how an overall pattern emerges from the data for the entire sample of participants.

6.2.4 Aggregate Distance

Whilst the previous section focuses on individuals' movements, in the following, the patterns emerge from the sample as a whole. It examines in detail how the activities compare and how patterns are maintained within the different sample groups, according to the different locations and the complete sample of GPS-tracked participants.

In Fig. 6.10 for the workweek and in Fig. 6.11 for the weekend, the complete set of participants are represented in a stacked graph showing the average distance

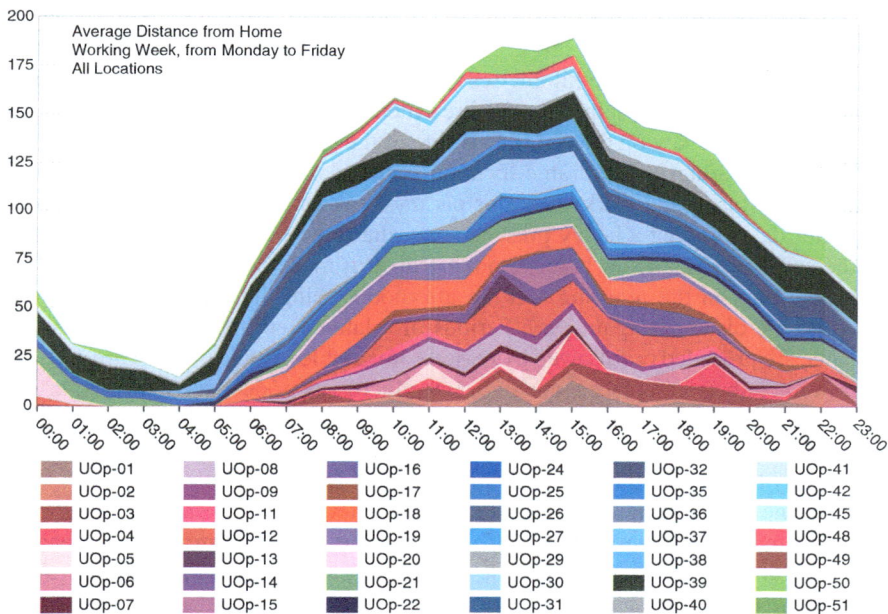

Fig. 6.10 Graph showing average distance in km from home over 24 h for all participants for workweek days Monday to Friday

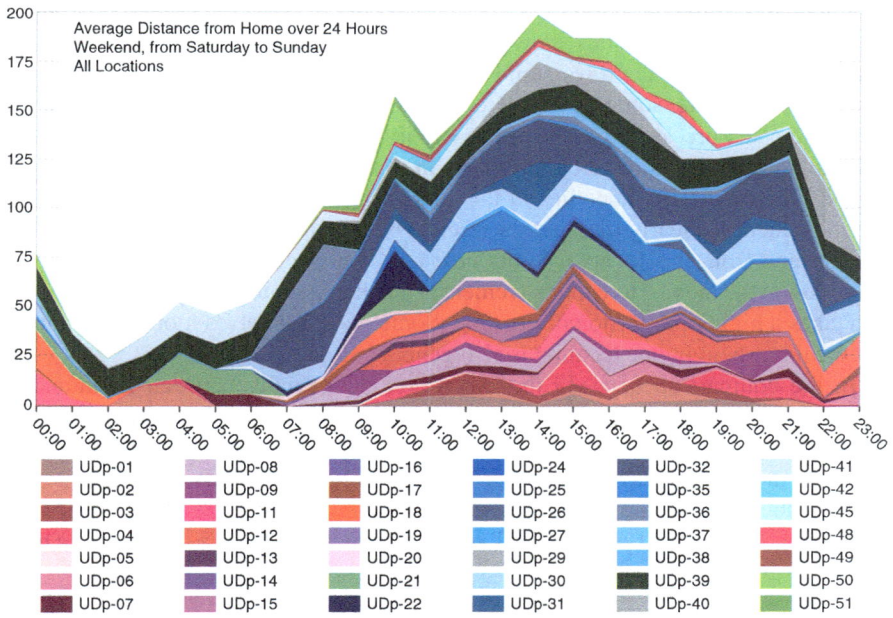

Fig. 6.11 Graph showing average distance in km from home over 24 h for all participants for weekends Saturday and Sunday

per hour to the home location over time. On the x-axis the time is represented as a day of 24 h, starting on the left at midnight and ending on the right at midnight. On the y-axis the distance in kilometres from the individual home location is shown vertically. In total 53 participants are represented, each with a different colour, listed accordingly in the key on the right-hand side of the graph.

Just as important as the home location is the work location. Especially, as to be expected during the workweek, this parameter is a fairly reliable constant, with day after day showing a heavy presence over the same time period. This is shown as the same distance from home. Looking at it in more detail, the work location is visible both in the morning and the afternoon, being interrupted by the lunch break where distances again vary. Periods of relative stability showing a constant distance are preceded or followed by periods of variable distance. These are sections of movement between places where distance increases or decreases. These periods of variables are the commutes. The graph in Fig. 6.10 shows how the distance from home increases on the way out. This is the morning window. The distance decreases on the way back home during the evening window after work.

Over the weekend the elements are the same: it is the configuration, the timing and the duration that are different. Commutes start later and are slower, meaning that these do not rise as steeply as their working day counterparts. The location away from home is also less stable, often only showing up as a peak, but not sharing the characteristic plane of a fixed activity, where the distance stays the same for a longer period. And the journey back home takes place about 5 h later than on weekdays. It is less directed and distance reduces more slowly, probably due to stopovers for shopping and similar activities.

Overall the characteristics show distinct patterns between weekdays and weekend days. In general the days simply start later on weekends than on weekdays. However, there are some common characteristics too. Between 3 and 5 a.m., all participants are at home and little activity is visible. This refers back to the assumption that the home location plays an important role as a hub over the 24 h period. The main difference lies in the characteristics of the movement. Whereas the movement is clearly directed on weekdays to a specific location for a distinct activity and from this location to the next one, the weekends present a pattern of activities on the move. Here the movement is much more integral to the activity itself and takes a higher proportion of the time. Sometimes it involves more destinations, but the time spent in one location is also generally less.

In the comparison between the urban areas, especially between London and Basel, a distinct time shift can be seen. Whilst the pattern and the sequence are the same, the timing is different. There is a 1 h difference between the same activities. In Basel most things take place approximately 1 h earlier. This is not due to the fact that there is 1 h time difference because of the different time zones. London is BST and GMT, whilst Basel is CEST or CET,[6] which equals as a constant one hour

---

[6]BST stands for British Summer Time and GMT is Greenwich Mean Time. Whilst GMT is essentially UTC during summer, the UK observes summer time which is 1 h ahead of UTC. Similarly in Basel the CEST stands for Central European Summer Time which is 1 h ahead of

6.2 Urban Diary Times

/+01:00 difference (ISO/TC154 2004). In the data this change has been accounted for all times are shown as local time. On workweek days, in Basel the commute starts as early as 5 a.m., ahead of London where 6 a.m. marks the beginning of the commute. The distance starts to level and stabilise at a consistent value, arriving at a destination in Basel around 8 a.m., compared with around 10 a.m. in London. The lunch break takes place at 12 noon in Basel, but 1 p.m. in London. This then has a knock on effect for the start of the afternoon, even including the time participants start to commute home. In Basel this is 4 p.m. and in London it is 6 p.m. The end of the day in Basel is around 12 midnight and in London around 1 a.m.

Also on weekends this occurrence of time shift is similar. Basel is in general 1 h ahead of the same activity in London. Even though, as described above, the weekend shows overall a generally slower pattern, a time shift between Basel and London of roughly 1 h is present overall.

In the introduction to this section, the programmatic assumption for the home location to have a strong influence directing a pattern is now being illustrated. The patterns of returning home every day as an assumption is confirmed by the results visualised here. The pattern shows that, early in the morning, the distance to the home location is in most cases zero and this gradually increases to a peak, from where it drops back down to zero, as participants return home. This is reflected in the average numbers calculated from all recorded track points according to urban area and participant.

By ordering the recorded data according to a strict time raster, the information can be compared. The examples and visualisations above have shown from different angles how activities fit into the time raster.

To organise the data according to clock time is the most obvious sorting criteria for this data. However, as outlined in the introduction to this chapter, there are other aspects to time. It has been shown that time can add a dimension to the resulting visualisation and places and dots turn into activities and sequences of action. What emerges is a picture of continuity rather than isolated events. Already from the graphs showing average distance from home, the ebb and flow and the changes in distance indicate the connections between events. Furthermore in the individual examples plotting distance over 24 h for individuals, as, for example, in Figs. 6.4–6.9, it is illustrated how one action leads to another and how travelling in space, indicated by the distance, is directly connected to travelling in time. Through this abstraction, it becomes possible to extract the essence of individual movement and highlight the importance of continuity as the defining parameter.

Continuity, as a concept of constant and uninterrupted events, is essential from an individual perspective. There is no point in the day or in the week when activities do not connect, paths do not match or times do not line up:

---

the Central European Time (CET, observed during winter). This is outlined in one of the standards set out by the International Organization for Standardization ISO in the document ISO 6801:2004 (ISO/TC154 2004).

> Of equal importance is the fact that time does not admit escape for the individual. He cannot be stored away for later use without complications for himself or society. As long as he is alive at all, he has to pass every point on the timescale. (Hägerstrand 1970, p. 10)

Continuity roots in the aspects of practice are thus at the heart of the discussion on routine. Practice, as argued in the introduction to this chapter, is the spatial dimension of activity in the manner of de Certeau (1984) and consists itself of travelling, as outlined here. The practice of travelling opposes directly any planned fixtures or the fixation of the map, as argued by Crang (2000), hence rendering any effort to disjoint the practice powerless.

Often clock time is taken as a measure to break the continuity into smaller pieces in order to conveniently take elements apart to judge and evaluate items separately, before regrouping them. This, however, can be misleading, since arguably there is no more important or less important time of the day at any given moment. Clock time is per se value-free, a framework to count and not to evaluate. There are, as demonstrated above in the examples, times for specific tasks and activities. These, however, are closely tied to each individual, group or even institution. Interestingly this links back to Hägerstrand's concept of constraints (Hägerstrand 1970) as critically discussed earlier, for example, in Sect. 2.2.

The discussion around the criticism of Hägerstrand's theory is illustrated to some extent by the graphs discussed in this chapter. On some levels, the arguments put forward by Hägerstrand hold up if elements are isolated and examined individually. As argued before, however, in the overall picture, beyond the materialistic and individual viewpoint, it can be argued that continuity is more important than the constraint model.

In the next section, other aspects of time will be discussed, as will the interpretation of the data under time aspects other than clock time frameworks. To complement the clock time as recorded by the GPS in the next section, we will be looking at how the individuals see themselves and how they organise their schedules.

6.3 Schedules and Planning

Planning ahead is a cultural behaviour, closely related to the concept of clock time and calendars. The idea of a string of events ahead is the basis for the plans recorded in the interview schedule sketches. Similarly, the concept of repetition is present in the fact that each scale or element, the day, the week and the year are repetitive and reoccurring. Furthermore they are interlocking and one directional and contained in one another. In other words, 7 days make a week and 52 weeks make a year, but 365 days make a year. This kind of counted repetition is a direct consequence of the linearity of clock time. In order to keep track of the count, larger units help with dimensions by creating time reference frames and keep the numbers at a range which

6.3 Schedules and Planning

can be grasped by the human brain. The reference frames help with orientation and segmentation. A 7-day rhythm is easier to relate to than a 168 h cycle (if a week were to be expressed in hours). A year broken down into 12 months is easier to navigate than a 365-day cycle, where 8 October would be day 281.

Part of the interview with each participant was a detailed discussion of the personal schedule on different time scales. This also included a table for them to write down the activities according to a time raster: see Fig. 6.12 for examples of filled-in questionnaires. The scales recorded are a 24 h day, a 7-day week and a 12-month year. Each category can be interpreted relational to the others if required. As such the 24 h day could be a day of the week and the week could be part of the year. Each category is rather general, and the participants were requested to give details of a typical schedule of activities. This can be of course an arbitrary description; however, it did work for most participants. It did not cause any complications, and participants were easily able to provide average times. In fact it seemed that every participant had a fairly good plan of their schedules ready in their mind, and the chosen categories resonated with units and reference frames they were using in everyday life.

The categories chosen were related to the recording period of 2 months. In this context it did not make sense, at the time, to ask participants to record the monthly schedule. The focus was on the day and the week. At the beginning, the assumption was that the longer the period, the less repetitive personal patterns there would be. However, as an experimental element, the yearly frame of reference was added as a third unit. On the one hand, this picks up on the question of scale and personal reference, but on the other hand it might highlight a transition from personal to collective in the type of activities recorded. The longer time scales could refer more directly to the cultural reference frame, or the constraint as Hägerstrand (1970) would put it.

The questionnaire intentionally did not specify the terms that were to be used. Participants were free to use their own wordings for activities. This approach was chosen in order not to preset the categories and leave enough room for each individual participant to express and work with her or his personal structure of reference. For the analysis, the recorded activities have been grouped under meaningful umbrella terms. However, for the cross-analysis, terms were simplified. This is partially because numerous terms do actually mean the same thing, but people have chosen to use one or the other term. Sometimes the same person used different terms for the same activity. Also the summarising in groups will show clearer clusters as opposed to lengthy rows of individual terms.

From the graphical recording method of the timetable, the data was transcribed to a tabular format. This included the details as well as a note regarding what scale was used to record the item. The scale is important so as to not mix up the recording methods. It can be noted from the recording sheets that participants have intuitively adapted to the scales provided, recording differently on a 24 h scale than on a weekly scale and again differently on a yearly scale. It seems that the time frame

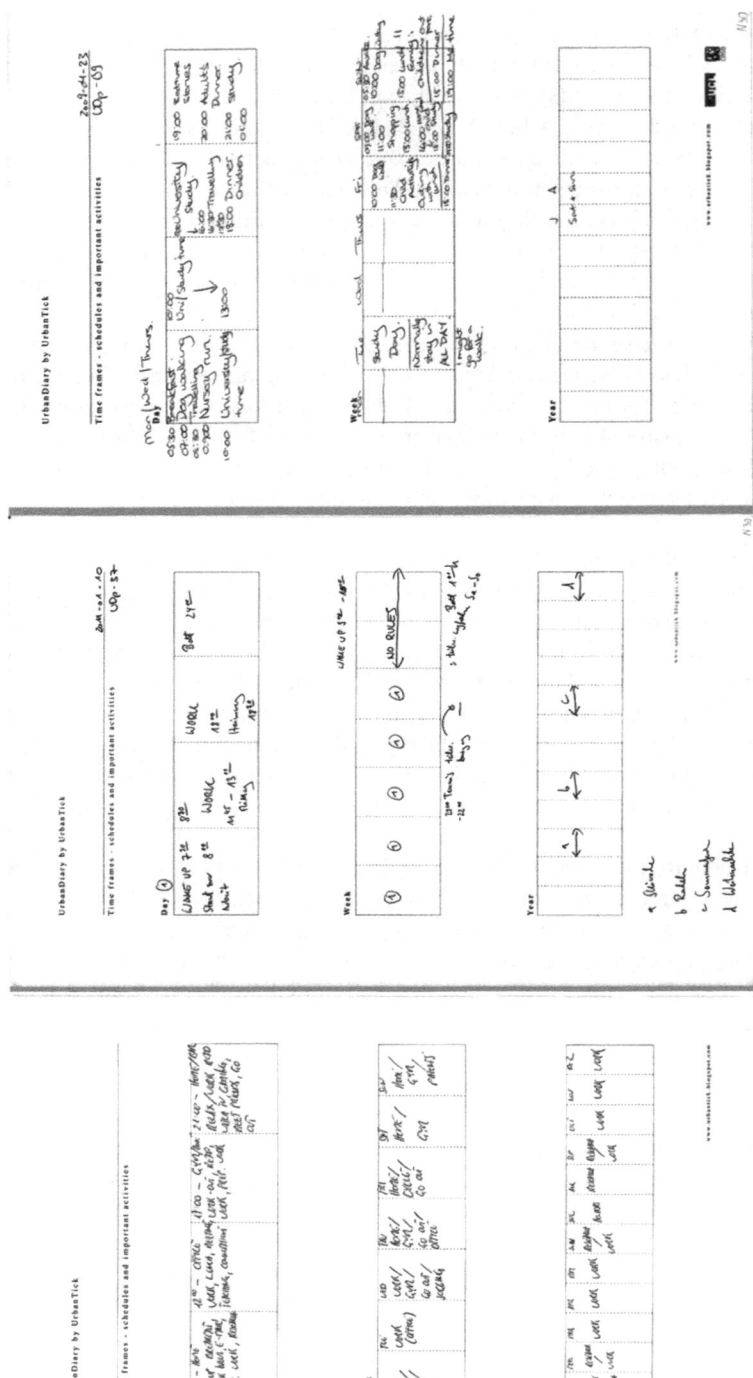

Fig. 6.12 Three examples of recorded schedules in the provided framework, from the top 24 h daily, 7-day weekly and 12-month yearly schedules. From the *left* the three examples recorded by the participants UDp-31 and UDp-37 (both Basel) and UDp-09 (London) (see http://www.flickr.com/photos/40984848@N04/8701250233/ for an online version)

6.3 Schedules and Planning

asked for does already preformat the answers given. Clearly on scale 01, the 24 h of 1 day, recordings are the most detailed. In most cases these recordings cover the entire day without any gaps. On scale 02, the week with its 7 days, the records are already jotted down sporadically with, in most cases, the assumption of the previous scale applying for most days of the week, especially the working day. This leaves weekends and evenings to be defined on scale 02. Scale 03, the year, is the least detailed, with notes on a monthly basis.

Overall, the scale applies directly in practical terms as a reference constantly linking between scales. It can be noted that there is no such thing as an independent daily schedule. It always coexists with the weekly schedule and for some details can only be presented in one or the other, but still have vital consequences for two or more scales. This topological problem came up during most of the interviews, and each participant imposed his or her own individual strategy to deal with this. On the one hand, the layout and presentation of the graphical recording sheet imposed this question rather directly by partially making a proposition as to how to separate between scales and what each scale should look like. On the other hand this framework makes it possible to characterise the responses against it as a base line.

It has to be noted that the data is to be understood as biased, owing to how it is recorded. As such the reference is clearly stated and so are the circumstances. The analysis can therefore cater for this fact.

The groups used in the analysis are mealtimes, *breakfast*, *lunch* and *dinner*; for the start of the day *get up*; travel is recorded as *commute to work* and *commute home*; the actual work time as *work* including "study", "seminar" and "meetings"; being at home as *home*, but this includes "taking a shower" or "watching TV", for example; sleeping has also its own category *sleep*; activities outside the house are *explore*, *sport* and *shopping*; *social* is a category summarising meetings with friends and family outside the house and includes the frequently used term "going out". A category *var* is used for specific activities of short duration. These activities are important for the individual but cannot be meaningfully grouped with any of the other overarching categories. Examples of this category are "walking the dog" or "playing music". A total of 14 categories are used to map out and analyse the activities.

6.3.1 The Day

There are clear distinctions between workdays and weekends. In all cases the work days are Monday to Friday and weekends are Saturday and Sunday. An example taken from the interview is shown in Fig. 6.13. Structurally the two categories are different. This is of course not unexpected since, in western cultures, in which all participants are embedded, a strict distinction is made between the working day and the weekdays. The activity of working is clearly one of the key structural elements

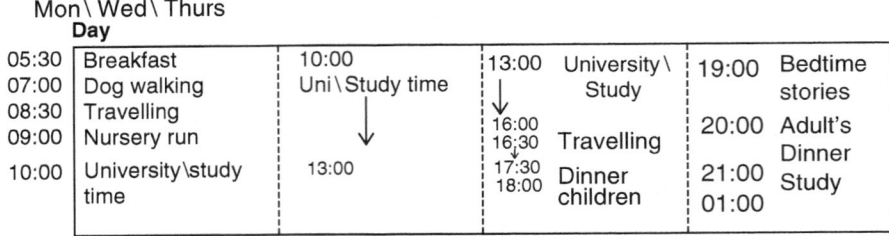

Fig. 6.13 An example taken from the interview with participants UDp-09. Listed are daily activities

as already seen in Sect. 6.2. It is important enough to influence the weekend through its absence, not least since the weekend is effectively tied to the activity of working as a time of rest.

A lot of the daily structure then is owed to natural needs such as eating and sleeping. These biological rhythms are the source for culturally standardised activities. As noted also, for example, by Hägerstrand (1970, p. 11), this is in connection to *constraints* in the form of the *capability constraint*, or Lefebvre (2004, p. 19), Elias (1992, p. 67) and also Zerubavel (1985). For example, the lunch break is one of the important elements, as it is placed at the division between the early part of the day and the later part. In the interviews this broader grouping was used frequently to place activities. In this sense AM and PM all of a sudden take on an additional cultural meaning.

It could be argued that the lunch break is a *capability constraint* in the sense of Hägerstrand's constraints concept (Hägerstrand 1970), limiting the range and duration of activities. On the other hand, the lunch break can also be interpreted as a *coupling constraint* since lunch is a social event where groups or families meet. And finally the lunch break could be an *authority constraint*, for example, looking at factory schedules of working hours and fixed lunch breaks.

Can an event as a result fit all three categories? Without going deep into the discussion around the constraint problem, it can be noted here, and will be demonstrated in the results section, that there is much more to something like lunch than simply being a *constraint*. It is much more of a social event, meeting people and participating in an activity with a group or family. The conditions in this case and many others are many and directed by a complex mixture of influences, guided both by time (moment) and space (location) in the sense of the German word *Zeitpunkt*, translated as point in time. It features a distinct spatial reference due to the specific use of *point*.

The weekday activities can roughly be put into three rank groups based on the number of times they were listed overall (for an illustration, see Fig. 6.14). The top ranking activity is *work*, with 40 listings. After this comes a group of activities all listed between 16 and 20 times. These are according to the rank *home* (21),

6.3 Schedules and Planning

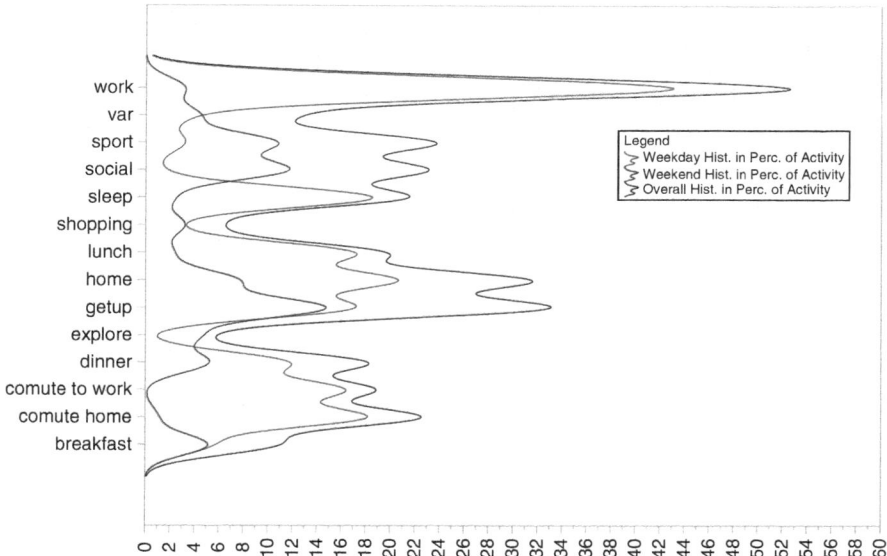

Fig. 6.14 The graph shows the number of listings for each activity category in per cent by weekday (*blue*), weekend (*red*) and overall (*black*)

sleep (18), *commute home* (18), *lunch* (17), *get up* (17) and *commute to work* (16). Activities in the third group are listed only few times, ranging between one and four. Two activities are sitting in between group two and three. These are breakfast (6) and dinner (12).

Activities listed for the weekend are in two rank groups only, roughly corresponding with rank groups 2 and 3 of the weekdays. The dominating category is missing. Higher rankings are activities like *sport* (11), *social* (12), *home* (8) and *get up* (15). The second group contains *work* (3), *var* (4), *sleep* (2), *shopping* (3), *lunch* (2), *explore* (4), *dinner* (5), *commute home* (1) and *breakfast* (5). The listing of *getting up* is high on weekends. It seemed important to participants to mark the difference from the working day by getting up late. Sport as an activity is important for a number of participants on the weekend, as is the activity group *social*. Meeting up with friends and family is also connected to the activity *explore* away from home on weekends. Still, the category at *home* is also high up on the list and people spend a fixed amount of time at home over the weekends.

Over 24 h the day is clearly structured into three main segments of activity. If looking at the beginning of a new activity, there is a morning block, a midday block and an evening block. All three blocks are illustrated in Fig. 6.15. This structure over the weekdays is again the pattern that would be expected. The activity gap between 2 and 5 a.m. is characteristic for an average day.

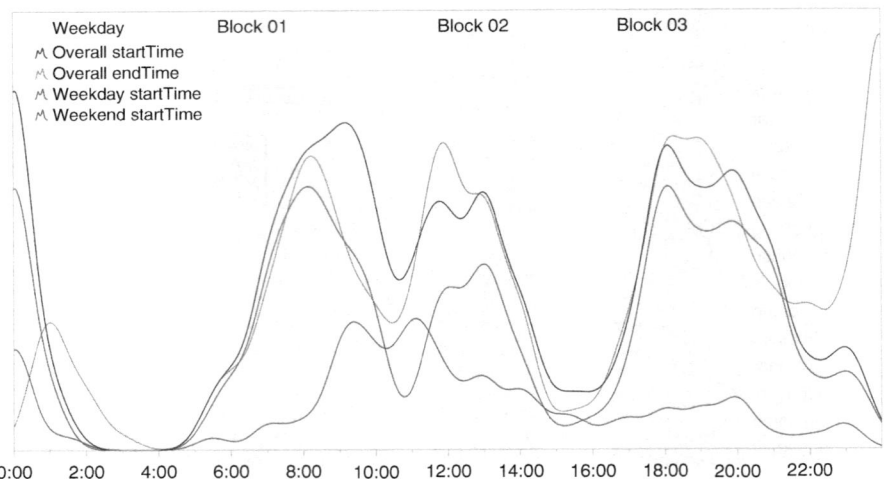

Fig. 6.15 Over the course of 24 h horizontally, the graph shows the number of listed activity start and end times for all locations. Colours indicate *black* = overall, *blue* = weekend, *red* = weekday

Looking at the three blocks individually, the changes of activity create separate segments within the blocks. The detailed blocks are illustrated in Fig. 6.16. The first block in the morning roughly is 5.30–10.30 a.m. It consists itself of three parts. The first is getting up and getting ready in the house. The second part is leaving the house and commuting to work, and the third part, also the most articulated one, is starting work. Over these three batches, we have a shift from one place to another, from home to work. This shift or commute is often a direct move, but can sometimes be a multi-stop journey: for example, it might include the school run. It is interesting how important these short stops are. The length of the activity does not explain its importance. Compared to other much longer activities, they feature prominently.

The second block, roughly from 11.30 a.m. to 2.30 p.m., is essentially the lunch break. It is built of two parts in this graph representation, as it contains the start of the lunch and the start of the afternoon work activity. All participants indicated a lunch. There are of course different ways in other cultures or locations, but no one in this sample claimed to work throughout the day: there is always a midday break. Sometimes it is also used to change activity: for example, participants who work part time often use the lunch break to change activities. As such it is an important point of reference in the daily schedule.

The pattern is backed up by a view on the activities listed for this time frame. There are only two: *work* and *lunch*. For this sample the activities

The third block, 4.30–10.30 p.m., comprises after-work and evening activities. A number of different activities are summarised here. The block starts with the end of work and the commute home. Sometimes shopping and picking up the kids becomes part of this commute. The activities are changing with *work* coming to an end. *commute home* and *home* as well as *dinner* become the main activities.

6.3 Schedules and Planning 157

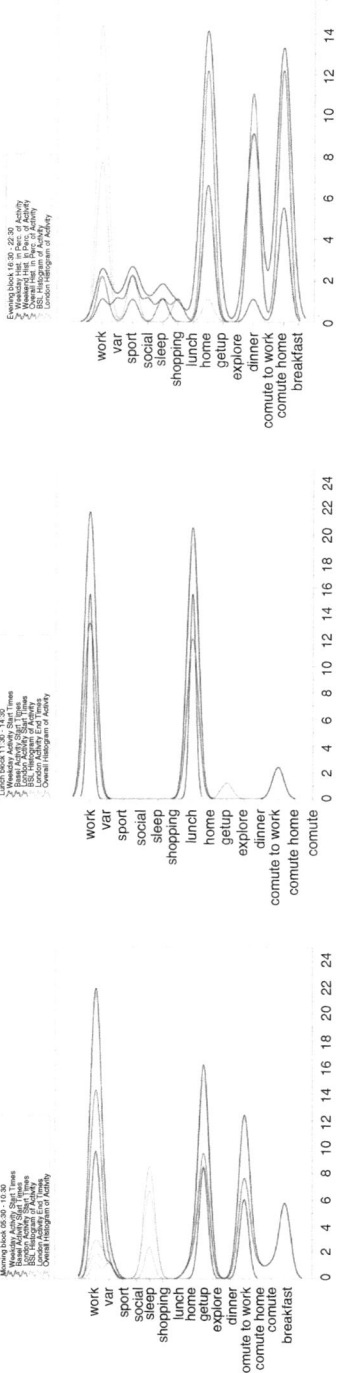

Fig. 6.16 The three graphs show the morning, lunch and evening blocks of a weekday, listing the distribution of activities. From *left* to *right*: block 01 morning, block 02 midday and block 03 evening. Lines show the start and *dotted lines* show the end times of activities during the corresponding block's time frame. Colours indicate *black* = overall, *blue* = London, *red* = Basel

However, as the evening progresses, it becomes a container for a whole range of activities circling around *social*, *sport* and mealtime as *dinner*.

Compared to the morning block however, the evening block is more variable. Times are less strictly organised and a number of participants show a high degree of flexibility. It is really only the times that are flexible. The routine or the sequence remains the same. For example, work end times fit in a time frame of about 1 h. This means that end times are not as accurate as start times, and thereafter activity start times vary accordingly. This can develop both ways, either earlier or later. Daily changes, however, do not normally influence the following sequence.

In this sense there is a small element of on-the-spot decision encapsulated in block 3. This flexibility is marginal, but compared to the rest of the day quite extraordinary. The earlier two blocks are tightly packed and stringently organised. The flexibility recorded here can be anything between 30 and 90 min. None of the participants have identified this time frame as a possibility in the sense of a window to plan in extra activities. The adjustment is managed for each unit of scheduled activities in an ad hoc way.

It is interesting to note that such opportunities are not to be utilised for additional or completely different activities. Flexibility appears to be a more complex option than it seems. The reasons for this might be many, but definitely include additional constraints such as location (as discussed earlier the location plays an important role in what activity can actually be performed and many actually require participants to relocate), preparation (the right gear or facilities might not be at hand), social context or economics like costs involved.

The core of each block can be identified around a mealtime. Over a 24 h period these are breakfast, lunch and dinner, corresponding to blocks one to three. The three meals are not equally weighted in the records, with lunch definitely getting the most attention, followed by dinner and breakfast being the least important. This might reflect the actual practice, where lunch is on a normal day definitely the celebrated mealtime and therefore of higher cultural importance. It is visible, though, from the background of participants that if they have family mealtimes, these are valued differently, and the family meals and/or the mealtimes for other family members are of importance too.

The graph in Fig. 6.17 on page 159 shows the times for both locations in colour, London in blue and Basel in red. Both participant groups were given exactly the same questions and framework to record the information. The results do broadly match, showing a similar overall structure. Interestingly though, looking at the scheduled times given for activities, there is roughly 1 h time difference between the same activities in Basel and London. This same effect was noted earlier in section *Clock Time* Sect. 6.2, especially in the subparagraph *Aggregated Distance* Sect. 6.2.3 based on the times recorded using the GPS tracker. Now, 1 h might not seem like much, but it is consistent across all three activity blocks, especially on weekdays. Whilst there is a time zone difference for technically measured times, in the case of the interview, times are always noted as local time by the participants.

6.3 Schedules and Planning

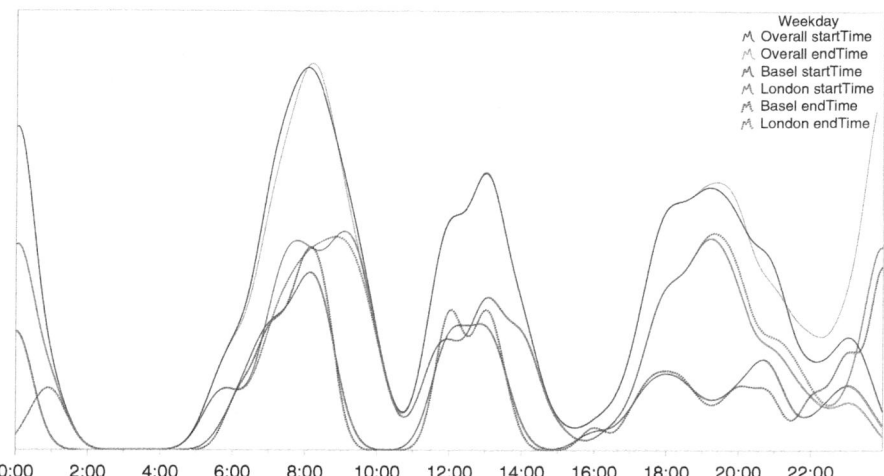

Fig. 6.17 The graph shows the activities over a general day as counts of start times and counts of end times. The graph is however as much to be read underneath the line as the change of activity as above the line where it represents the consistency and duration of activity

The midday event *lunch* during the working day in Basel (and most of Switzerland) is generally held between 12 and 1 p.m. In London (and the UK in general) however, it takes place between 1 and 2 p.m. The same 1 h delay has been discussed in Sect. 6.2. In the GPS data, it is visible in the morning and the evening, with start times roughly deferred by 1 h. Such a difference in the timing of one event, such as lunch, could simply be an anomaly. However, having observed that the 1 h difference persists throughout the day, the pattern could be down to cultural differences.

It is not immediately obvious why such a difference exists. There are different strands to start explaining this phenomenon. Partially this might be explained culturally as a local form of organisation. The natural aspect would be the daylight factor, which is actually different, with sunrise and sunset earlier in Basel. The answer might be a combination of different aspects.

In other areas the two graphs show much the same pattern. Both behave in the above-characterised three blocks in terms of start and end times. The overall structure appears to be almost identical. There are some minor differences as to how strong one aspect weighs over another: for example, the evening block is slightly differently structured. Where in the London graph activities peak after work, just after 7 p.m., in Basel this block is structured as a flatter sequence of three similar activity peaks at 6 p.m. (commute home), just after 8 p.m. (evening) and one at around 11 p.m. (sleep).

This suggests that participants have given more detail about this time period, and this time frame is filled with some three different activities. Given that the graph showing end times (dotted line) behaves in a similar way, it can be assumed that if one activity is in many cases followed by another, they are synchronised. Given the additional detail, the period after work in Basel appears to have greater importance. It is not only just about

> ...chilling out... (Interview UDp-30)

as one of the London-based participants put it. The activities *sport*, *var* and *social* are clearly more used in the evening in this weekday evening block. On the other hand, dinner is the activity that was put down as definitely more important to London-based participants.

Regarding an interpretation of the data collected from participants' responses, it can be said that, in general, people seem to have no problem listing the activities in the given time frames. However, it seems that most people do not so much think of the activity as reoccurring. If asked about a pattern, participants usually have to think about the pattern, as if they have to piece it together, before giving an answer. On the other hand, if asked about an activity or time, the answer comes instantly. The schedule is clear to the participants, but works rather like a calendar. Activities are planned out according to their time, not in terms of their recurrence.

Shopping was not as prominent as one would expect, especially on weekends. The listing of *getting up* is high on weekends. It seemed important to participants to point out the difference to the working day by getting up late. Similar as expected the dominating activities are *work*, *sleep*, *home* and mealtimes, in particular *lunch*. These activities are not only relevant for feeding in the sense of a natural need but furthermore have a cultural and especially an organisational dimension. The cultural aspect is locally rooted and can take on different forms. These in turn influence the daily structures tremendously. In this study the two locations, London and Basel, both share a similar culture. Moving on to other climate zones in particular, cultural agreements on mealtimes and importance vary dramatically.

6.3.2 The Week

The records of activities across the week were recorded using a 7-day grid to provide the participants with a framework as in Fig. 6.18. They were asked to record reoccurring activities other than what has already been noted for a general weekday. Clearly the weekend, with both Saturday and Sunday, are active. The other weekdays more or less operate according to the schedule recorded in the previous day section.

The weekend pattern over Saturday and Sunday is, as indicated in the introduction to this section, structurally different from the pattern of the working day. In general the activities differ, and the scheduling of activities is organised in a different fashion. The *work* activity is not present, and with it a highly dominant and structuring element is missing on weekend days. If you want, the elephant is not in the room, which leaves a lot of space for everything else. There is space for more activities, and the timing of activities is much more flexible. During the week, *work* pushes everything to the side, leaving only small windows for other activities.

6.3 Schedules and Planning

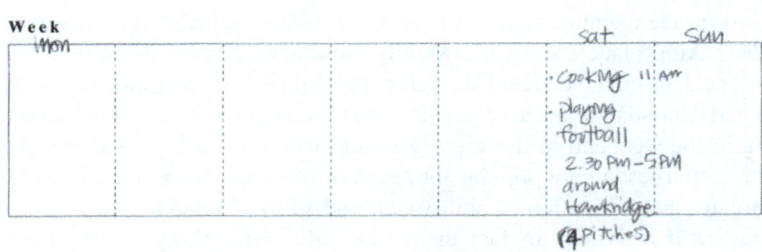

Fig. 6.18 An example taken from the interview where participant UDp-07 has noted down the various weekly activities that are repetitive

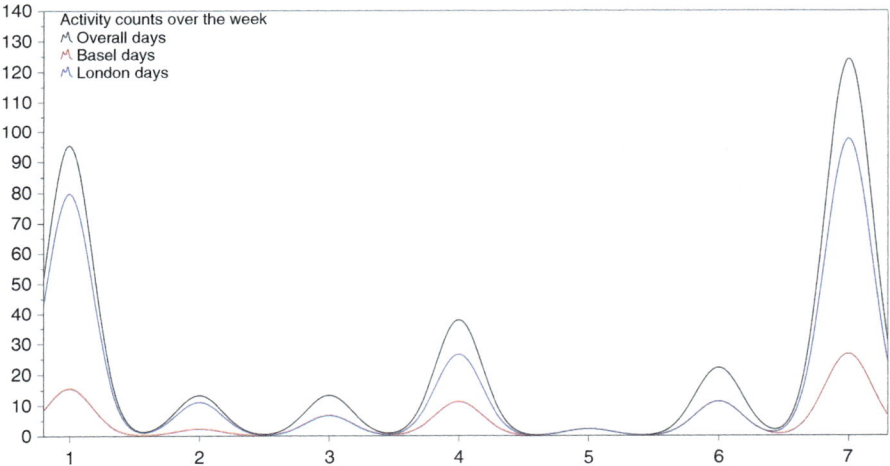

Fig. 6.19 The graph shows the activity count over the week for each day. This graph starts with Sunday (1) and ends with Saturday (7). Overall numbers are represented in *black*, and the locations (Basel = *red* and London = *blue*) are shown as separate lines in this graph

On weekends there is now more flexibility with the timing of activities too. The overall results are summarised in the graph in Fig. 6.19.

Generally this flexibility is most visible in the later start of the day. Activities only start to pick up between 8 and 9 a.m., with a first, and major, peak around 10 a.m. This activity block lasts right across the middle of the day into the beginning of the afternoon. The first low point in terms of new activities is around 15:00, before it starts to pick up again for an evening block of activities. This second block peaks between 7 and 8 p.m., with a rather quick drop off to nearly zero recordings of activity start times between 9 and 10 p.m. The third block of activities on the weekend is pushed over into the morning of the next day, with activities slowing down around 2 a.m., flowing into the dead zone of the morning hours starting between 2 and 3 a.m. and lasting until after 5 a.m.

Activities which are additional to the general daily schedule are mainly concentrated on the evening block. All weekly activities scheduled on a weekday are recorded taking place after work. Mainly these activities are listed under *sport* or *social*. The days covered are all similar, with Wednesday being more active than the others and Thursday less active. Friday, as expected, is in its evening block already leading in the weekend in the activity *social*, which is an expected characteristic. Monday and Tuesday are popular for *sport*. Wednesday, being the midweek mark, accumulates a combination of both *social* and *sport*. Thursday is the day with the least recorded activities. In fact there was only one activity recorded across all responses. See Fig. 6.19 for details of the summary.

The different structure of the weekend in comparison to the weekday is, as we see from the description given above, rooted in the fact that activities are more dynamic. Plans are usually made for the full day and often activities drag on. The lack of activity changes during the afternoon can be explained with the observation that participants usually organise activities on weekends that span the blocks of time. For instance, a visit to family or friends might start around 11 a.m. and continue into the evening, ending as they arrive back home. This umbrella of visiting family and friends consists obviously of a number of activities, but it is organised mentally by the participants as one activity. The same is true for days out, where participants go out *exploring*.

The week does show an internal pattern, consisting of pronounced weekends (Saturday and Sunday) and a sequence of similarly structured weekdays. One additional element is a pronounced midweek on Wednesday. The distinct separation between weekend and weekdays was expected, and also Wednesday as a popular day for activities is no surprise. Having Thursday as an outlier with next to non-activities is surprising. However, few activities are listed for the other weekdays, so this should not be rated too important.

6.3.3 The Year

The reference frame of the year appeared to be the least immediately graspable unit for the participants. It appears to be difficult to generalise such a long time span whilst recording personal events. An example is shown in Fig. 6.20. In the context of a whole year, individual activities often seem not relevant, as expressed frequently by participants, for example UDp-41:

> ...but then I am not sure if I will do this next year. (Interview UDp-41)

As a consequence the participants recorded information only sparsely. Only 20 out of the entire sample of participants actually responded to this category, most only after repeated prompting for details during the interview. Three main topics stand out from all the responses. Holidays are the most cited reoccurring schedule item across all month and participants. Often these are specific types of holiday, such as

6.3 Schedules and Planning

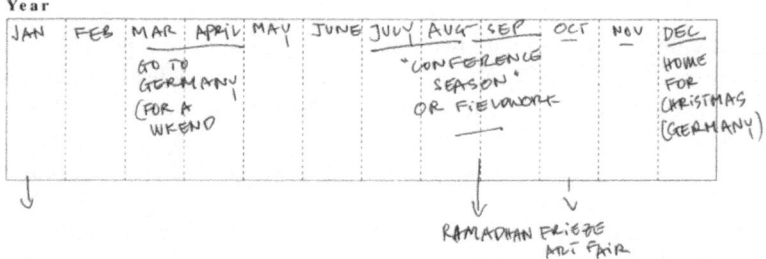

Fig. 6.20 Example of activity listings for the yearly time frame taken from the interview with participant UDp-17

ski holidays, summer holidays, bike holidays and so on. Holidays as a scheduled activity were recorded throughout the year except for June and November. Most popular are summer (July and August) and winter (December and January, over New Year) holidays.

This category is embedded in a wider context. For one thing, booking holidays in advance can be cheaper. Tour operators, airlines and hotels often offer cheaper prices if booked well in advance. Also, holidays have to be booked with the workplace in advance, to allow the company and clients to plan ahead. Secondly, holidays carry strong cultural meaning and social value. It is an often talked about topic with friends and family and often serves as an indicator of status.

The second topic is cultural festivals, which participants noted down as yearly points of reference. December with Christmas and New Year was the most noted month for cultural events. However, Ramadan, Diwali and the Thai New Year were also raised as recurring annual events of cultural importance. This also reflects the diverse religious and cultural background of the sample of participants in the study. It should be noted that the London sample is more diverse than the Basel sample.

Here the cultural dimension of temporal frameworks is most notable. In this context it could even be argued that the two things – annual festival and unit – go hand in hand and reinforce one another. The cultural festivals become essential for the identity of the unit and act as time markers. The next Ramadan or previous Ramadan is fixed units, acting like landmarks, which individuals can refer to and communicate about with others.

The third category that is worth mentioning from the interview results is birthdays. A number of participants have mentioned these recurring days as manifestations of a yearly pattern. Usually the birthday is that of a close friend or family member. These focal points represent a personal connection to the time framework and are contrary to the collective nature of the cultural festivals focused on the individual. It is interesting, however, how these two patterns manifest a concrete context perfectly balanced between the individual and collective.

6.3.4 Discussion Schedules and Planning

On the individual front, there are two other categories that are repeatedly recorded. One is the reoccurrence of cultural events such as music festivals, concerts or carnivals. These of course attract a larger audience and stand for a peer group forming a collective, organised for the individual. The second one is the planning of personal events and goals such as work- or education-related upcoming milestones such as conferences or exams.

It is here where cultural festivals become the essence of the repetition and allow for a densification of the temporal organisation. As argued earlier in Chap. 2 the literature review in Sect. 2.2, time-space is in reference to Reichholf (2008) who argues that cultural institutions such as places and sites for rites where initiators form permanent settlements through the permanence of repetition they provide. Reichholf argues that the time framework provided by the rhythm of the practice created permanence desired for settling. This theory stands against the commonly believed theory that permanent settlements started to form following a change in practice from hunting to farming. Reichholf, however, argues that due to a rich provision of resources at the time, it is more likely that permanent settlements had another origin and the change in practice to settlements and farming was developed in conjunction or as a consequence.

If this is translated into the urban context, interesting consequences for the relationship between space as place and space as temporal framework can be observed. It would mean that the meaning of a place principally has a rhythmic definition and is manifested in practice: its physical existence is only a second step. The cultural aspect of place is moved to the foreground with the definition of the temporal framework through cultural practice as the defining element.

The cultural practice essentially relies on the activity, and in this context the annual cultural festival is vital. For one thing it defines the temporal framework, but on the other hand it acts as a reference point for the collective, acting as a social compass. In this context annual cultural festivals are much more than pure celebration, family meals and presents. They stand for a fundamental temporal practice producing the spatial framework: *spacemaking*.

With these lines of thought in mind, it is a small step to the perspective of Jane Jacobs, particularly in the *The Death and Life of great American Cities* (Jacobs 1961). There she argues for the bottom-up approach and illustrates how the people grow the city. With the cultural practices being at the heart of *spacemaking*, this is easier to understand as to how the people make the city. The physical environment is a result of everyday practices and is constantly reinforced and rebuilt each time the same activity is practised in everyday life, just like the rhythmic activity patterns discussed and illustrated earlier in this section. The planning and design Jane Jacobs is talking about emerges directly from this lively source of practice, and in this context it could be the habitus. Such a highly decentralised approach can explain the diversity observed in cities, the multitude of solutions to the same problem, but

6.3 Schedules and Planning

foremost the adaptability of the system as argued for by, for example, Batty (2005). As Batty interprets Jacobs in the introduction to his book *Cities and Complexity*:

> ...what she (Jacobs) was arguing was for a new understanding of cities in terms of how individuals behave and the processes that they use to develop their environment. (Batty 2005, p. 3)

In this sense, the results shown in this section on Clock Time can be seen. It was shown how individuals travel rhythmically and repetitively through their individual part of the city. It can be argued that through these actions the individual area of the city is maintained.

In this context, and the current discussion around large and megacities with regard to structural problems, the question is to what extent can these cultural aspects guide the framework of the city and continue to guide it. If spatial practice based on habitus is the defining element, how does it relate to size and organisation? Urban planning has been concerned with this problem for almost a century, and various proposals have been put forward. One approach was to cut down on size and scale the city accordingly. This was planned and implemented by the garden city movement proposed by Ebenezer Howard (Howard and Osborn 1965) (Fig. 6.21). If this proposal is interpreted from the cultural perspective of habitus as the producer of spatial consequences, the size of a garden city, at around 30,000 inhabitants, becomes more important. In the diagram *The Three Magnets*, Howard lists the attractions of his proposal in the context of the attractions of the existing context, and he does list cooperation as one of the benefits of his proposal. It produces

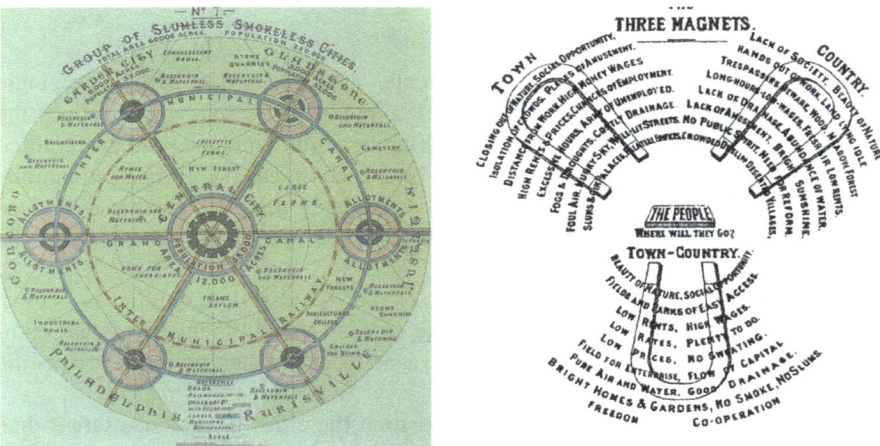

Fig. 6.21 Ebenezer Howard's *Garden City* diagrams as printed in *Garden Cities of To-morrow* 1902. The diagram on the left shows an abstract plan of how the garden cities are positioned around a central or core city. The diagram on the right shows three different forms of living as identified by Howard, in town, in the country and in his proposed garden city here titled as the ideal combination of the previous two, town-country. The diagram lists the positive aspects of all three forms of living and is entitled *The Three Magnets*

Fig. 6.22 The London communities map 1942, drawn by Arthur Ling and D. K. Johnson in 1943 as part of the comprehensive 1943 *County of London Plan*, which was the responsibility of Patrick Abercrombie. The map shows a simplification of the communities and open space survey undertaken as part of the investigation for the county plan and shows the existing main elements of London. The communities are in *brown*, the green spaces in *green*, the main industries in *dark blue*, the shopping centres as *light blue dots* and the town halls as *red dots*. This plan can arguably be identified as the origin of the description of London as a collection of small villages with their own identities

a manageable community which theoretically could be governed by one single cultural reference. This would result in a less complicated setting and reduce the amount of clashes and conflicts between different cultural interests and the required amount of negotiation. Essentially it could be argued that the idea was to streamline communities into monoculture settlements. Arguably this was not the conscious intention. From this perspective, however, comes a possible interpretation of the proposal.

Similarly this applies to Abercrombie's plan of London communities, as shown in Fig. 6.22. This would illustrate how the city is indeed organised around communities, with each of them defining a separate pattern and creating a specific identity.

This line of argument would clearly position the cities as artefacts rather than naturalistic organisms. The cultural basis of *spacemaking* describes the settlement as an intellectual act and embraces the human power of creation as the driving force for the building of cities. The habitus leads to the form and identity of the city as a consequence, not a fact in itself.

Even though we stated at the beginning of this section that the results are not promising in terms of the data collected, it turns out otherwise. The results demonstrate a number of points, for example, they illustrate the relationship between repetitive patterns of activity and the differences between the personal and the collective manifestation of a rhythm. However, it is clearly not one or the other; it is all that shape the identity that is appreciated as a result of the cyclical repetition in the context of the physical environment of the city.

The traditional field of time-use research goes back as far as the early years of the twentieth century. Especially in Northern America and Russia, time-use studies became popular together with studies into biometrics and ergonomics as described, for example, by Pentland and Harvey (1999). The quantification of the everyday was of interest, and over the last 100 years, many studies approached this field. One of the main criticisms is the fact that time-use studies mainly focus on a macro-level (Steinbach 2006).

An important point is to what extent personal cognitive capabilities are required to plan time use. It is suggested that sub-competencies such as structuring, planning, dividing, deciding and so on are required to successfully schedule activities (Steinbach 2004; Freericks 1996).

The three categories for recording the details have provided a framework conceptualising the information. It is clear that the setup of the framework always is a factor that influences the results, particularly in this case, where essentially the participants were asked to generalise the information. This is insofar as the question was not only to note down a series of particular days but to give information about how they plan their time. As such it can be noted that this is not a pure time-use study, as it aims to capture orientation and planning rather than pure activity.

In this sense, the actual activities are to be understood as what they are, processed details of what the participants identify as important to their planning. They can be looked at as the cornerstone of the schedule. There is no claim for a complete representation of the nitty-gritty of everyday activities. However, the date provides a good picture of how participants manage their time and activity space as well as how well they identify with the routine inscribed in the activities.

Each scale category has shown an individual rhythm: in fact, all have shown several modes of organisation. Especially interesting are the influences across scales. It has been, for example, difficult to separate the day from the week. There are many dependencies and entanglements creating conditions influencing the rhythm down the line. This made it difficult for participants both to record the information and to analyse the information provided.

It has become clear that the categories and especially the activities are influenced by additional factors, and time is not the primary factor here. Schedules are not organised according to the required slice of time available. Clearly planning ahead does not primarily orientate on the time as a finite entity, like a monetary budget (Steinbach 2006). Rather, organisational parameters are drawn from the context of each activity and the purpose. None of the participants have expressed a reading of time as a limited resource that has to be divided into smaller segments. The driving force of the context has a regulatory character and relies on a structure of repetition and experience rather than individual cases. Patterns are, as expected, related to several different factors. These include scale, activity and context. For example, an activity can have a strong influence on one scale in the context of several others, but take on a secondary role on another scale.

In regard to spatial consequences, the presence of such strong practices inscribes the degree of social mobility, in the sense of both class and area. As seen from the examples and the conclusion, social ties play an important role. They are a time and space constraint, to quote (Hägerstrand 1970), according to group and class. As the data demonstrates, a handful of locations dominate both the time and the space pattern. The nature or the class association of these can be expected to strongly influence the class of the actor. This would mean that the habitus could play a role in hindering social mobility.

As discussed earlier, the dominant aspect is continuity over the unit or time frame. For example, commuting is a dominant feature of the daily routine and a defining activity in itself. However, on a weekly scale, commuting is a necessary secondary activity that is often implied rather than actively planned. Nevertheless, it is important to note that the scales are a simple and effective tool to structure the information, but in detail impose boundaries difficult to account for in the results since most activities cross them more than once.

From the results, it has become clear that the time frames for mobility and interactions in the urban environment between stationary locations of activity are rather limited. Besides the commuting times, there are few concrete activities engaging with people's surroundings. Some activities such as socialising and exploring, mainly focused on weekends, do include various degrees of movement. Mainly however, activities focus on a specific location which is convenient for a specific task. In this context, infrastructure is quite an important aspect of the schedule, with mobility being secondary.

Fundamental to the results presented here is that the context of the activity is the structuring element to the schedule and not the planning according to time budgets. This explains partly how cultural and local specifics can influence the schedules, creating different organisational patterns over the same periods containing similar activities. Even the local influence of strong activities such as *work*, for example, impacts the schedules in such a way. The importance of *work* is much stronger than the actual percentage of time it occupies. Similarly the fact that participants commute further every day to go to work cannot be explained by Londoners having more time to spend commuting. It is a combination of local practice, spatial and economic conditions. Of course in the end, other activities will not fit or have to be shortened, since in both locations participants have the same time budget, but the time commuting occupies is not the primary factor.

Moving from the individual level to a collective level and looking at the aspects of activities in time, the next section of this chapter will analyse Twitter data collected as part of the New City Landscape project. Activities are here no longer assigned to an individual or a specific task. Rather the analysis picks up on the findings discussed above and specifically looks at places and times in the context of the aspects of activity patterns outlined. Does the urban area as a whole follow similar patterns of activity, and how do the cultural differences influence the appearance of local structure?

6.4 NCL Times

Looking at the time information stored as part of the Twitter message metadata collected through the New City Landscape project, a similar time-use and activity analysis to that in Sect. 6.2 has been undertaken. However, in comparison to the Urban Diary data, here we have an overall view of the urban area as described previously in Chap. 5. Essentially the perspective is shifted from the individual and personal activity to the collective buzzing of action across the entire city. Times and activity measures are no longer based on strings of sequences, but are now representing the ebb and flow overall.

Each message sent on the Twitter platform is tagged with a time stamp, indicating the exact time and date it was sent. The Twitter service is based on the idea of a continuous feed of information, ticking in messages and feeding them to clients and platforms in real time. The sequencing of the messages is primarily based on the time stamp of each message. By default there is no thematic or user-based grouping of messages. The basic user interface is designed around a vertical timeline, where new messages come in at the top and old messages are pushed down at the bottom. The time stamp therefore is one of the prime features of interaction with the Twitter service. As such the service is promoted in the present, clearly directed at current trends with little interest in the past. In fact there is no real history to the messages. At least Twitter does not archive anything beyond the most recent 2,000 messages. There are, however, other projects that log Twitter messages and store them for different purposes. There is the Library of Congress,[7] where the entire Twitter archive is being stored. For some time Google ran a service called Google Realtime where the Twitter archive could be accessed. However the deal between Twitter and Google that was set up in 2009 ended in June 2011, and Google is currently not providing this any longer.[8] In this context, time is used in different ways, serving as the technical sorting indicator in a linear fashion, as well as the basic idea the service implies on an experiential level. The massive flood of individual messages, however, lends an ephemeral appearance to each tweet, since the individual message is rather quickly superseded by many thousands of new *public shouts* as the 140 character message was labelled in the early days of Twitter.

The total number of tweets sent per minute, however, does not allow the time stamp to be a unique identifier.[9] The time stamp is a mere approximation. Time with minutes and seconds as a framework simply is too broad to cope with the amount of data. This extends beyond the sequencing capacity to the reference of a tweet.

---

[7] In 2010 Twitter donated the entire archive of tweets, since the beginning of 2006, to the Library of Congress. It was jointly announced on the Library of Congress Blog (Raymond 2010) and the official Twitter Blog (Biz 2010).

[8] The deal was announced in October 2009 at the Web 2.0 Summit. The ruling contract for Google to access the Twitter archive expired on 2 July 2011 according to Barnett (2011).

[9] By the end of 2011 (December), a report on the blog The Big Picture (Ritholtz 2011) shows that Twitter receives about 98,000 messages a minute. This is staggering 1,650 messages per second.

Identifying when an individual message was sent is not straightforward. The time shown usually corresponds to the local time of the device. The sender however might be in a different time zone. As a unique identifier therefore, the tweet id number can be used. This however is not a standard feature of the tweet message. It can only be found in the tweet's URL or the metadata. Most of the mobile or specialised Twitter platforms do not show either of the two which makes it difficult to uniquely identify the message.

For the purpose of this research however, the provided time stamp information is sufficient. The sequence is not of high importance. What we are looking at is the number of tweets per time unit. The time stamp therefore can be utilised for this project to access the data under temporal aspects. Using this time tag information, a detailed timeline of the messages recorded in real time for the New City Landscape project can be created and analysed. To do this, Time Rose Diagrams are used to visualise the flotation of the vital activity in different cities from around the world.

6.4.1 Time Rose Diagrams in Detail

The invention of rose diagrams is widely associated with Florence Nightingale (1820–1910) in her famous *Diagram of the Causes of Mortality in the Army in the East* printed in the publication *Notes on Matters Affecting the Health, Efficiency, and Hospital Administration of the British Army: Founded Chiefly on the Experience of the Late War* (Nightingale 1858). See Fig. 6.23 for illustration. The diagram[10] depicts the data collected by Nightingale during her time as a nurse serving with the British Army. Over time it shows the casualties for each month over the period of 24 months. By plotting the time around the circle, Nightingale assigned a section of 30°. Within this section the data values are plotted outwards. The circular arrangement, with its similarity to the familiar clock face, makes the time dimension comprehensible and, with its implied repetition pattern, indicates that a rhythm is inscribed in the data.

The diagram is also used in other areas. For example, rose diagrams are commonly used in geology. Here the rose is mainly used spatially in the manner of a wind rose with north, east, south and west orientations. In this context the branches are giving directions. This refers to shifting direction of stone pattern and suchlike. There is also a use in meteorology for wind directions with a similar spatial orientation of the rose.

---

[10] For a detailed discussion of the diagram, see, for example, an animated version of the diagram on http://dd.dynamicdiagrams.com/2008/01/nightingales-rose/ [accessed on 2011-08-30], or the documentation on the BBC at http://www.bbc.co.uk/programmes/p00chk4w with the title *Florence Nightingale's Rose Diagram* as shown on BBC Four [accessed on 2011-08-30].

6.4 NCL Times

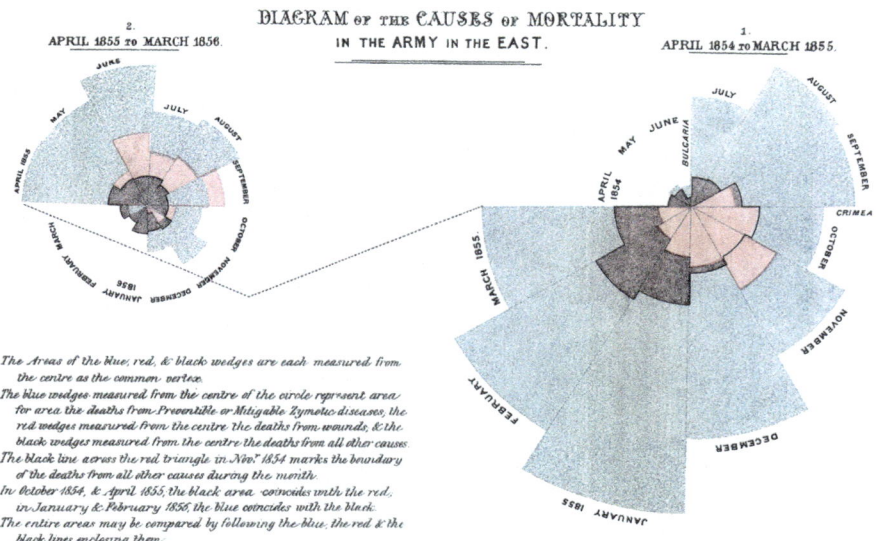

Fig. 6.23 *Diagram of the Causes of Mortality in the Army in the East* printed in the publication Nightingale (1858). This graphic illustrates the number of deaths that occurred from preventable diseases (in *blue*), those that were the results of wounds (in *red*) and those due to other causes (in *black*). The legend reads: The areas of the *blue, red and black* wedges are each measured from the centre as the common vertex. The *blue* wedges measured from the centre of the circle represent area for the deaths from preventable or mitigable zymotic diseases, the *red* wedges measured from the centre the deaths from wounds, and the *black* wedges measured from the centre the deaths from all other causes. The *black line* across the *red triangle* in Nov. 1854 marks the boundary of the deaths from all other causes during the month. In October 1854, and April 1855, the *black* area coincides with the *red*, and in January and February 1856, the *blue* coincides with the *black*. The entire areas may be compared by following the *blue*, the *red*, and the *black lines* enclosing them (See image on Wikipedia http://en.wikipedia.org/wiki/File:Nightingale-mortality.jpg)

The Time Rose Diagram used here however is, as its name indicates, based on the circular clock framework, plotting continuous time sequences clockwise around the centre. Similar to a spider diagram, data values and numbers are plotted outwards for each time unit. Different elements can be added, such as multiple values or multiple repetitive time axes, building outwards as individual rings. Such an example is shown in Tufte (2009, p. 72) by Antonio Gabaglio (1888) in *Teoria Generale della Statistica*. See Fig. 6.24. Another example is developed by Fisher (1995) as a Time Rose Diagram. The illustration Fig. 6.25 is an example taken from his book *Statistical Analysis of Circular Data* (as before) and shows arrival times at an intensive care unit over the period of 12 months.

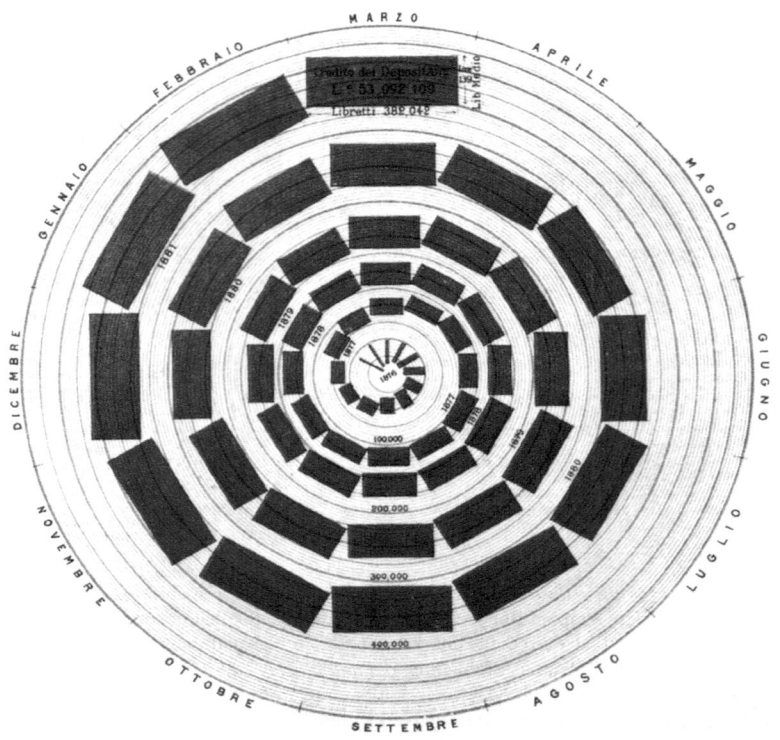

Fig. 6.24 Diagram by Antonio Gabaglio, illustrating low-dimensional multivariate data with time spiralling from the centre outward, the period being 1 year

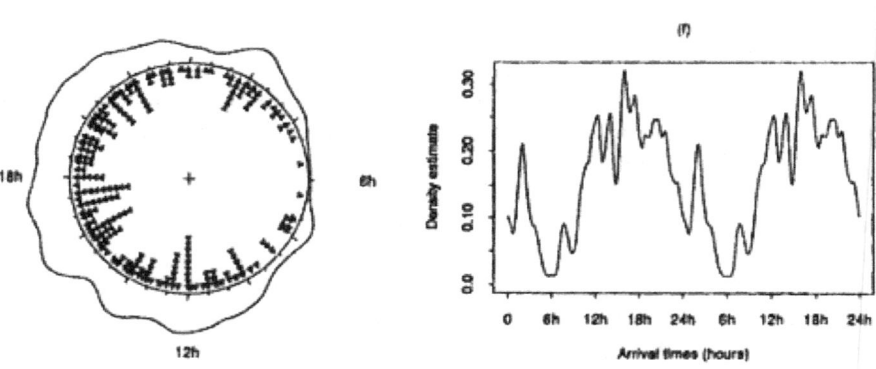

Fig. 6.25 Diagram by N. I. Fischer, taken from his book *Statistical Analysis of Circular Data* (Fisher 1995). It shows the arrival times of patients at an intensive care unit over 12 months. On the left the data is plotted clockwise on a 24 h face as stacked units with an additional ring showing smoothed values as a curve. On the right the same data is shown as a smooth curve linearly

6.4.2 24 Hour Activity in Urban Areas

Using Time Rose Diagram to visualise the data collected with the NCL project, the ebbs and flows of Twitter activity are visualised, purely focused on the time aspect. This has shown that the fluctuations between different times of day or different days of the week, for example, are significant and clearly show a pattern that can be related to activities. Coming from a rhythm analysis perspective, the emerging patterns are not surprising and seem to be in line with the general everyday real-world activities already observed in the earlier sections of this chapter: see Sects. 6.2 and 6.3. For example, during the early morning hours, between 3 and 5 a.m., there is a clear dramatic dip in numbers of messages sent on a 24 h cycle. This characteristic appears across all samples visualised to date. This is the same drop discussed above, at times of little to no activity. Furthermore there are clear peaks for morning and afternoon activities. Nevertheless, the different cities have their individual characteristics, related to the amount of activity at a particular day of the week or time of the day. Cultural aspects become visible quickly in this data set.

The Time Rose Diagram is a circular plotting of the 24 h day unit allocating a 15° slice for each hour. It is the same principle as a spider graph. Each hour is counted clockwise from the top starting at midnight on a 24 h face. Midday or 12h00 is located opposite at the bottom with six, 18h00 respectively on either side. This circular visualisation puts a strong emphasis on the repetitive character of the day unit. Reading from the Time Rose Diagrams of the different places analysed, a number of observations can be noted. In the following results, two samples of cities from around the world are discussed. This allows comparing the influence of location and cultural aspects. The first sample includes the following cities: Dubai, Istanbul, Los Angeles, Boston, Atlanta, Manila, Mexico City, Cairo, Bogota, Sydney, Paris, Munich, New York and London. The second sample includes 20 cities as described in Chap. 5 in Sect. 5.6: Amsterdam, Bangkok, Barcelona, Bogota, Buenos Aires, Calgary, Den Haag, Denver, Dubai, Hong Kong, Lagos, London, Mexico City, Moscow, Mumbai, New York, Paris, San Francisco, Seoul, Singapore and Tokyo. The analysis focuses on each hour of day, specifically on peak hours and low volume times during 24 h as well as weekday differences.

The data collected for Sydney is plotted as a Time Rose Diagram in Fig. 6.26 and thus reflects a more or less perfect kidney shape. The kidney shape is due to a dramatic drop in activity during the early morning hours, with otherwise a more or less steady activity flow throughout the day.

In this Sydney example as the kidney type, the day starts around 9 a.m., just as the curve starts to reach its daylight average. During the daylight hours, this will stay more or less the same with a slight increase around early afternoon, 15h00, and another one between 7 and 8 p.m. After 8 p.m. the volume reduces slowly first to 10 a.m. and just after midnight drop steeply to its lowest counts between 3 and 4 in the early morning hours of the day. If we now compare other cities to this *kidney-type* shape of Sydney, the following differences between the cities can be noted. It appears that Munich shows an early morning peak, about 1 h before London and

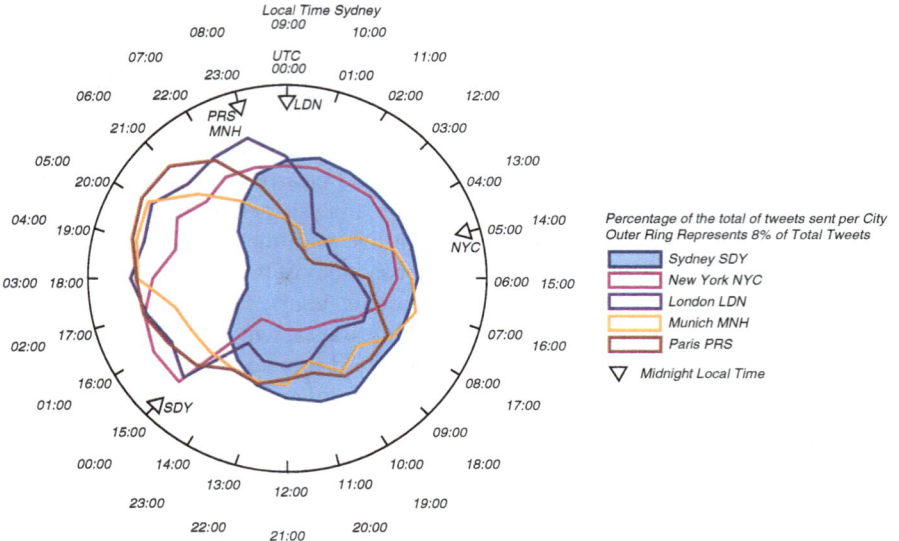

Fig. 6.26 The Sydney Kidney, a time activity plot in a Time Rose Diagram comparing Sydney to New York, London, Munich and Paris. Each city is plotted at the same UTC time showing how the peaks vary over a 24 h UTC period

up to 2 h before Sydney or Paris. On the other hand, New York and London are the cities that *never sleep* with a low dip in numbers of tweets around 2 h after midnight. Such a low dip stands in contrast to Munich where the slowing of the activities starts already after 9 p.m., Paris after 10 p.m. or indeed Sydney at around midnight.

The temporal aspects show variations between the different cities. As a different mode of classification, a sample of nine cities have been ordered according to activity preferences (see Fig. 6.27 for an illustration). Depending on whether they are more active in the morning, in the evening or all day round, they are ordered in line from evening (on the left) to morning (on the right). Some cities are clearly more active in the morning and others in the evening. Dubai and Istanbul, for example, are clearly more active in the late hours, where on the other end Cairo and Bogota are early birds and tweet a lot more in the morning. The US cities Boston and Atlanta have both a peak in the morning and in the evening.

Looking at the sample of 20 specific cities in detail, similar patterns can be observed. Depending on the characteristics, several groups emerge that share key aspects. What is being looked at in detail is the low point and the peak time, as well as the periods of rise and fall of activity. The average times across all 20 cities give a good first impression in terms of activity across 24 h.

The main characteristic, as noted above, is the dip in activity during the early hours of the morning. Between 3 and 6 a.m., there is a clear low point. No matter which location, the early morning is the time people are asleep. On the other hand, in the average plot, activities are only marginally rising towards the evening. Mainly

6.4 NCL Times

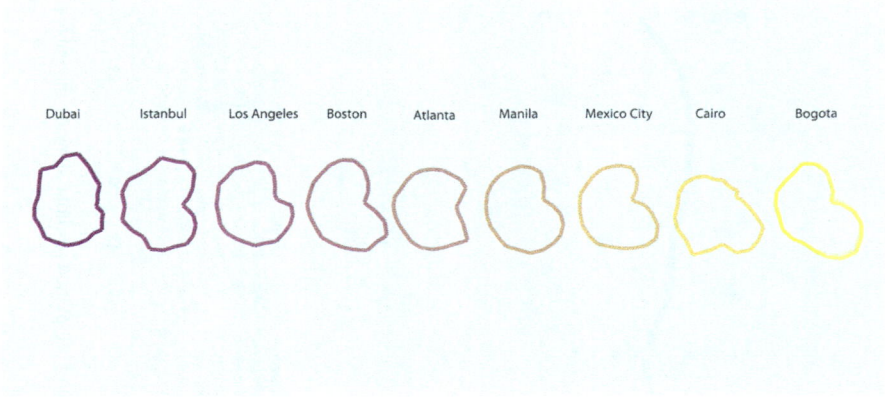

Fig. 6.27 Ordering nine cities from around the world according to the Twitter activity by hour of the day. *Far left* represents evening; *far right* is the morning. Cities in the centre have less particular occurrence of activity at a certain time of the day

they are steady between 10 a.m. and midnight. The period of decline is showing between just after 11 p.m. and 3 a.m., and the uptake of activity starts after 6 a.m. until around 10 a.m. See Fig. 6.28 on page 176.

Comparing the 20 cities with reference to this average plot, the characteristics are drawn out as below average, above average or similar to the average. After this, the cities are grouped according to their similarities. For a summary of peak and low point times, see Fig. 6.29.

There are two cities that have a late low point where activity is at a minimum. These are Tokyo and Dubai. Both cities record the least activity on Twitter at around 6 a.m. Both also record a slow start and remain below average for the rest of the day. It is not only until the following early morning hours when both overtake the average and remain above average activity, until they hit the low point around 6 a.m. again. It is also important to note that the low point for these two cities is not as low as the average. Activity still remains relatively high, at least half a per cent point above average.

The cities recording low points around 5 a.m. are Buenos Aires, Hong Kong, Moscow, Mumbai, Seoul and Singapore. They display clear similarities as to how they drop down, along average lines, in activity from after midnight and how they rise, but remain below average in the morning hours. All except Seoul remain a good half a per cent point below average at the low point. Similarities are then suspended throughout the afternoon and evening, with each city following its own patterns.

On average the low point is at 4 a.m. The cities which share this characteristic are Amsterdam, Bangkok, Bogota, Denver, Lagos, London, Mexico City and Paris. All except London remain below average at the low point. The rise of activity after the low point is rather steep (between 5 and 8 a.m.), along the average line. London is

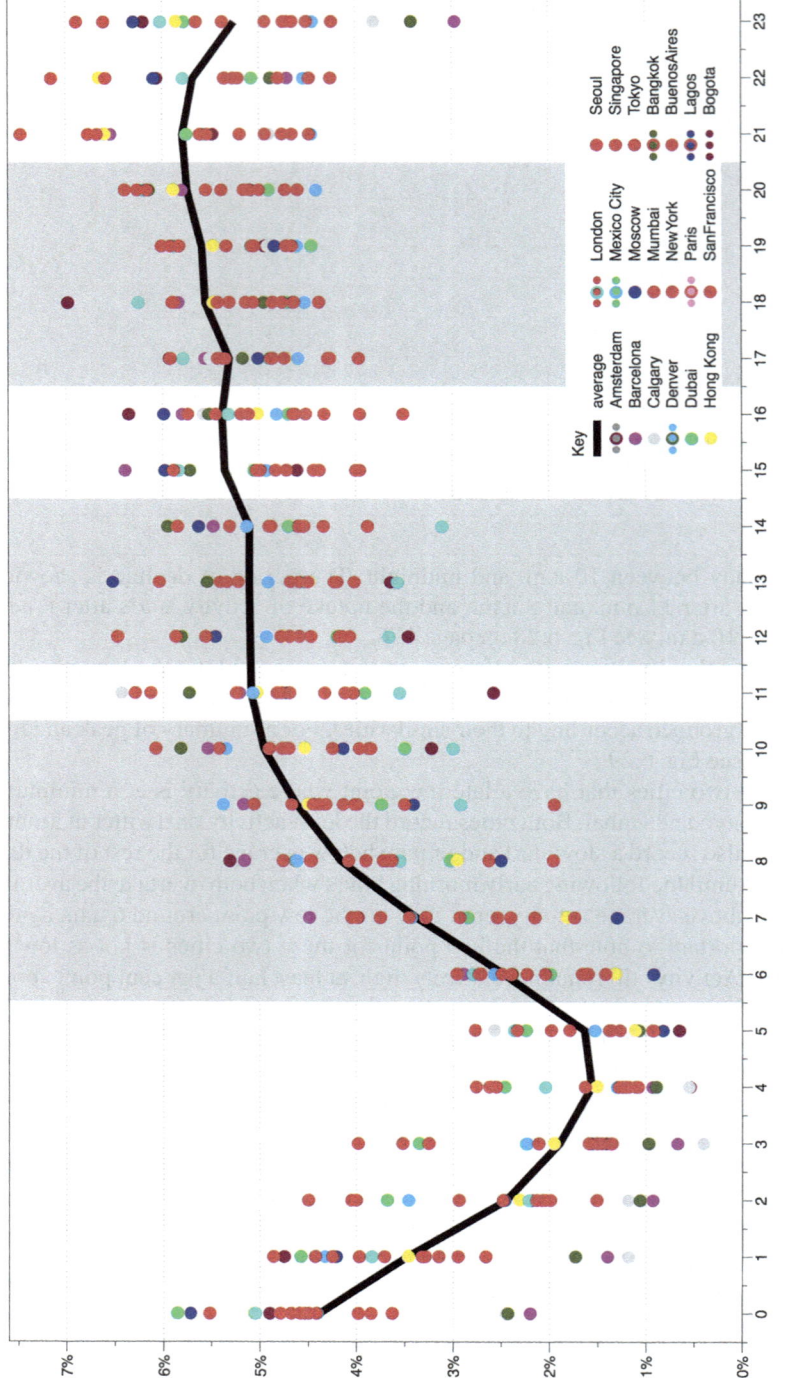

Fig. 6.28 The plot shows the average activity curve in overall per cent over 24 h (*black line*). The individual cities are plotted per hour showing the range of activity amount

6.4 NCL Times

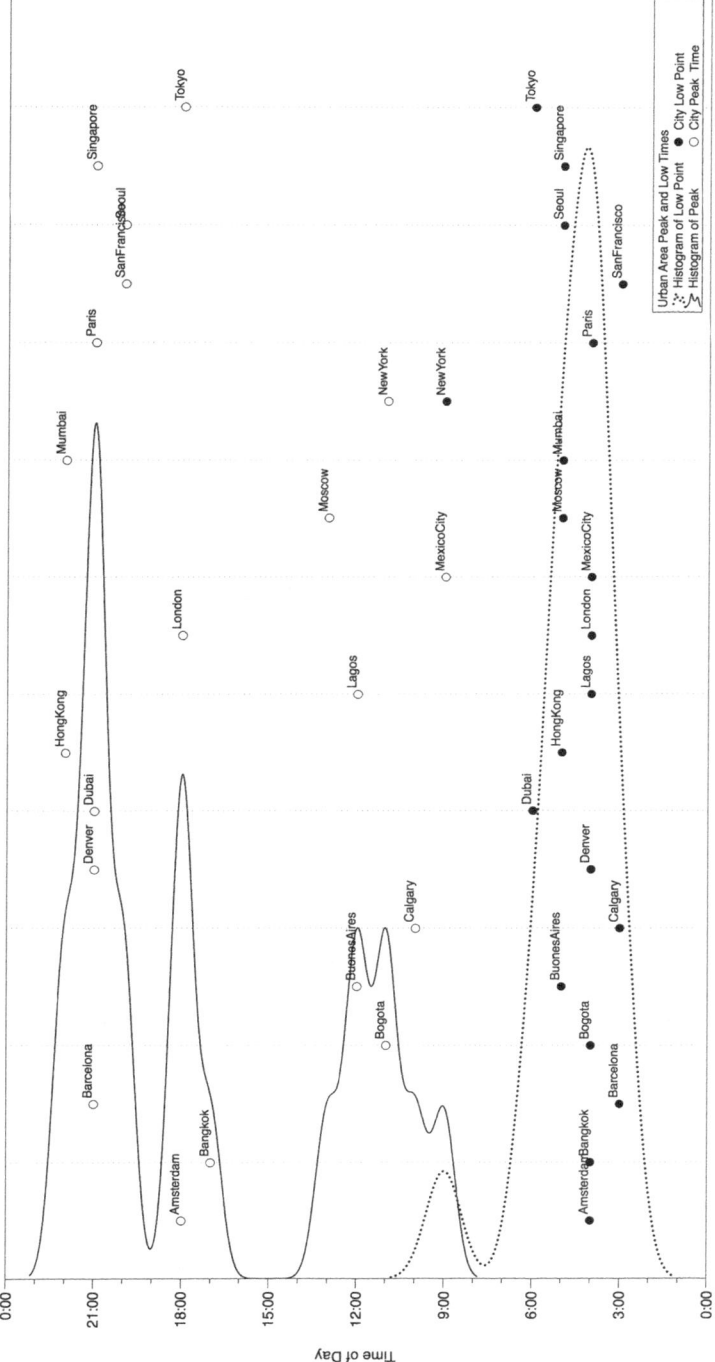

Fig. 6.29 The plot shows the time of day for both peak activity (*black line*) and low point (*black dotted lines*) for all 20 cities. The *circles* show the position of the city's peak times and the *dots* the low points, according to the 24 h scale on the y-axis

again the outlier here. Throughout the rest of the day, there is little similarity within the group before they fall in unison to the low point, after midnight.

Early cities to record a low point are Barcelona, Calgary and San Francisco. They share again a rapid rise in activity to 5 and 7 a.m. where they rise from around 1.5 % to around 4.5 % points. All three are amongst the earliest cities.

The observations can be summarised, that an early low point in this data set results in an early start the next morning. Similarly a late low point results in a slow morning.

Regarding the peak times of activity, the 20 cities are divided into two groups. One group of cities reaches their peak between morning and midday and the other between late afternoon and evening. Seven cities represent the morning group with Mexico City, Calgary, Bogota, New York, Lagos, Buenos Aires and Moscow, ranging from 9 a.m. to 1 p.m. The latter group has 13 cities spread between 5 and 10 p.m.

The analysis of the 24 h time activities has shown that there is a clear pattern in cities from around the world, characterised by a regular dip of activity in the early morning hours and a rise of activity throughout the daylight hours. Whereas there is only slight variation across the sample in the dip phase, where activity drops to a minimum, there is greater variety in the ways different cities are active during the day. No clear pattern has emerged from the data, but there are two main groups. One group of cities is more active in the morning, and the other one has its activity peak in the evening hours. In some cases specific tales or sayings about cities are reflected in the data; for example, New York is the city that never sleeps, which can be an interpretation of the NYC activity pattern as seen here.

The way cities are active for 24 h and especially during the night and early morning hours is being investigated as the night-time economy, picking up on suggestions made by Jacobs (1961). In the past 10 years, the literature suggests a dramatic change has taken place, transforming the urban landscape of activity with extended opening hours and services, as, for example, discussed by Hadfield et al. (2001). It can be assumed that now, 10 years on, a shift towards even later activities can be observed. Nevertheless, it also is notable that the period of low and high activity is somehow linked, and cities with an early low point, at around 3 a.m., also show an earlier start in the morning, as compared to cities with a late low point around 6 a.m., such as Tokyo, which has a slow uptake of activity in the following morning.

6.4.3 Weekday Preferences

Similarly there are preferences regarding the weekdays (see Fig. 6.30). Not all areas tweet on the same day. The early weekdays Monday and Tuesday are generally less active than the rest of the week. Manila clearly prefers the weekend, where Cairo,

Fig. 6.30 Comparing the same nine cities as in Fig. 6.27 over the period of the week per day. The weekdays are drawn in *green* and the weekend days in *purple*

Istanbul and Mexico City prefer the end of the week, Thursday and Friday. Dubai and especially Bogota have the least differences between the weekdays with similar numbers of tweets throughout the week. These patterns can be attributed to cultural aspects. For example, Thursday and Friday are the weekend in the Islamic countries, which explains peaks in Cairo and Istanbul on these days. Manila on the other hand is about 83 % Roman Catholic, responsible for the strong Saturday/Sunday activity, as shown in Fig. 6.30.

6.4.4 Topics and Keywords

Other indicators of temporal patterns are the keywords and general content of the messages sent. The word cloud generated from the content captures the topics on discussion in the area. Twitter is quite responsive and reflects wider trends in almost real time. Twitter has mutated from a geek tool to a mass-curated news platform. A number of events in the past 2 years have shown how quickly information spreads on the platform and how it has overtaken established media channels (Lerman and Ghosh 2010; Kwak et al. 2010). One of the first events to spread over the Twitter network, surprising established channels, was the Iranian Election in 2009

(Leavitt 2009). A number have followed since, including the earthquake in Haiti, the earthquake in Japan in early 2011 and the London Riots in 2011 (Burn-Murdoch et al. 2011) to name a few.

Monitoring of keywords and tags, with sophisticated software systems, is in place as a result. It is still challenging to run full-text analysis in real time on such a flood of individual messages, not to mention the interpretation of the content. Several systems rely on a way of aggregating by focusing on tags, as the words preceded with a # are called, or keywords, as top-mentioned words based on a numeral ranking, for example, the mood mapping project (Mislove et al. 2010) focusing mainly on the United States as discussed in Chap. 5 on page 97. This project rated the content of the tweet based on ranked indicator words for which the messages were scanned. The ranking covered a range from good to bad, giving an idea as to how the nation feels.

The content however is not necessarily Twitter based. The tool ties in closely with other media channels. A new topic can then be boosted by a range of users jumping across to the network, streaming from its source. This could be in response to other media activities, such as a TV programme or an Internet rumour. The Twitter platform has become a vital part of the tools for information distribution throughout the Internet. Information can spread quickly and become *viral* if the right Twitter users pick it up. This is visible in the London data set, as shown in Fig. 6.31, collected during the UK Leaders Debate on British television (streamed in spring 2010). The hype of the third and last live discussion before the national election was reflected in the data on the day before and the day after the live discussion. The hashtag #leadersdebate was at the top of the hashtag league table throughout the week, but clearly led it around the hours of the actual live debate. This reflects a dominating interest across the network.

This phenomenon is, however, localised. In the data set of other cities collected at the same time other topics dominate. Whilst in London the #leadersdebate is top,

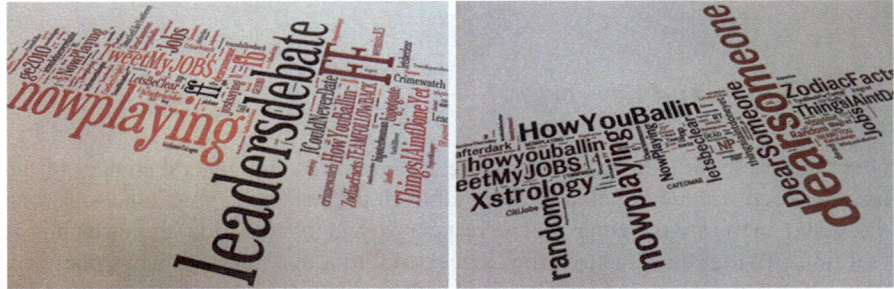

Fig. 6.31 Tag cloud showing top-level keywords for London on the left and New York on the right. Data was recorded simultaneously over the period of 1 week as part of the New City Landscape project

in New York social tags such as #dearsomeone or the music-related #nowplaying[11] predominate. Another popular tag frequently in the top ranks is the #ff used for something worth looking at. This tag features in all the large European cities and is usually related to a link sent in the same message. All these elements of the tweets tie in directly with the temporal flows of information and interests. It reflects how news travels the city and where it peaks.

Trends can sharply rise and dominate over other topics. However, similarly steep trends can collapse and disappear. As we could observe from the example of the tsunami resulting from the devastating earthquake in Japan in early 2011 crossing the Pacific, the lifespan of such a topic is short lived even, or especially, in a place that is directly involved (see also Fig. 6.32 for illustration). After the earthquake the keywords, earthquake, wave and tsunami in Honolulu, Hawaii, which lay in the direct path of the tsunami, have steadily increased. They did peak in sync with the arrival of the wave on the shore of the islands of Hawaii. Within hours of the passage of the wave however, the topics disappeared again from the platform.

At the same time the topic provides an additional link between activities, interests and the place. This place can be the current location or a destination still to be reached. The ephemeral nature of the message is reflected in the movement of the user. The here and now is at the same time reflection and prophecy. It is the activity in the making and as such fundamentally different from any other mapping exercise focusing on the objective and true nature of the representation. In this case the temporality takes over and merges the different steps, tasks, achievements and plans into one large stream of activity.

6.5 Rhythmic Time

The concept of a rhythm as a pattern of repeating structure in time is present from the start in the conception of clock time as built in the form of repetitively added units. There are, as shown in the examples above, also rhythms and patterns in time as an experience or a social continuum. The structuring aspect is not only present as a counter but in many different ways as a point of orientation, organisation and in the long term as a navigation aid.

From the example of clock time, it becomes clear on how the seconds, minutes, hours, days, months and years build the structure of orientation for both past and future events. The units provide a simple grasp of duration and brief scale of

---

[11]It has to be mentioned that some of these social tags are put into the Twitter stream through service providers. For example, the #nowplaying tag is put in by music streaming sites to promote the service through the consumer's personal social network. Nevertheless, it still reflects the activity pattern of users. It describes what they were doing since the automatic tags are based on usage.

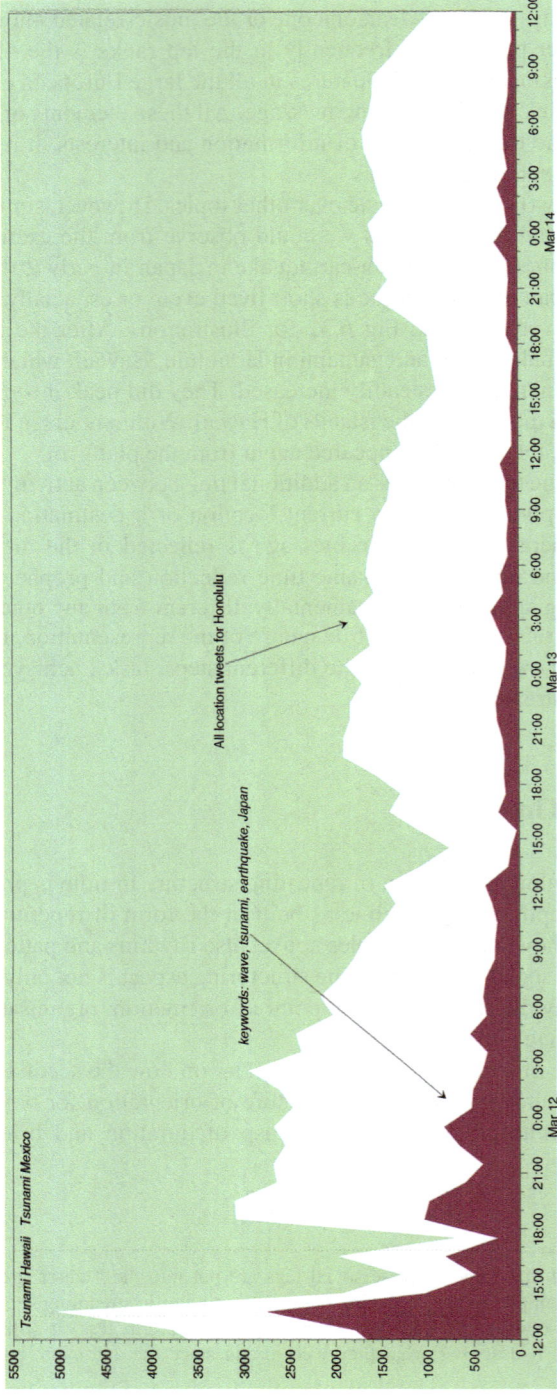

Fig. 6.32 Graph plotting the number of tweets in Honolulu, collected as part of the New City Landscape project, over time on the horizontal axis. It starts just after the earthquake in Japan, showing the arrival of the tsunami wave in Hawaii at the *first vertical line* and the arrival of the wave in Mexico at the *second vertical line* from the left. Overall tweets are in *white*, and in *purple* the number of tweets mentioning the keywords wave, tsunami and/or earthquake

comparison. As discussed earlier, these mathematical units do not lend themselves to all levels of experience. Quickly the units become either too short to express reality or too long to achieve any exactitude.

Clock time, as Elias (1992, p. 118) argues, can be read, mainly in the urban context, like masks in pre-urban societies. It is clear that masks, just like the clocks, are made by people, but they somehow represent an extra-human existence, a myth. Masks appear as embodiment of spirits. Clocks appear as embodiments of *time* (Elias 1992). In other words, there is much more to the clock than the mechanical count of units. It conveys a social message and has a specific presence. Through this, a clock can be just as spatial as the *here and now*.

A number of natural phenomena do visualise this referencing character of the clock time in connection to experience. The tide continuously ebbing and flowing at the shore is one such event beyond the human capacity of which to recognise detailed changes. Over approximately 6 h the water level drops by 3–5 m along the shore of the British Isles, depending on the position of the sun, the moon and the earth. Depending on the morphology of the location, this results in a steep and abrupt drop, in rocky and steep landscapes, or a wide stretch of flat land emerging and disappearing. The changes are visible if referenced and compared; however, if witnessed in real time, it is impossible to recognise as it happens. Even though the process is steady and happening at a large scale, it is hidden from the human eye.

It is impossible for the human rhythm to relate to such a large scale. The duration from high tide to low tide is a large chunk of an individual's daily waking time, if we set aside 8 h for sleep. Within this time frame, the individual will have performed a number of tasks, including eating, moving and so on. The attention span will not allow us to grasp such large-scale change.

Every now and then however, according to a reference, it will be possible to make out the progression of the change, as the water swallows this rock, grabs a stick or causes the sandcastle to collapse.

Examples shown relate to this in a number of ways and offer, amongst other things, insight as to how the orientation scope of human time is structured in order to navigate the everyday world of the urban environment. In the context of actual activity tasks, such as getting to work, taking the children to school, arranging for lunch or meeting with people, it is not enough to handle each activity by comparing it as progress. The whole task is managed in one piece, structured in a way tailored to the individual capabilities of rhythmic time.

As argued above, the single event is not outstanding, but it is the continuity that lies at the heart of the practice. The doing is more important than the achievement, and through doing, each activity finds its place in the daily schedule according to the wider context and cultural condition.

The various examples have been used to illustrate the aspects of the data recorded for this study. Overall the repetitive aspects have been implicit in the graphs and visualisations, but explicit in the activities of the recorded events. Routine plays an organising role in the everyday lives of participants. This has been shown in the interviews but also here under the aspects of clock time as GPS recordings and time-space as recorded plans and schedules. Interestingly, rhythms do play a role across scales and groups of individuals. Whilst the routine might be an obvious element for a single person, the larger effect of synchronised routines for a group and the wider context might be less noticed. The examples above have demonstrated that, beyond the individual, rhythms persist and remain consistently similar.

In some ways, it could be argued that routines only make sense if they are to be synchronised with other patterns, hence the requirement to retain a study pace. In the context of spatial planning and urban organisation, this aspect of the topic is of course of major importance. A city being more than the sum of its parts relies on the compatibility of these parts. Taking these observations to such a general level, the suggestions would be that the collective rhythm is reflected in the individual routine and vice versa. The connection is the cultural convention expressed in spatial practice.

Overall the findings have not been surprising. There have been a few oddities in terms of how different individuals or different places diverge slightly on rather specific aspects. Some places tend to be later, some earlier, and some have a particular peak or downtime that is more articulated than others. However, overall the similarities have been a really strong feature of the patterns tracked in this research. Arguably the rhythms do produce similar patterns and a strong set of core activities. Whilst social convention might be responsible for such a synchronisation, the strength of the similarity, the convergence in fact, is stunning.

Whilst the investigation and the probes focus on the differences, the strong result is in the similarity that is shown as the sort of negative image of the research presented here. This aspect of the rhythms found in the time data is something often too obvious to be looked at. Many aspects of the everyday are, as earlier outlined in the introduction, normal enough to be taken for granted.

That we have a 5-day working week and a 2-day weekend is not so old an invention, but nowadays it can hardly be imagined differently. Only a generation ago, the normal case was a 6-day week with 1-day off for the weekend, and two, possibly three, generations ago, there was no such thing as a weekend known to the general public (Burke 1995).

The same is true for the activities. Whilst holiday and leisure activities on the weekend are *expected* nowadays, this is a relatively recent phenomenon. The tourism industry is barely 100 years old (Burke 1995). This is also true for working hours. Where nowadays working hours are assumed to be roughly 8–9 h between nine and five, in the early days of the industrialisation, shifts were common and labourers as a collective worked 24/7.

All these examples illustrate the point that rhythms do change and they are not static at all. At the same time, cyclical patterns seem overwhelmingly dominant and enforcing for the individual. On numerous occasions, the participants expressed that in general they are not thinking about their routines. In particular they do not think of their schedules in the sense of routine as discussed in the introduction to this chapter in Sect. 6.1. Surprisingly there is still enough irregularity and instant decision making involved to mask the fact that the person sitting in the next seat on the bus is basically following the very same routine.

One of the possible explanations for this is that in regard to time, the capacity for humans to capture and process slow and long-term units is limited. Thinking of watching the tide come in on a sandy beach illustrates this argument from an experiential point of view. Intellectually it is known and a fact that the water level is rising. The tide timetable tells the exact time of low and high tide for any particular day. It takes a total of roughly 6 h for the process, and the sea level changes depending on the location and time of year between 2 and 7 m. The facts are there and still it is difficult to *see* the change. The rise is slow enough and over a long-enough time span that it is not visible to the human eye, and only when the waves suddenly wash over the beach towel is it physically clear that the tide is rising and the party has to be relocated further up the beach.

The everyday situations and the habitus in the city are constituted similarly. It may be more complex, in the sense of a dramatic hustle and bustle of millions of other individuals around following their own individual, but pretty much identical, routine. Arguably a constant stream of *in the moment* decisions takes up a lot of attention, preventing the individual from focusing on the bigger picture.

Whilst in the middle of all the activity unfolding, the habitus does not present itself. This reinforces the earlier argument in regard to the importance of continuity. The fact that everything is happening now does not distinguish between the past and the future nor the beginning or end of an activity. As a consequence the cycle has little relevance in the individual moment. However, for example, regarding any social organisation, the background structure of the cycle enables flexible but organised possibilities.

In many ways, this is where time directly feeds into the urban context of the city. As suggested earlier, cyclical arrangements are vital to provide the spatial conditioning that underpins, for example, meeting for trade. Market places and also religious sites are most likely even older examples, demonstrating this entanglement between time, location and activity. Aspects will be to extend these points in terms of permanence and temporality. One specific point of interest is the resulting space in a Cartesian sense. In Chap. 7 on *Space*, these conditions are examined using the same data sets analysed in this chapter. The main focus will be not on the time information, but on the spatial dimension.

Bibliography

Andersen HK, Grush R (2009) A brief history of time-consciousness: historical precursors to James and Husserl. J Hist Philos 47(2): 277–307
Barnett E (2011) Google realtime search suspended after Twitter deal ends. Telegraph.co.uk. http://www.telegraph.co.uk/technology/google/8617985/Google-Realtime-search-suspended-after-Twitter-deal-ends.html
Batty M (2005) Cities and complexity: understanding cities with cellular automata, agent-based models, and fractals. MIT, Cambridge
Batty M, Longley P (1994) Fractal cities: a geometry of form and function. Academic, London

Biz (2010) Tweet preservation. http://blog.twitter.com/2010/04/tweet-preservation.html
Brand S (2000) The clock of the long now: time and responsibility. Phoenix, London
Burke P (1995) The invention of leisure in early modern Europe. Past Present (146):136–150
Burn-Murdoch J, Lewis P, Ball J, Oliver C, Robinson M, Blight G (2011) Twitter traffic during the riots. http://www.guardian.co.uk/uk/interactive/2011/aug/24/riots-twitter-traffic-interactive
Cowan HJ (1958) Time and its measurement; from the stone age to the nuclear age. World Pub. Co., Cleveland
Crang M (2000) Relics, places and unwritten geographies in the work of michel de certeau (1925–86). In: Crang M, Thrift NJ (eds) Thinking space. Critical geographies. Routledge, London/New York, pp 136–153
de Certeau M (1984) The practice of everyday life. University of California Press, Berkeley
Elias N (1992) Time: an essay. Blackwell, Oxford
Fisher NI (1995) Statistical analysis of circular data. Cambridge University Press, Cambridge
Freericks R (1996) Zeitkompetenz: Ein Beitrag zur theoretischen Grundlegung der Freizeitpädagogik. Schneider Hohengehren, Baltmannsweiler
Glennie P, Thrift NJ (2009) Shaping the day: a history of timekeeping in England and Wales, 1300–1800. Oxford University Press, Oxford
Hadfield P, Lister S, Hobbs R, Winlow S (2001) The '24-hour city': condition critical? Town Ctry Plan 70(11):300–302
Hägerstrand T (1970) What about people in regional science? Pap Reg Sci 24(1):7–24
Howard E, Osborn FJ (1965) Garden cities of to-morrow. MIT, Cambridge, MA
Hughes DO, Trautmann TR (1995) Time: histories and ethnologies. Comparative studies in society and history book series. University of Michigan Press, Ann Arbor
ISO/TC154 (2004) ISO-8601 data elements and interchange formats – information interchange – representation of dates and times. Standard 8601:2004, International Oranization for Standardization, Geneva
Jacobs J (1961) The death and life of Great American cities. Random House, New York
Kwak H, Lee C, Park H, Moon S (2010) What is Twitter, a social network or a news media? In: Proceedings of the 19th international conference on World Wide Web, WWW'10, Raleigh. ACM, New York, pp 591–600
Leavitt A (2009) The iranian election on Twitter: the first eighteen days. Web ecology project. http://www.webecologyproject.org/2009/06/iran-election-on-twitter/
Lefebvre H (2004) Rhythmanalysis: space, time and everyday life. Continuum, London
Lerman K, Ghosh R (2010) Information contagion: an empirical study of the spread of news on digg and Twitter social networks. In: Proceedings of 4th international conference on weblogs and social media (ICWSM), Washington, DC
Longstaff GB (1893) Rural depopulation. J R Stat Soc 56(3): 380–442
Loua T (1885) Les déplacements de la population en france d'après les derniers dénombrements. Journal de la société de statistique de Paris 26:118–129
Lynch K (1972) What time is this place? MIT, Cambridge, MA
Miller CR (1992) Kairos in the rhetoric of science. In: A rhetoric of doing: essays on written discourse in honor of James L. Kinneavy. Southern Illinoise University Press, Carbondale
Mislove A, Lehmann S, Ahn Y-Y, Onnela J-P, Rosenquist JN (2010) Pulse of the nation: U.S. mood throughout the day inferred from Twitter. http://www.ccs.neu.edu/home/amislove/twittermood/
Nightingale F (1858) Notes on matters affecting the health, efficiency, and hospital administration of the British Army: founded chiefly on the experience of the late war. Printed by Harrison and Sons, High Wycombe
Noulas A, Scellato S, Lambiotte R, Pontil M, Mascolo C (2012) A tale of many cities: universal patterns in human urban mobility. PLoS One 7(5):e37027
Pentland WE, Harvey AS (1999) Time use research. In: Pentland WE, Harvey AS, Lawton MP, McColl MA (eds) Time use research in the social sciences. Springer, New York
Portugali J (2004) The mediterranean as a cognitive map. Mediterr Hist Rev 19(2):16–24
Rämö H (1999) An aristotelian human time-space manifold from chronochora to kairotopos. Time Soc 8(2–3):309–328

Ravenstein EG (1885) The laws of migration. J Stat Soc Lond 48(2):167–235

Raymond M (2010) How tweet it is!: library acquires entire Twitter archive. http://blogs.loc.gov/loc/2010/04/how-tweet-it-is-library-acquires-entire-twitter-archive/

Reichholf J (2008) Warum die Menschen sesshaft wurden: Das größte Rätsel unserer Geschichte, 2nd edn. Fischer, Frankfurt

Richards EG (2000) Mapping time: the calendar and its history. Oxford University Press, New York

Ritholtz B (2011) 60 seconds: things that happen every sixty seconds. http://www.ritholtz.com/blog/2011/12/60-seconds-things-that-happen-every-sixty-seconds/

Smith JE (1969) Time, times, and the 'Right time'; "Chronos" and "Kairos". The Monist 53(1):1–13

Smith JE (1986) Time and qualitative time. Rev Metaphys 40(1): 3–16

Steinbach D (2004) Ideale temporale Muster als kognitive Wissensstrukturen über den Umgang mit der Zeit: Eine qualitative Studie am Beispiel engagierter Ausdauersportler. PhD thesis, Deutsche Sporthochschule Köln Deutschland, Köln

Steinbach D (2006) Alternative and innovative time use research concepts. World Leisure J 48(2):16–22

Stouffer SA (1940) Intervening opportunities: a theory relating mobility and distance. Am Sociol Rev 5(6):845–867

Tabboni S (2001) The idea of social time in norbert elias. Time Soc 10(1):5–27

Tuan Y-F (1977) Space and place: the perspective of experience. University of Minnesota Press, Minneapolis

Tufte ER (2009) The visual display of quantitative information, 2nd edn. Graphic Press LLC, Cheshire

Whorf BL (1950) An American Indian model of the universe. Int J Am Linguist 16(2):67–72

Zerubavel E (1985) Hidden rhythms: schedules and calendars in social life. University of California Press, Berkeley

Zerubavel E (1989) The seven day circle. University of Chicago Press, Chigago

Chapter 7
Structuring Space

Abstract This chapter presents the results from the perspective of space, again including both, Urban Diary and New City Landscape data sets. The starting point is Cartesian space. Discussed are the aspects of space on this basis, questioning its relevance for the different data sources. In this context, the focus is put on the implications for the interpretation of the data regarding the comparison if this spatial system is used as the reference. The sections in this chapter are used to test the aspects of space in various settings from examples of globally mapped out tracks and traces, as well as cities, to the animated visualisations of time-based mapping. These can be discussed under the heading of space for both visualisation but also for the interpretation of space. Furthermore in section Social Space, space looked at as it is described and registered by the individual. Here the data is sourced from the mental maps, using the experience of space as communicated by the participants. This leads to the discussion of urban spaces as self-territory, with an individual perspective composed of personal cities, and how this links the spaces from both individual and collective perspective. With Urban Islands, the focus shifts to the idea of how the important locations act as anchor points and relate to one another in analogy to the Naked City as proposed by Debord in 1957 and the creation of different types of spaces according to travel patterns.

This chapter presents the results from the perspective of space, again including both, Urban Diary and New City Landscape data sets. The starting point is *Cartesian space*. Discussed are the aspects of space on this basis, questioning its relevance for the different data sources. In this context, the focus is put on the implications for the interpretation of the data regarding the comparison if this spatial system is used as the reference. The sections in this chapter are used to test the aspects of space in various settings from examples of *globally* mapped out tracks and traces, as well as cities, to the animated visualisations of time-based mapping. These can be discussed under the heading of space for both visualisation but also for the interpretation of space. Furthermore in Sect. 7.5, space looked at as it is described and registered by the individual. Here the data is sourced from the mental maps, using the experience of space as communicated by the participants. This leads to the discussion of urban spaces as *self-territory*, with an individual perspective composed of personal cities, and how this links the spaces from both individual and collective perspective. With *Urban Islands*, the focus shifts to the idea of how the important locations act as *anchor points* and relate to one another in analogy to the *Naked City* as proposed by Debord (1957) and the creation of different types of spaces according to travel patterns.

7.1 Cartesian Space

For the best part of the last 200 years, science was built on an objective understanding of nature. The natural sciences especially have strictly implemented the elimination of subjective views in science. As Daston and Galison (1992, p. 1) argue that the credo was:

Let nature speak for itself.

This goes along with a strong emphasis of the image over the description. Photographs of *facts* and mechanically generated curves are valued more highly than the words and expressions used to describe the phenomenon. This was not only about accuracy and precision in scientific experiments and data analysis but about morality itself (Daston and Galison 1992).

The rise of strict objectivity in spatial terms can be seen in this context. Cartography and the surveying of land gained importance within trends to mechanically generate images whilst evading subject-introduced mediation:

Where human self-discipline flagged, the machine would take over (Daston and Galison 1992, p. 1).

In geography, as one of the sciences concerned with space, the *Cartesian system* consequently implies a demand for *objectivity*. The reference system of coordinates spanning the entire globe with latitude and longitude (see Fig. 7.1 for illustration), providing values for every possible location on our planet, is the manifestation of this dream, achieving objectified space. Ultimately GPS is the electronic implementation of this concept. A tiny gadget calculates the present position on the spot in reference to the virtual grid spanning the world.

The survey of land across the USA is one of the biggest projects to implement the strict *Cartesian systems* as a physically organising framework. It goes beyond the spatial aspect and deep into a cultural understanding of *objectivity* and is, to some extent, the continuation of the colonisation of America. It is the act of man ruling over nature. Manhattan's grid system was planned and laid out in the nineteenth century (Rose-Redwood 2002). Beyond the implementation of a logical system to enhance real estate values and economic efficiency, the designers of the grid had in mind a disciplined society, as argued by Rose-Redwood (2002). This of course would suit the needs of the emerging capitalist economy of the American Empire.

Today objective spatial understanding has moved on from only governing the plot of land to taking over the modern modes of transport. The moving object is

7.1 Cartesian Space

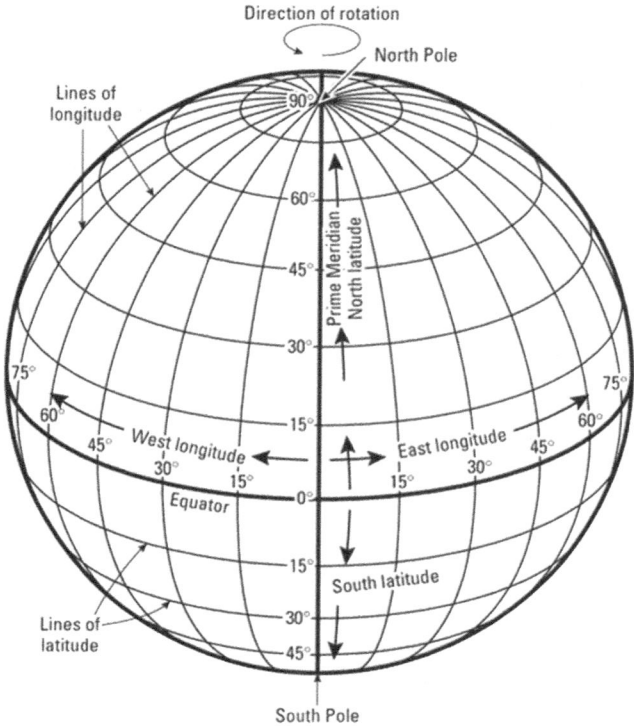

Fig. 7.1 The virtual grid as it spans the globe as latitude and longitude lines, starting from Greenwich in London and from the Equator, respectively (Image taken from Ocean Drifters)

being integrated in the static spatial framework through a sequencing of positions. In the following, the transformation of the city into a functioning object, an Urban Machine, took place. With this shift, objective usage of space has entered everyday life. No longer is it the scenic route or travel through a familiar terrain that guides the choice of route, it is the objective selection of the *best option* from the Sat Nav box installed on the dashboard, using GPS, thus instructs the driver what route to take.

The rivalry between *subjective* and *objective* in a range of disciplines is currently trending in favour of *objectivity*. Here geography of course has traditionally been at the forefront, advocating in favour of *objectivity*. There have, however, been strands of geography which question pure objectivity.

Both human geography and time geography, with their strong focus on the human subject, constantly have to question *objectivity* as a result of their field of interest. However, together with the social sciences, they work hard to be as *objective* as the natural sciences. As Bisbee (1937, p. 371) says:

> To be scientific is to be objective.

This is a more general statement in the context of the paper, she continues:

> This is a sine qua non in the social scientists' efforts to keep their research on a par with those of the natural scientists. (Bisbee 1937, p. 371)

To argue for the position of human geography, Ambrose (1969) discusses the term *environment* at length in his introduction to *Analytical Human Geography*. He states:

> One may conclude that the human geographer's interest in purely physical aspects of the discipline may decrease as non-physical phenomena become increasingly important in influencing man and as, in any case, the realisation grows that man is acting in a subjective not an objective environment. This is not to say that we have reached a stage where a comprehension of the physical landscape, and the processes affecting it, are unimportant to a human geographer. (Ambrose 1969, p. 4)

Whilst the data from the fieldwork is visualised using the Cartesian space concept as a reference in geographical tradition, wherever possible it is the intention to specifically highlight and represent aspects of subjectivity.

The analysis of such data is challenging in numerous ways, and the interpretation of individual subjectivity cannot always be detached from the subjective view of the researcher. However, with the methodological setup of the data collection, subjective data was specifically and systematically collected. These elements of the data will allow comparison to position *subjectivity* and *objectivity* as equal aspects of the research material and subsequent analysis.

7.2 Urban Diary Spaces

As part of the Urban Diary project, subjects' movements have been recorded using GPS technology, as discussed in detail earlier in Chap. 4. The GPS tracks are used here to map participants' activities. The data recorded by the device is effectively only locations, as a series of points. Every 7 s, the GPS tracking device logs the current location. The trajectory map results from connecting these points with lines (see Fig. 7.2 for illustration and explanation). However, in many ways this linear logging comes close conceptually, and often graphically, to the *path* concept, proposed by Hägerstrand (1982), discussed at the beginning of this book.

These trajectory lines represent what is assumed to be the line of movement. If the recorded points are close together, that is, recorded at frequent intervals, this is believed to provide an accurate representation of the participant's movement. With less frequent location points, the resulting *path* is a more abstract representation of the movement. In this sense it is a question of resolution, and, coming back to the *Cartesian system* of space, its application to the moving object, the sequential recording of absolute locations can be in itself a contradiction.

The aspect of *scale* is important in this context since, as indicated in Fig. 7.2, it notably influences the interpretation of the spatial nature of the GPS data. Whilst it might be reasonable to interpret the trajectories as lines of movement, in the example given as shown in Fig. 7.2, at a smaller scale, it might not be. If analysed at the level

7.2 Urban Diary Spaces

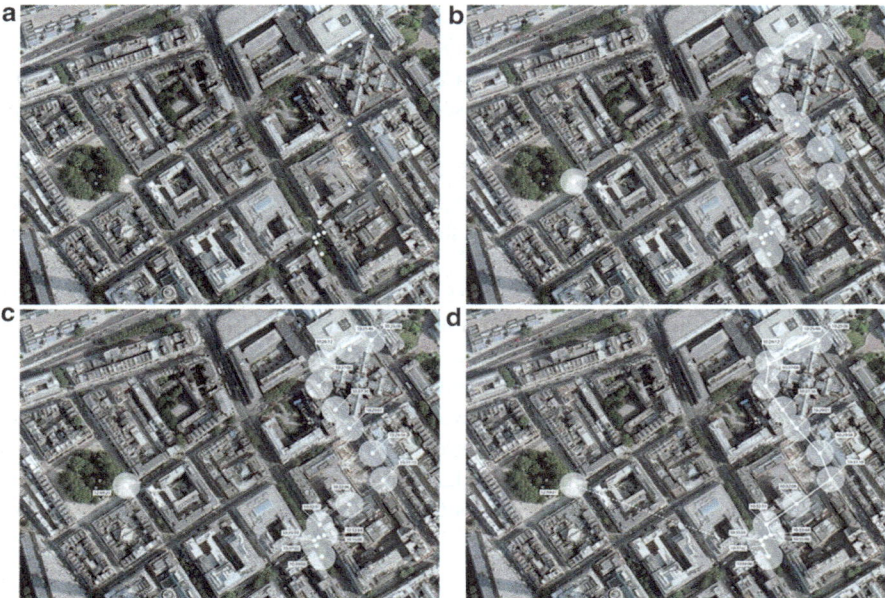

Fig. 7.2 The map (**a**) shows example GPS points as accurate latitude and longitude positions. In map (**b**) we have added an area for the possible GPS error, resulting in a *spatial area* of possible positions for each point. Map (**c**) shows the time stamp associated with each location, introducing a sequence. The sequence is then connected in map (**d**), in order to show the assumed path. At this scale, the points are relatively close. This allows the assumption that the path is following the roads. The last point on the far left has moved a considerable distance over a longer period. The sequence is clear, but the path and spatial connection cannot be interpreted from the data

of the street, questioning which side of the street the movement took place, whether or not the zebra crossing was used, no clear answer can be provided, and the data must be interpreted as fragmented. On the other hand, at a larger scale, on the level of the neighbourhood, the data, including the far point on the left, could make perfect sense.

Taking these aspects of influence into account leaves us with a subjective interpretation of the data. It is experience and intuition that strongly influence the results and the visualisation. Whilst resolution and sequence are to some extent two descriptive attributes, they are stylised as natural aspects of the *objectified* problem. Often such descriptive attributes, or in this case value attributes, mask the fact that the observation, the mechanically or more modern digitally reproduced observation, produces a mere abstraction in the sense of simplification of the real-world phenomena studied.

The sequencing in GPS tracks, as discussed in Chap. 4, is based on the time information recorded together with the location information. In the absolute time concept of clock time, each point is positioned in time with a clear before and after, implying a string of points resulting in the sequenced path.

194   7 Structuring Space

The data recorded by the participants of the Urban Diary study captures the *spatial extension of their everyday routines*. As such, the visualisation of the data points, or indeed the tracks, is an illustration of where they have been, but at the same time how they got there. In this sense, the sequence is to some extent to be understood as a process of the activity.

Although the raw data is only point information, for visualisation, mainly 2D mapping purposes, the recorded location points are used to render a line representing the movement. The main reason for this is the better sense of continuity and sequence this provides. An individual's map of everyday movement recorded over a longer period of time generally shows strong patterns (for a visualisation of Urban Diary tracks, see Fig. 7.4). It is obvious that there is no randomness in the way participants move about in everyday life. For each participant, the recorded location forms an individual pattern, strongly influenced by characteristic key locations or *anchor points*. As predicted in the 1970s, for example, by Golledge (1978), the location of these anchors structures the shape and orientation of the overall map (see Fig. 7.3 for the original illustrations). Whilst Golledge mainly focused on the build-up of a map from scratch, an individual learning about a new environment, what we can see from

Figure 1 Skeletal node–path relations   **Figure 2** Nodes, paths and neighbourhoods

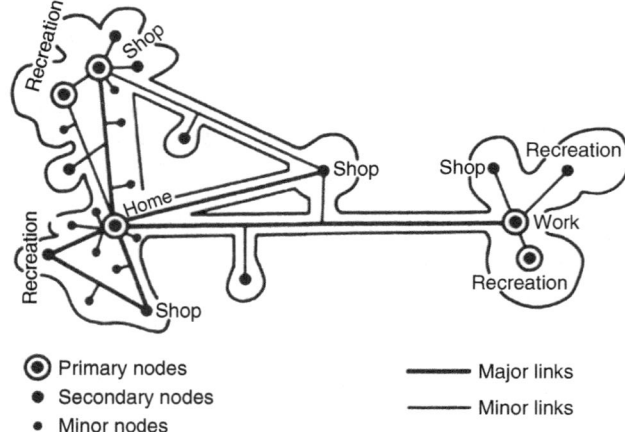

Figure 3 Nodes, paths, neighbourhoods and linkages

Fig. 7.3 Illustration taken from Golledge (1978) showing key or anchor locations and linking connections between them. The original figures 1–3 illustrate the building up of a spatial cognition of the environment which is continuously expanded

7.2 Urban Diary Spaces

Urban Diary records is a similar thing. Participants did not learn about the city from scratch during the period of this study, but as the recordings illustrate, they showed a similar *sense of exploring*. It could be argued that the element of time and the repetition are clearly contributing to this sense of building up knowledge of space.

Over the period of the tracking, the map that built up started to show patterns, which mainly represent the repetition that occurred in the participants' behaviour. By using the thickness of the lines, tracks start to accumulate on the daily routes and mark out the personal transport arteries of the city. For an illustration, see Fig. 7.4 showing London with all 20 participants' journeys recorded over the period of 2 months each.

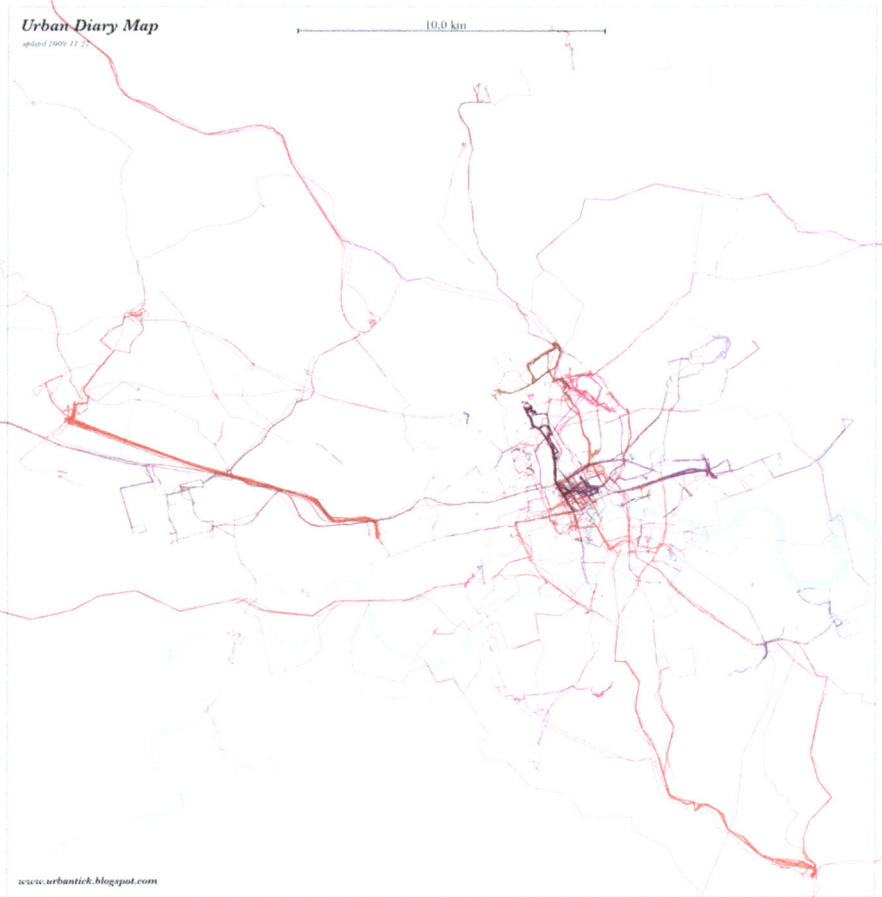

Fig. 7.4 London as drawn from the 20 participants' movement paths in different colours for each individual as shown earlier in Fig. 4.5. The river Thames has been added to the map for orientation purposes

Most of the participants rely on the public transport network, and therefore they are channelled into the routes simultaneously used by thousands of fellow travellers. These personal arteries become collective arteries and start to represent transport corridors. The major overall pattern that starts to show up from early on is the London-like characteristic of a centralised radial structure as discussed earlier in Chap. 4. The trajectory map starts to look like a star or a tree, with the majority of the tracks drawn from a peripheral location into the centre. Based on the transport for London zone plan, most participants lived in zone 2 or 3 but travelled into zone 1, which is the centre, for work. This travelling takes place in a predominantly straight line, pointing at a virtual centre point. The radial pattern that emerges from collective activity is therefore, it could be argued, related to the structure of the London infrastructure layout and consequently to the city's morphology. Compared to records of other cities, as shown in Fig. 7.10, this characteristic may be individual to each city and is determined mainly by the morphology, transport network and citizen behaviour.

There are cases in the recorded data where the device has lost its signal, leaving it unable to determine the required position over a longer period due to environmental factors. Some of these have been discussed in Chap. 4 and explained in the technology section.

Lost signals are, however, generally picked up at a later time. This can also be the case if participants are travelling underground, using the tube. In this case the underground station is the point of disappearance and another station the point of reappearance. This allows an observer to make sense of a number of islands of activity, by tying them back into the overall picture. However, we have not added any additional data to fill these gaps.

The *anchor points* for each participant can easily be detected on the map where numerous journeys lead to and from a place, and a thicker layer of lines indicates more activity. Whilst one-off trips are common, they leave less of an impression on the map. In Fig. 7.5, we show the anchor locations for three participants. These are *home*, highlighted with a larger circle, and *work*, with a smaller circle. They are connected with a dotted line to indicate the relationship. All three examples are similar in distance, and the mode of travel is public transport. For two, which are showing continuous travelling lines, these are mainly bus journeys, and the one disconnected example is the underground.

Whilst the *home* location is important throughout the week every day, the work location is only a key reference during the working week. During this time, the connection between the two locations is of great importance spatially. As we will be showing later on, for example, in Sect. 7.4, the local area is more important at weekends.

All three examples show the path between *home* and *work* is not just a line. There are variations in choice of route in either direction. UDp-06 varies the route only little, with one or the other tube station to catch the train. It is more of a choice for the other two. For participant UDp-02, variation is practised for shopping reasons. This participant buys dinner on her way home from a local shop, whilst the route

7.2 Urban Diary Spaces

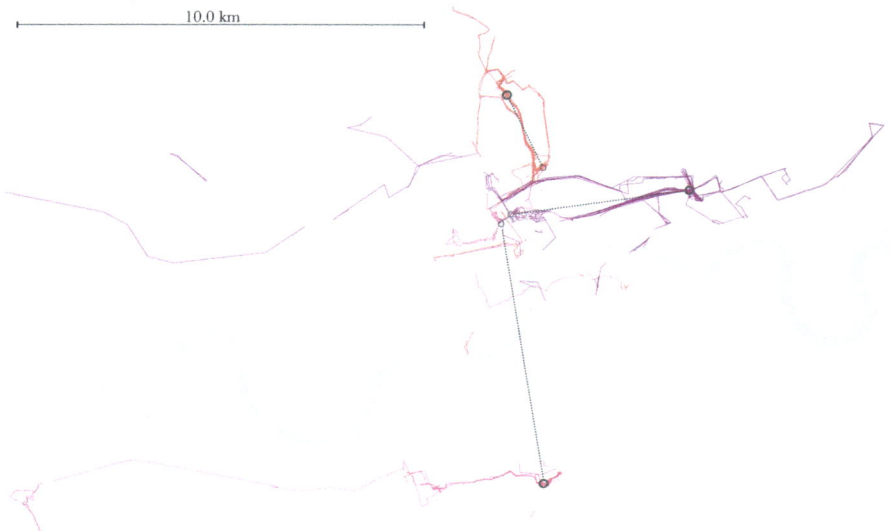

Fig. 7.5 The figure shows central London with the data of three participants, UDp-02, UDp-06 and UDp-09, and indicated *anchor points* home and work

taken in the morning does not pass any shops at all. For participant UDp-09, the rerouting between the way in to work and back home is related to the arrangements for childcare and pick-up times.

Whilst all of this is concerned with spatial representation of location, the *narrative*, as discussed in Chap. 3, is of importance. The story involving the sequence and the activity is tightly connected to the spatial activity. The participants build in purpose to the trips, design them according to needs, shaping a practice. This practice establishes itself over time, manifesting itself as a spatial *habitus*.

It has to be noted that, over the recording period, all participants have only established a few anchor locations. In fact there are hardly any locations to be identified from the GPS data as an *anchor point* besides *home* and *work*, hence the focus on these two in Fig. 7.5. Two aspects of the data have to be taken into account regarding the anchor locations captured in the Urban Diary data. One is the fact that the *home* and *work* are so dominant, standing out more with a presence of 5 days a week. Also, the recording period, 2 months, is a relatively short period to activities apart from the normal working patterns are often only recorded four times or fewer. Compared to the regularity of the two main locations, discussed in the previous Chap. 6 with regard to the time pattern, these are minor counts.

However, coming back to the schedule recordings as discussed in the previous Chap. 6, on the weekly scale, only few reoccurring activities were recorded that would require a specific destination which would here be identified as an *anchor point*. In most cases found, this is a weekly training session, a club, hobby or a

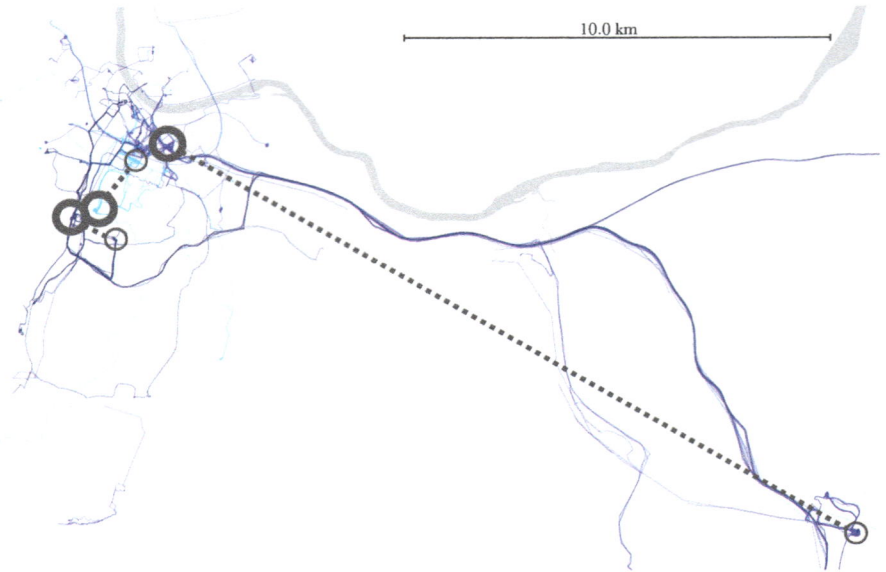

Fig. 7.6 The figure shows Basel with the data of three participants, UDp-30, UDp-36 and UDp-37, and indicated *anchor points* home and work

cultural activity. Whilst in theoretical models of everyday activity, these activities feature prominently, spatially, in this study, there was little evidence recorded.

What becomes obvious in comparison between the participants is how little ground each individual covers. From the individual maps, but especially on the maps showing multiple participants, one gets a sense of how tight the spatial extent of the covered area is. See Fig. 7.5 for a sample from London and Fig. 7.6 for Basel. Spatially this is tied to the anchor points, but in general this is a spatial feature of the *habitus* and its practical limitations.

Each individual of course draws a distinct shape, as we will discuss later on in Sect. 7.2.1. Overall, however, each map has a strong direction introduced by the *anchor points*. The spatial extent is strongly influenced by the *anchor points* and results in a specific perspective on the city. It can be described as a professionalisation of spatial usage. Each individual creates a unique spatial combination of city spaces according to their interest, needs and experience. This represents the individual spatial rhythm. Within this specific area, they are familiar with the spatial arrangements and spatial organisation. Such a spatial specialisation is again based on the routine, which in turn is anchored around the key points. This results on the other hand in large areas of the urban environment being *black spots*. Here, individuals have little knowledge about the detailed organisation. These observations have been confirmed by the participants and were discussed using the sketches of cognitive maps.

7.2 Urban Diary Spaces

These observations draw a different picture of the city overall. Generally, one thinks of the city as being one large artefact. On the level of the individual, we can see that this is not entirely true, in the sense of actually being in touch, visiting or experiencing all parts of the city. Especially in London, a large city with a population of over seven million, inhabitants specialise around selected locations, building their own city. However, the same specialised focus of spatial coverage is observed in Basel, which is considerably smaller, with a population of 170,000. This focus is highly selective and can be interpreted as the manifestation of the *agency* of the individual.

There are a lot of decisions involved in shaping the individual *habitarium* in the city which results in the spatial coverage we are observing here. *Central Place Theory* (Berry and Garrison 1958; Ullman 1941; Mulligan et al. 2012) discusses how economic constraints to a great extent limit this choice of focus on the city as well as the distance and direction of individuals' movement. Such economic and peer group constraints are packed alongside the previously discussed physical and time constraints and help define the spatial extent of the *habitarium*.

As we have seen in the previous Chap. 6, there is on the time side, little scope for experiments and variation in path choice due to constrained travel times. Spatially there is similarly little room for experiments and explorations of the urban environment within the everyday routine. It can therefore be noted that the rhythm has a strong spatial significance.

7.2.1 What Shape Are You?

The individual perspective on the city is not only limited spatially overall as shown above; it is structured individually. Whilst a strong routine creates the patterns, as shown in the previous Chap. 6, the spatial configuration *shapes* the everyday movement.

Each individual creates a unique *shape* out of *path* lines. This overlies the city, just like a city scale drawing. These *paths* describe the urban environment the individual has experienced and claimed as individual *territory*. With the specialisation, the focus is on where in the city the activity took place. With the idea of the *shape*, the emphasis of the observation is on the structure of the movement itself. See Fig. 7.7 for an illustration of all London sample *shapes*. The map in each square is drawn at the same scale and over the same recording period of 2 months.

Numerous aspects and personal preferences influence the structure of the individual *shape*. For instance, there are the *anchor points* as discussed above, especially their relative location to one another. Whilst in some cases, for some participants, they might be close together, with short travel routes between them, in other examples one or two might be quite some distance away, leaving the participant to travel longer distances between them.

Fig. 7.7 A shape drawn by each of the 20 individual study participants from the London sample (See http://www.flickr.com/photos/40984848@N04/8702380112/ for an online version)

7.2 Urban Diary Spaces

Fig. 7.8 Three *shapes* representing the three size categories. On the *left* the smallest category A with 1–2 km, in the *middle* the medium group B with 4–8 km and on the *right* the large category C with 10–20 km (See http://www.flickr.com/photos/40984848@N04/8702383814/ for an online version)

Another aspect is the mode of transport chosen to connect the different locations. There are great differences between modes of public transport, most notably between travelling overground or underground. And of course, with individual transport, we have different potential routes to shuttle between locations.

From the 20 London participants, the individual shapes can be categorised roughly into three groups according to size (see Fig. 7.8 for an illustration). The size refers to the core area of the drawing, which is the routine movement. This is based around the *anchor points* of home and work in most cases. Whilst the largest group C covers travel distances of up to 20 km between anchor points, the medium group B shows travel routes between 4 and 8 km, and the group A with the smallest shapes only travels routinely between 1 and 2 km.

In group C we have five representatives, about 1/4 of the entire sample. Group B summarises the large chunk of the sample with 12 individual *shapes* and drawing the smallest *shape* in group A are just three examples. The conditions of the primary *anchor points* are close proximity of home and work for the examples small of UDp-05, shown in Fig. 7.8 on the left. UDp-07, the middle of Fig. 7.8, shows a strong routine with frequent travelling of approximately 8 km between home and workplace. This is supported by the fact that the same route of public transport is used on each occasion. This results in a sort of backbone route for the entire drawing. The example of a long commute shown in Fig. 7.8 on the right is UDp-12, again on a public transport route, thus is a trip of around 20 km into the city centre.

Whilst these distances are calculated only for the routes that are travelled repeatedly, most *shapes* show a number of less frequent or single journeys. This feature is here interpreted as a second characteristic of the movement. A similar distribution in three categories can be distinguished from the sample.

It is these fainter lines of the *shape* surrounding the core element that let it appears as if glowing. This characteristic describes to what extent the individual travels outside the daily routine. These fainter *path-lines* are guided by secondary

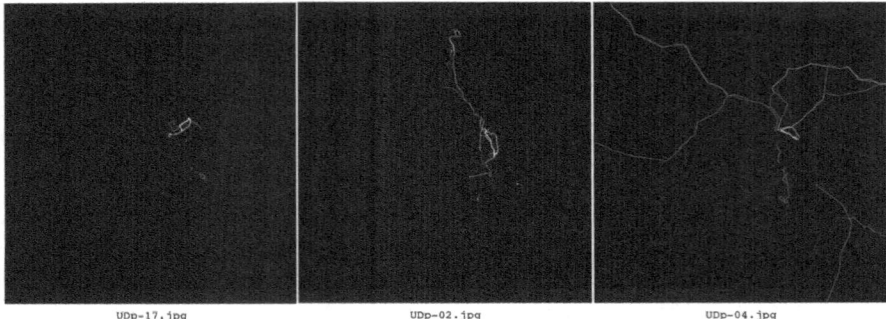

Fig. 7.9 Three *shapes* representing the three glow categories. On the *left* the smallest category *D*, in the *middle* the medium group *E* and on the *right* the large category *F* (See http://www.flickr.com/photos/40984848@N04/8702384442/ for an online version)

anchor points. Here we have, for example, trips to visit friends or family as in the example of UDp-02, shown in Fig. 7.9 (middle), who live in a different part of town. There may also be weekend trips out of town as with participants UDp-04, shown in Fig. 7.9 (on the right). The example *D* is a constraint in movement on one hand by the proximity of primary and secondary *anchor points* and a demanding working schedule leaving little time for long-distance journeys.

7.2.2 Self-Territoriality

The variety of shapes *drawn* by the study participants illustrate the range of influences and conditions everyday movement follows. The *constraints* as Hägerstrand (1970) terms them and as discussed earlier in Chap. 6 can be found here described in spatial terms of travelling. It however also describes the *agency*, the inherent property of individual decision making. Arguabnly, both aspects being present means the characteristics run deeper than simple context *constraints*. This partly is described by the individual type but is foremost perpetually reinforcing if, as presented in the introduction Chap. 1, routine functions exist in the sense of a *habitus*. The recurring parameters, revolving around the same pattern, will lead to a similar characteristic every time. The shapes observed here can be expected to persist over longer periods and represent true characteristics.

These characteristics have been summarised in the field of individual and collective territoriality formation. For example, Scannell and Gifford (2010) and Manzo and Perkins (2006) discuss the different influences shaping *place attachment*. Their model is based on the three factors of *person, place and process* as *PPP*. They specifically highlight a number of functions *place attachment* benefits. As a not exhaustive list, Scannell and Gifford list (in order): survival and security, goal support and self-regulation and continuity.

What we visualise in the *What Shape Are you?* maps above is partially how this *place attachment* manifests spatially. Whilst place and person are clear elements, some aspects of the process are obviously invisible along with the details of time. Scannell and Gifford (2010) list three aspects of the process part of *place attachment*. These are *affect, cognition and behaviour*. For example, Tuan (1990) describes *affect* as an important aspect of something as a *sense of place* with the term *topophilia* or *love of place*. The term *place* was earlier described in more physical terms as *anchor points*.

Interestingly, Scannell and Gifford (2010) argue for *place attachment* as distinguished from *territoriality* as argued for, for example, by Altman (1975), Altman and Chemers (1984), and Brown and Werner (1985). The main aspect of difference put forward is related to aggression. These authors suggest that *territoriality* demands action, more specifically aggression and force, whereas *attachment* is presented as lovingly self-focused and inclusive.

It could be argued, however, that the *shapes* represented here are a representation of the *self-territory* created consciously as a result of a series of activities. There is not necessarily any aggression or reinforcement whatsoever involved. Nevertheless, the physical presence of the individual lays a claim on the spaces, not least in the sense of experience. In this context, using the term *territory* as opposed to *place* seems to make sense since it usually includes several *anchor points*. These are of importance, since the creation of identity arguably is closely connected to the identification with the context or environment (Proshansky 1978). As stated by Fredrickson and Anderson (1999, p. 22):

> For it is through one's interactions with the 'particulars' of a place that one creates their own personal identity and deepest-held values.

Being, and having been places, is therefore essential, and as speculated in Sect. 3.1 *body*, this spatial experience could be interpreted as a *body extension*. With this, *territory* is physically important. Conflicts of interest are inevitable, but need not result in aggression or force. They are a form of negotiation and navigation in time and space.

Of importance in this context is the momentary nature of the territory. The claim is dynamic and closely tied to presence as well as individual memory. As discussed in the previous Chap. 6, this is an essential condition where the physical parameters of both *clock time* and *Cartesian space* break down, being unable to account for these aspects. For both these concepts, if something cannot clearly be identified as one or the other, it cannot be included, leaving *self-territory* undescribed.

Here, however, we have arguably a condition, and the above maps visualise this, where *self-territoriality* is a feature. It can be described as proposed by Scannell and Gifford (2010), but also, and in this context more importantly, in terms of the routine nature of its occurrence. The repetition, or the cyclical aspect of the presence, has a reinforcing aspect which arguably supports *territoriality*, but in a nonviolent sense. Being present is not principally an act of violence but an expression of *agency* which is in part based on negotiation in a social context.

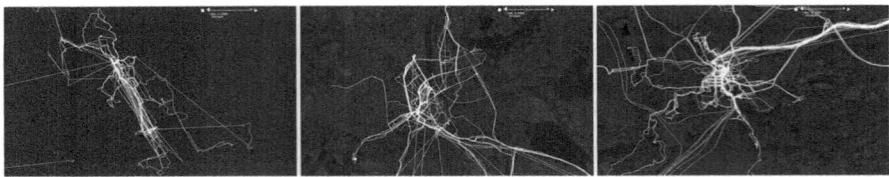

Fig. 7.10 Qualitative comparison between three cities based on path records. On the *left* we see London, in the *middle* Basel and on the *right* Plymouth (See http://urbantick.blogspot.co.uk/2008/12/comparison.html for an online version)

7.2.3 Path Lines and Urban Morphology

The individual paths of residents between Basel and London show similar characteristics at different modes and distances. It is interesting to compare overall characteristics. This view of a city's *path morphology* is of interest as it shows how locations are shaped by the routes connecting places.

A simple comparison of the path collections shows the morphological differences between cities. We illustrate in Fig. 7.10 the differences of records between London, Basel and Plymouth. Whilst this is a qualitative look at morphology, the characteristics of the individual city are being picked up and will be discussed below.

It is interesting to compare how individual movement responds to the urban surrounding. The three cities shown in Fig. 7.10 have distinct urban patterns. For example, Plymouth was completely replanned almost from scratch after it was destroyed in the Second World War by Nazi Germany due to it being a major UK Navy base. Patrick Abercrombie developed the plan for the reconstruction of the city centre. He also presented ideas for the reconstruction of London after the Blitz. Basel, on the other hand, is a similarly sized city in a different setting, with its original growth patterns structuring its appearance. London, our third example, is a world city with a single dominant core but a large collection of partially independent local subcentres.

To explore how these characteristics influence the interaction with the built environment in terms of *path morphology*, Fig. 7.10 shows tracks overlaid onto a structural city map showing the morphological characteristics of the cities. For Plymouth (Fig. 7.11), this is the original Abercrombie Plan.[1] The tracks redraw the orthogonal pattern laid out. Movement is confined to an either north-south or east-

---

[1] The Abercrombie Plan was first proposed by Patrick Abercrombie and James Paton Watson as *A Plan for Plymouth* in 1944, although it carries the date 1943, and consequently it was revised several times before implemented in large chunks (Mosley 2012). To a large extent, the key features are still visible in Plymouth today. Some key changes are currently being implemented by the local council. The political leaders have decided the city needs to outgrow the tight structure. In many ways however, it can be argued that to this day Abercrombie remains largely misunderstood in terms of what his plan is trying to achieve for Plymouth and how it should be linked to the wider context of the city centre.

7.2 Urban Diary Spaces

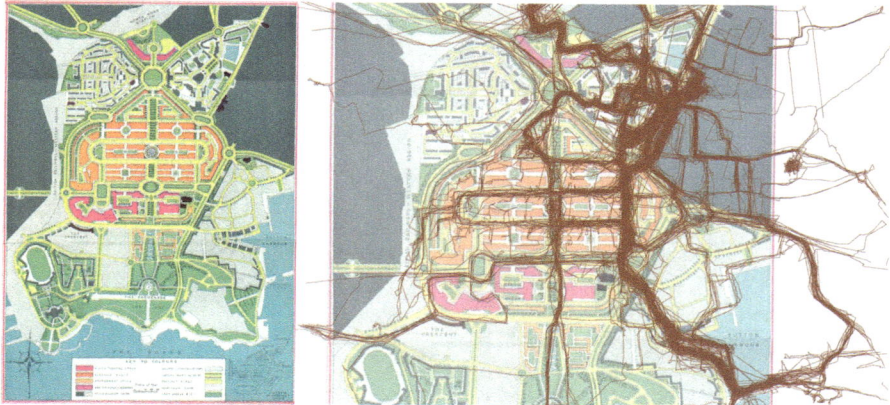

Fig. 7.11 The *Plan for Plymouth* as presented by Patrick Abercrombie in 1944 on the *left*. On the *right* showing only the centre with overlaid path data retracing the core structure

west orientation. This strict layout leaves no room for experiments and channels movement clearly in special arteries (Fig. 7.11).

In Basel, by comparison, the historic structure has been preserved although it has been reinterpreted in terms of usage. The medieval city walls have been demolished and replaced by major arterial roads. In this sense the current structure can be described as *grown*. In Fig. 7.12 this is represented by path tracks and area colours indicating the urban growth over a period of roughly 100 years. During this period, the walls were extended and then in a second step abolished to accommodate the growth of the city. In Fig. 7.12, the area in dark brown shows the old medieval town surrounded by walls dated around 1860. The area coloured in beige is the extension, around 1875, still surrounded by a new and extended wall. The area shown as a black outline is the extension of the city around 1926 but is mainly the present extent. Basel-Stadt (the city of Basel) is a political unit, a canton. The city has grown and completely filled its political boundaries at this point. Growth after this happened as densification and of course as sprawl into the suburbs.

It is important to note that, after the walls have been demolished, the freed-up space has been used for major infrastructure projects such as roads, but also as open spaces. This means that, in addition to the link roads, from the centre outwards there are a number of ring roads (on the ground of the former walls) that tie in well with the rest of the network. Moving circumferentially is quite simple, hence the great number of tracks recorded on these routes.

In London, by comparison, it is much more difficult to travel circumferentially as it has a strongly centralised transport structure. Roads mainly lead into or out of the city centre. This is represented in the London track log as shown in Fig. 7.10. It is strongly linear, and this represents exactly this centrality as the line points towards the centre. This is enforced by the public transport routes again having a strong focus serving the centre. For example, all underground lines but the Central Line

Fig. 7.12 A map of Basel with path tracks overlaid on a growth map. Indicated are the growth periods 1860, 1875 and 1926 which corresponds to the present state in terms of the overall expansion of Basel-Stadt

are arranged radially. The same is true of the bus routes, although minor tangential bus routes do connect locations parallel to the centre.

The morphology of the urban tissue is connected to the movement patterns and paths of everyday life activities. These visualisations demonstrate how tightly knit and interwoven with activities the urban fabric is. However, the question of the detailed relationship between the urban fabric and the activity remains unanswered.

In her paper *Urban morphology as an emerging interdisciplinary field*, Moudon (1997) presents the beginning of a separate field of urban morphology in the early twentieth century, mainly led by M.R.G. Conzen (b. 1907) and Saverio Muratori (1920–1973). The two schools of urban morphology started in Italy and England. For the three basic aspects, Moudon (1997, p. 7) lists:

1. Urban form is defined by three fundamental physical elements: buildings and their related open spaces, plots or lots and streets.
2. Urban form can be understood at different levels of resolution. Commonly, four are recognised, corresponding to the building/lot, the street/block, the city and the region.
3. Urban form can only be understood historically, since the elements of which it is comprised undergo continuous transformation and replacement.

It can be argued that the pressures of urban migration and everyday movement, as a result of activity patterns in connection with land use, are major factors often referred to as *continuous transformation*. As such, the force of cumulated individual paths, as a result of *coupling constraints* (using the term introduced by Hägerstrand 1970), means that the infrastructure is put under pressure during usage.

This is observed in the local park, where, despite having beautifully laid out pathways, informal routes as *shortcuts* emerge across the grass. However, major routes of travelling and trading have often informed strategic settlements. Whilst some settlements were built to stabilise the routes, others rose to become major centres because of trade.

7.3 New City Landscape Spaces

The Twitter data collected in cities from around the world was analysed in the previous Chap. 6 in regard to its pattern in time, specifically over 24 h. In the following section, the same data is discussed according to its spatial dimension using mapping techniques.

From the data collected on the social networking platform, 3d landscapes of message densities were created—as introduced in the earlier Chap. 5. These virtual city landscapes, as they are termed, represent the city uniquely on the basis of location, activity and interaction in a virtual world of real places. We term them *New City Landscapes* (NCL).

A selection of individual landscapes will be discussed to draw out some of the specific topics observed from a spatial perspective. Subsequently the focus will shift to the role of activity and mobility in generating a distinct pattern. On the one hand, these are infrastructures and other real-world elements that play an important role in the virtual world of online networking. On the other hand (discussed in a third part), these are emerging morphologies, compared across the locations analysed.

7.3.1 Characteristics: Features and Comparison

As a link between the virtual nature of the New City Landscape maps and the real world, the textual descriptions of places enable such a connection. The drawn-up New City Landscape appears unfamiliar: the landscape features we are navigating by, if we are familiar with the place, do not correspond with these peaks, hills and plains in this graphical version. However, on closer inspection, some of the main landscape features can actually be found as determining elements of the new visualisation. This is not true in a primary sense, but they appear to be defining and shaping indicators of the resulting virtual landscape.

For example, large natural features such as bodies of water, bays, seasides, or rivers can be clearly identified from the map as these are often non-tweet areas. Cities like Barcelona, San Francisco, Hong Kong or Singapore are dominantly structured by the real-world water features and coastlines, and this can clearly be identified in the NCL maps of these places. The two sides of the bay in San Francisco (see Fig. 7.14) are the main feature in the NCL map of the urban area. In other urban areas, for example, London, Seoul or New York, the rivers as structural elements influence the outcome as secondary elements. Often as, for example, in the case of Seoul, we can observe a split of a core centre into two centres on opposite sides of the real-world river feature. This goes hand in hand with the way this feature influences the built up area, as in the case of New York City where it clearly defines the island of Manhattan and reflects the population density.

The high rising landscape features of the virtual NCL maps are the hot spots of Twitter activity, the peaks. Here the morphology varies dramatically between the urban areas. How the Twitter density is structured is unique to each urban area. Similar to the *path morphology*, discussed in the previous section, we can here talk of a *tweet morphology* uniquely tied to the specifics of each city constellation including natural setting, land use, morphology and accessibility. There are however some characteristics that can be pointed out. The three different groups summarising the major characteristics are described below.

The *centre* is the type where one main location dominates as a single major peak in the whole urban region. The *island* category has different hot spots appearing as mountains and hills scattered across the area surveyed. The third type, *feature*, is where one or more features draw out as shapes, groups of peaks or ridges.

For the *centre* group, some of the examples are London (see Fig. 7.13, first on the left), Barcelona, Mexico City and San Francisco; see Fig. 7.14. One peak pinpoints

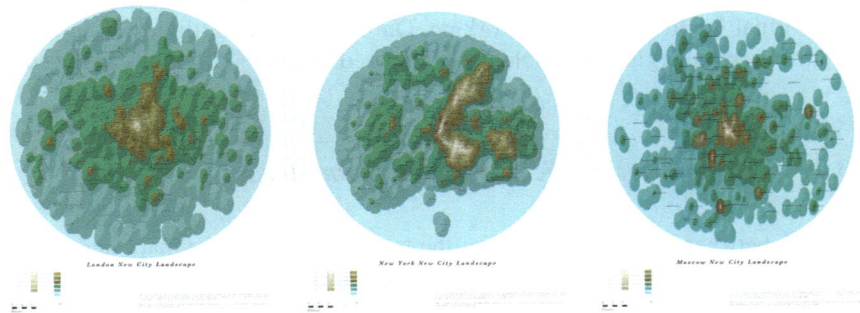

Fig. 7.13 An example for each of the three morphology types. From left to right the centre type represented by the London-NCL map (See http://www.casa.ucl.ac.uk/urbantick/maps/london_ncl_100628.html for an online version), in the *middle* the feature showing the New York-NCL map (See http://www.casa.ucl.ac.uk/urbantick/maps/newYork_ncl_100628.html for a large online version.), and on the *right* the island, with the example of the Moscow-NCL map (See http://www.casa.ucl.ac.uk/urbantick/maps/Moscow_ncl_100814.html for an online version). See appendix for larger prints of all 20 city maps

7.3 New City Landscape Spaces

Fig. 7.14 Three examples representing the *Centre* type of landscapes. These are from the left, NCL-Barcelona, NCL-Mexico City and NCL-San Francisco (See http://www.flickr.com/photos/40984848@N04/8701298525/ for an online version)

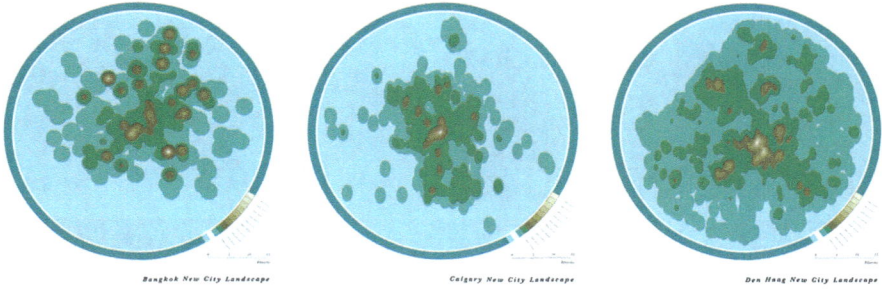

Fig. 7.15 Three examples representing the *Island* type of landscapes. These are from the left, NCL-Bangkok, NCL-Calgary and NCL-Den Haag (See http://www.flickr.com/photos/40984848@N04/8702420256/ for an online version)

an area of dominant activity. Since it is based on kernel density, this represents not just one location but a region from which many messages are sent. Together these messages add up to the peak. There can be other significant hills, but in this type they are clearly lower than the one major area.

Examples representing the *island* group are Moscow as shown in Fig. 7.15 on the right and Bangkok, Calgary and Den Haag as shown in Fig. 7.15 from the left to the right. Here a number of peaks reach similar levels. These different locations compete for height but remain at similar level. Sometimes, as in the case of Bangkok, trails can be traced between these peaks, suggesting that users travel between the locations on popular routes.

Representing the *feature* group are New York (see Fig. 7.13 on the right) and also Dubai, Mumbai and Singapore, illustrated in Fig. 7.16 from the left to the right. Here a whole group of linear features are active and as a result produce not a single point of high activity but a ridge or a shape. Shopping streets, boulevards or strong natural features can be the cause of this.

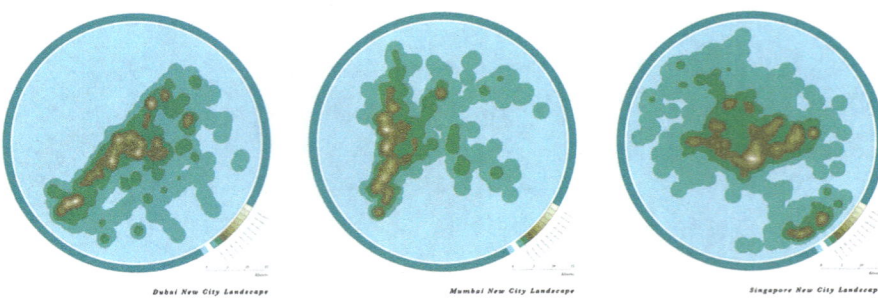

Fig. 7.16 Three examples representing the *Feature* type of landscapes. These are, from the left, NCL-Dubai, NCL-Mumbai and NCL-Singapore (See http://www.flickr.com/photos/40984848@N04/8701298177/ for an online version)

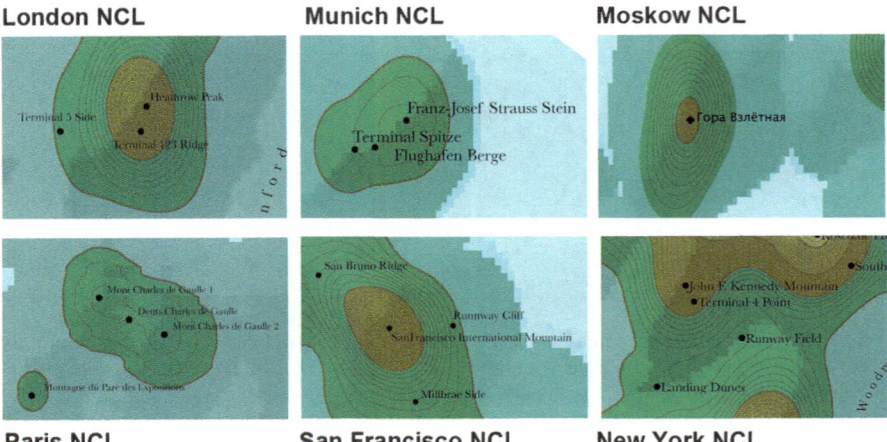

Fig. 7.17 Twitter surfaces showing a range of airports as seen on the New City Landscape maps

Natural features define the virtual map to a great extent. It can be speculated that the density of messages is connected to accessibility; hence, inaccessible areas show fewer tweets. This is illustrated in the maps showing costal centres. Buenos Aires and Lagos show a clear *cut-off* line, where the water limits accessibility. Other features are mountainous landscape, which influences the morphology of Bogota or Mexico City. A third type is the unlimited city growing in the plainness, such as London, Paris, Denver and Calgary. Here space is not as restricted, and activity can spread more easily to desired locations.

Beside natural elements, other physical features shown are the major infrastructure installations. Two examples illustrate these. The airports (see Fig. 7.17) are examples of intense activity. Almost all the NCL maps show a peak in the vicinity of airports. Parks, on the other hand, are an example of city infrastructure that manifests through the absence of activity, in the form of Twitter activity. Whereas at airports, tweeters seem to be bored waiting for the plane or excited to just have landed, people in the parks are most likely engaged in physical activities and therefore leave these locations empty on the Twitter maps. A key example is the Central Park in New York, a virtual Twitter activity desert, whilst the area nearby, and in Manhattan generally, is a high tweet area. The same phenomenon is observed in Singapore, where the large green space in the centre of the island is left empty of tweets. The same phenomena can be observed in Calgary, with its large green space at the northern fringes of the city showing little online activity.

Airports show up, as mentioned before, as activity hot spots on all maps. There excitement and boredom go hand in hand and turn airports into major urban features in the virtual urban landscape. This creates transport hot spots where individuals are caught in transition, highlighting the importance of these connections both as a spatial feature and as a destination. The aspect of destination is supported on the one hand by accessibility, as speculated before, but land use also plays an important role. Activity tends to intensify with the *trendiness* of the urban neighbourhood in many cases. Land used for leisure, shopping and recreation facilities attracts large numbers of active users with the time and mood to tweet.

The maps in many cases show the accessibility in economic terms. Especially in the outskirts of the cities in lesser well-off areas, the usage of service clearly drops off. Here the luxury of using a handheld gadget and mobile Internet is simply not available. The data shows the division between rich and poor in the form of the accessibility of the technology.

Mexico City is a good example of this aspect being clearly visible. It shows a strong east-west separation with the eastern parts of the urban area clearly being less wealthy, where access to Twitter is limited.

Here the New City Landscape maps are discussed as picking up the urban morphology of the actual places showing the characteristics of the city such as density, accessibility and land use. However, the perspective is clearly that of the individual and the activity that leads to the location. Although shown as collectively produced patterns of the urban area, it is the interaction between the *place* and the individual that makes the map. The attractiveness of the location is represented here with regard to how Twitter inspires the various locations.

In other words, the maps visualise how the urban environment influences the activities and the choices made in an everyday context. These are as much the natural features of the place as the infrastructure and the actual morphology of the urban framework. Like massive magnets, these aspects of the urban area pull and push, influencing individual decisions and shaping collective usage patterns.

7.4 Timed Space

The aspect of time in the production of maps is not often represented as such. Cartesian space is exclusively mapping the static positions of objects as they present themselves in relation to one another and to an objective point of reference.

To visualise time as an aspect of the information, other techniques have to be introduced. Next, the spatial dimensions are discussed in relation to spatial changes over time. To illustrate the potential and the problems, a range of different examples and approaches are discussed.

An often-used method involves adding time as an attribute of the represented feature. This can, for example, be achieved via a form of coding, using attributes such as icon, colour line type or transparency. The method of applying colours to represent time opens up a dimension in that it uses colour to show different units for each data point, such as days or hours. This method has its limits as the colours are more often used to distinguish between different individuals, but it was successfully tested in a single-participant environment as shown in Fig. 7.18. Five colours have been used to code the party of the GPS track corresponding to morning, midday, afternoon, evening and night.

Fig. 7.18 A GPS track colour coded according to time, starting with red at midnight fading into *yellow*, *green*, *blue* over the day and turning into purple at night again (See http://www.flickr.com/photos/40984848@N04/8701345067/ for an online version. See http://urbantick.blogspot.co.uk/2009/03/ud-aquarium-example.html for an online version)

7.4 Timed Space

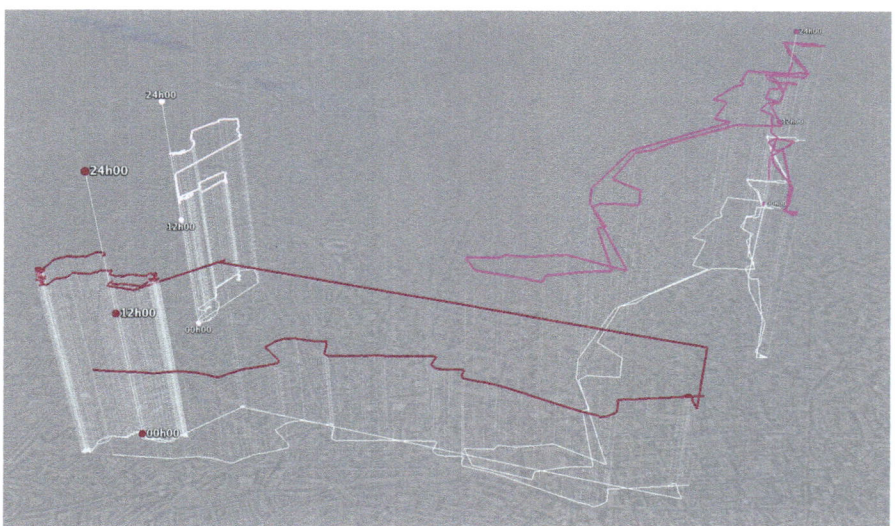

Fig. 7.19 Space-time aquarium, after Hagerstrand, plotting three participants in the Urban Diary London context. The model represents 1 day of 24 h

Another method to represent time and location simultaneously was developed in the 1970s by Hägerstrand (1970). The model (Fig. 7.19) produces a 3D visualisation of the data, using the x- and y-dimensions to refer to the spatial location of events and the vertical z-dimension to plot the passage of time. Whilst time passes, the z-dimension increases. This space-time *aquarium* (Carlstein et al. 1978) comfortably merges the two different types of information. However, it does simplify space to a great extent, and details such as topography are lost. Hägerstrand originally used the term *prism* to describe this phenomenon, but the metaphor of an *aquarium* perhaps captures more accurately the internalised, situated experience of the participant. The approach was implemented successfully in work done by Kwan (2004), summarised in *GIS Methods In Time-Geographic Research*. Although the readability of the object can be tricky, it works well in an interactive 3D environment but can be confusing when used as a static 2D print with a lot of contextual information. Patterns of repeated activities do emerge clearly with this method, as illustrated in Fig. 7.20.

The commercial application *GeoTime* has turned this method of representation into a functioning software programme, including a user interface (Kapler and Wright 2005). From a set of data points, it can build the space-time aquarium and offer a set of tools for analysis. During all processes, the software keeps the representation of the aquarium flexible, and the viewpoint and data displayed can be altered at any time. This maintains a welcome flexibility and helps with the reading of the data. GeoTime has been used to run analysis on the UD data set. Specifically the *Meeting Finder* proved to be of use. With this tool, points can be identified where

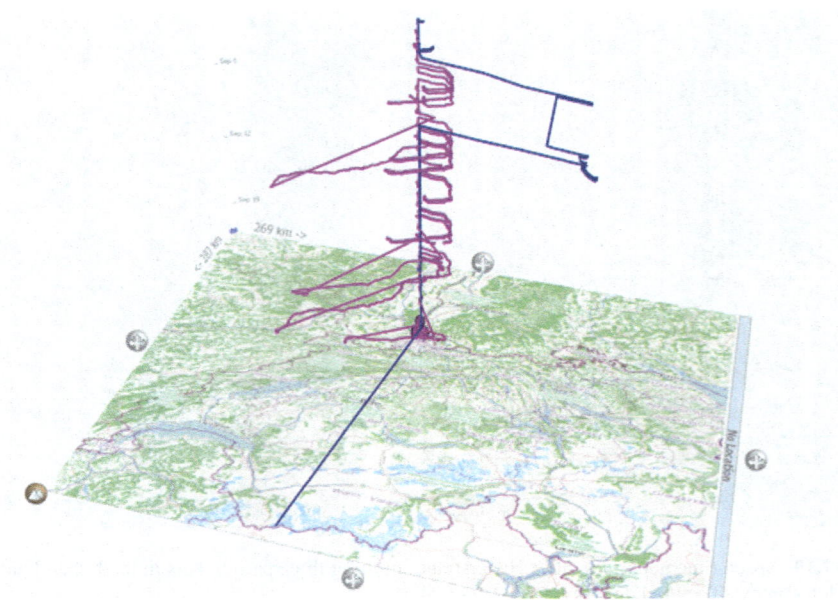

Fig. 7.20 Showing the movement patterns of a single person over the course of multiple days where the time rises vertically using the GeoTime software

personal trajectories of movement intersect with one another in terms of the criteria of distance and time. It turned out that participants on the project have been in the same location at the same time without realising, since they do not know each other. This suggests that, perceptually, London is not so big after all.

A further method to map time-based information is by employing animation techniques. With this method, the passage of time can be represented through sequential frames. In the framework of Cartesian space, this is a likely option. It is basically a collection of slight changes in static image frames in sequence. A number of clips have been produced and continuously updated as the UD data log grows. In our case Google Earth was used as the visualisation platform. The built-in functionality to replay time tagged location information is simple to use and powerful. To achieve more clarity in pattern representation, the recording period was usually compressed and represented as a single day. This means superimposing all the days onto a 24-h period. Recurring events show up as accumulated activities, whereas one-off activities are represented as single lines. The Virtual London Model, developed by Batty and Hudson-Smith (2005), has been used to set the recorded locations into a spatial context (see Fig. 7.21 for an illustration). This setting is regarded as a first step to combine the time-based information with the morphological space of the city.

The New City Landscape maps, as discussed in the previous Sect. 7.3.1, are purely spatial. Nevertheless, the time pattern, as discussed in the previous Chap. 6,

7.4 Timed Space

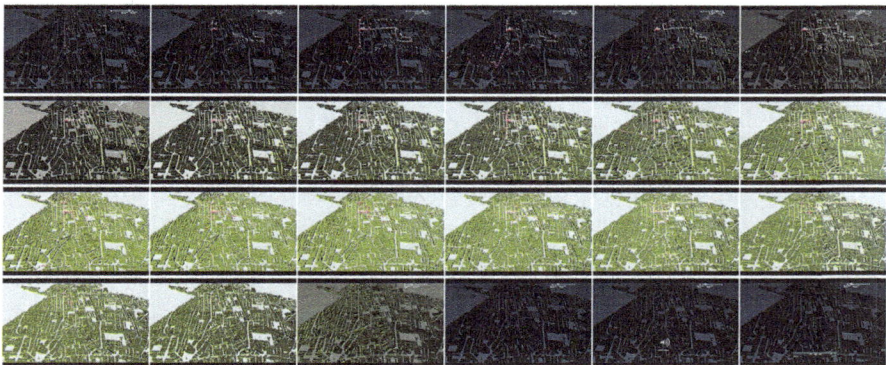

Fig. 7.21 Shows 24 animation stills taken from a 24 h day GPS trace visualisation in a local neighbourhood. The brightness gives an indication of the time of day (See https://vimeo.com/4469668 for an online version)

influences the extent of the map temporarily. Whilst the maps showing data are collected over the period of 1 week as a summary, there are short-term changes in real time where the activity shifts both in focus and intensity.

As an example, Fig. 7.22 shows six snapshots over the course of 24 h and how the activity landscape changes in its extent.[2]

In Munich there are some hot spots that are constantly active over the course of the day, like the airport terminals in the top right-hand corner and of course the centre. Interestingly, the centre remains strong throughout the observation period, and there are no clear shifts or jumps between locations as the day passes. There is notable fluctuation around the centre and occasional spots on the outskirts.

These examples illustrate the potential of integrating time as an aspect of space, whether as dimension, colour or slices. Whilst the Cartesian space does not change, its configuration does. It could be argued that the spaces are being activated with the usage, then light up to function in individual cases, whilst at other times, they remain dormant.

The spatial aspects are throughout examples of the dominant characteristic, and time is represented as an addition. As such, all representations are biased, and do not achieve a dualism of time and space in an equal sense. In this context, the analysis always has to account for this fact and is forced to favour spatial parameters over temporal aspects.

It is interesting to note that the *constraints* described by Hägerstrand (1970) in different ways related to the same problem. Either these are spatial *constraints* such as *only one body in the same space at the same time* or temporal constraints such as *no simultaneity at different places*. Whilst these parameters describe the abilities,

---

[2] An animated version can be accessed online at http://urbantick.blogspot.ch/2010/07/tweet-times-activity-over-24-hours-in.html

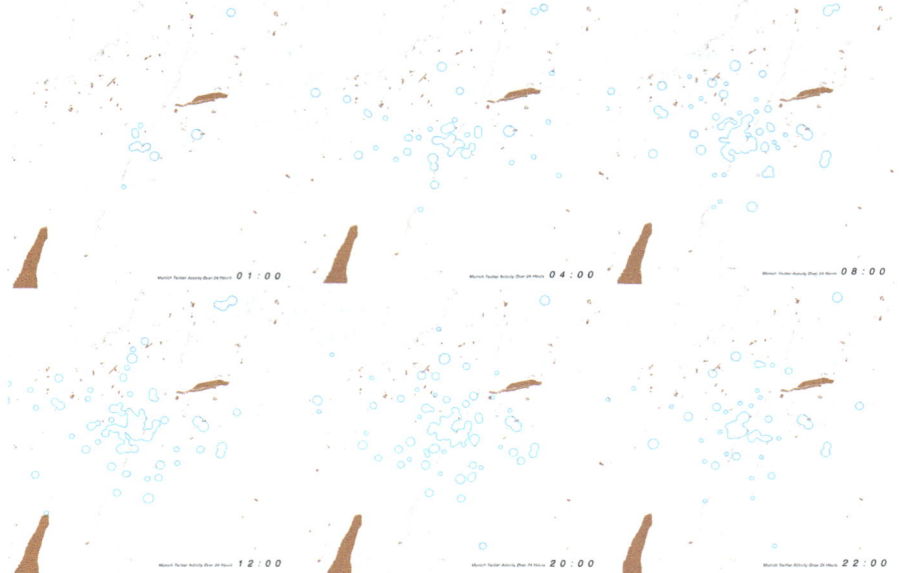

Fig. 7.22 The snapshots of the New City Landscape map of Munich over a 24-h period show how the activity pattern influences the spatial coverage. Tweeting areas shown in blue and green spaces are shown as points of orientation in *orange* (See http://urbantick.blogspot.co.uk/2010/07/tweet-times-activity-over-24-hours-in.html for an animated online version)

they also apply to the representation. As if the *constraints* were not complex enough, one or the other aspect guides the representation and thus makes this more problematic. In this sense, either the space or time constraint applies twice. This results in an extreme distortion of the facts in the representation that has to be accounted for in the analysis.

7.5 Social Space

The social aspect of space has so far only played a background role. However, as discussed under the aspect of time in the previous Chap. 6, the activity, and with it the social dimension, is a guiding factor as to how spaces are accessed and experienced.

Several aspects have been discussed in this chapter *Space* so far. One is the concept of *anchor points* and how individuals build up a collection of key locations through repetitive movement and exploration in the proximity of a location. Besides this *spatial* connection, there is the aspect of *place attachment* we have discussed in connection to *What Shape Are you?* and the build-up of an emotional attachment to specific places, referring to the specific concept of the *PPP* model by Scannell and Gifford (2010).

7.5 Social Space

These aspects stand in direct connection to the social aspect of space and the ways space allows one to interact and position oneself socially. Such a view recognises the limitations of Cartesian space and highlights traditional cartography as being a contested practice, as argued by Pinder (1996).

As discussed in Sect. 7.4 above on *Timed Space*, the tight time dimension connected to spatial presence is important in the social context. It is visible in the dynamic visualisation of the movement patterns of individuals, as shown in Fig. 7.21, but in the continuous changes on the collective level of the New City Landscape maps as to how hot spots of activity grow, shift and shrink, as shown in Fig. 7.22.

A whole range of specialised tools and instruments can be used to measure space in different ways, but how do we measure social space? As argued in the quote in the introduction to this chapter, how do we best let the studied subject speak for itself?

> In front of me, on the desk where I write, I've assembled a bunch of instruments useful in measuring the environment, instruments that I've found around the house. In front of me, on the desk where I write, I've assembled a tape measure, a yardstick, a stopwatch, a watch, a goniometer and an arm protractor, a clinometer, a map measure, a compass, a wall thermometer, a pocket thermometer, a percentage protractor, a level, a plumb, a light metre, a camera, a pocket scale, a postage scale, a barometer, a measuring cup, a set of measuring spoons, a pedometer, a stud finder, and a passel of questionnaires. Some of them, like the pedometer, no longer work, but still I hold on to them. Others, like a couple of the questionnaires, never worked at all, but even these I am loath to throw away. All of them have told me, or promised to tell me, something about my world, and since the world is something I'm eager to know about, I'm not eager to part with these instruments, functioning, ïĆawed, or broken down. It's 84° F where I sit at 11:30:36 in the morning. It is nine minutes and 47 seconds since I typed the first word in this paragraph. There's another instrument in this room, and I am it. I would have said it was stuffy where I sit and that half an hour had passed since I started writing, although my stopwatch now says it's been 11 minutes and 38 seconds at, according to my other watch, 11:34 on the nose. I won't argue with my instruments. They're measuring different things than I. My thermometer knows nothing of the humidity oppressing me; my watches, recording the pressure of their drive springs, know nothing of the pressure of trying to say something with words. (Wood 2010)

The question as to which instrument is the correct one to measure the human geography of the city has several answers, but as Wood (2010) summarises in his paper, there is one obvious tool, the humans themselves. Several techniques to do this were proposed in the past with the Situationists' *derive* and Kevin Lynch's *Mental Maps* being the most prominent examples.

Whilst it obviously changes dramatically with the usage of virtual spaces and mobile communication, *place* is still a facilitator of social interaction. With the availability of real-time video calls and mobile Internet, the physically significant enabling aspect of social interaction of space and even more so of *place* is reduced. Interaction is possible almost anywhere on earth, at any given moment, in real time. The meeting *place* as a physical location no longer has to be the same for everybody participating in the same conversation.

Arguably, with this development, already underway for a good 20 years, Cartesian space and the visualisation in geographical maps thereof are losing its significance to some extent, in the sense that social interaction does not necessarily need to coincide with the Cartesian concept of space. Spaces can be, and in this sense are already practised, as individually defined.

The shrinking of space due to the invention of increased speeds of travel and subsequently new communication technologies was visualised by Buckminster Fuller (Krausse and Lichtenstein 1999). Fuller illustrates the graph of increased speed with the symbol of the world map, in this case a Dymaxion projection, becoming smaller and smaller, proportional to travel time.

The visualisation was composed in 1963, and since then, technological development has continued to increase sharply. Telecommunications and the Internet, in particular, have again transformed the sense of space in terms of the travel time dimension and presence in practice. Spiekermann and Wegener (1994) argue that whilst Hägerstrand (1970) predicted that increased speed may be transformed into a greater amount of free time or a larger action radius, empirical studies such as Zahavi (1979) have shown that the individual daily time travel budget is relatively constant. More free time leads thus to more travel or long-distance travel. As a consequence, increased speeds lead to a perceived shrinking of space.

These observations can be confirmed by comparing the findings from the Urban Diary data from London and Basel. The commuting times are significantly different, but this does not lead to more activity or a change of schedule. The structure of the day and its key routines are still the same in both cities.

Whilst travel speeds increase, the travelled route might lose significance as a *place* in itself, but the destination gains importance. This can be, for example, observed with the increased importance of faraway holiday destinations. Particularly for everyday travel, the speed of travel and the engagement with the route between destinations is of importance in the urban context. For the construction of something like a *self-territory*, the physical experience of space is key.

As an alternative to the Cartesian mapping of the urban environment, the Situationist International (SI) around Guy Debord developed the method of *dérive* (drift, drifting), an exploratory, destination-less wander through city streets, detecting and mapping *ambiences* in the city. These tactics were termed *psychogeographies*. The strong emphasis on the subjective perspective and the experiencing of space and context put the tactics in direct opposition to the established Cartesian space and its mapping.

Guy Debord examined and defined the phenomenon of the city islands as isolated and spatially disconnected areas in his Naked City text presented in 1959, challenging traditional ideas of mapping with the map of the same name (Sadler 1999), by dramatically departing from the grid and introducing a fragmented, subjective and temporal perspective (Fig. 7.23). This view and description of space as a personal perception and experience are summarised in the term *psychogeography*.

What Debord and the SI essentially talk about is how attractiveness and repulsion of urban elements can guide the experience in the city context, influencing decision and orientation. This is not necessarily a free floating *dérive* practice. Arguably, it is similar to what individual everyday travel entails, negotiating attractiveness and possibilities. However, the experience of the Urban Diary project suggests

Fig. 7.23 *The Naked City*, Debord (1957), Illustration de l'hypothèse des plaques tournantes en psychogéographie

that the emergent pattern is connected to people's personal preferences. Some of the participants would never use the tube for personal reasons, and others would always use the tube for the same reasons. Often the argument is about the sense of orientation that participants associate with the particular mode of transport, e.g. the *simplicity* of the London tube map or the *simplicity* of the surface bus route, respectively.

7.5.1 Urban Islands

The time patterns discussed in the previous Chap. 6, the *anchor points*, also strongly influence the way individuals perceive the environment and shape their personal city and their attachment to place. Depending on the participant's transport preferences, the emerging pattern of activity either draws a continuous track or starts to build up isolated and spatially disconnected areas. One major factor influencing this pattern of detached locations is the individual's chosen mode of transport, which in the case of London may involve using the underground.

What Debord (1981) describes as the *hinges of the city*, the key areas to influence the practice of the *dérive* in the urban context, can be to some extent reinterpreted for individual movements in the city. The Situationists focused on the *ambiences* of the place whilst wandering aimlessly in the streets. What attracted or repulsed them guided the direction of the derive. In the personal context of the individual, navigating through the streets of the city, not aimlessly, but similarly, in a web of demands and desires, creates an arguably similar effect. Certain aspects act as hubs directing the path.

The islands illustrated by Debord (1957) in the map, whilst resembling urban islands, are at the same time the hinges of the city as perceived on *dérive* tours. Similarly, the *anchor points* and the resulting movement maps of individuals create urban islands, between which individuals are going about their everyday business.

Mode of transport plays an important role in that the connections between locations change their characteristics. Whether the mode of transport is fast or slow, engaged or disconnected, over or underground, dramatically changes the way these islands are perceived and maintained. However, not all patterns end up as collections of disconnected locations. There is another group of participants that maintains a record of continuous traces of movement. Travel overground generates a continuous spatial corridor connecting places. This leaves the individual with an embodied experience of the transition. Such a movement practice creates a very different identification and orientation potential from traveling underground.

These two groups of connected routes, with a continuous spatial experience and the disconnected routes of isolated places, are illustrated in Fig. 7.24 with the

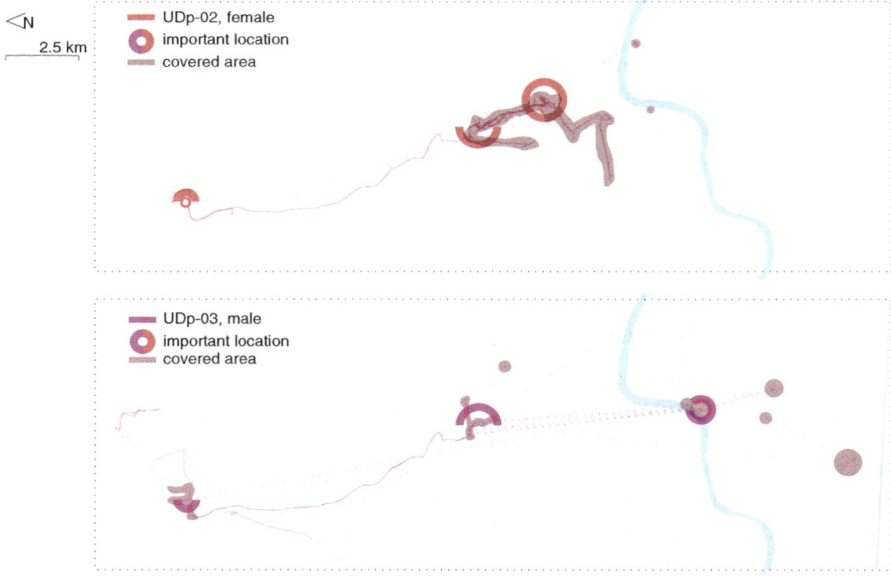

Fig. 7.24 City islands, the impact of mode of transport on personal psychogeography of the city. The *top map* shows a continuous spatial experience using the bus and walking as the mode of transport. In the *bottom map*, the main mode of transport is the underground, resulting in a collection of urban islands of spatially disconnected areas

movements of two participants from the Urban Diary London sample. Two of the participants, who happen to be a couple, are each representative of one group of movement patterns.

Whilst participant UDp-02 mainly uses the bus, participant UDp-03 mainly uses the tube, as shown in Fig. 7.24, top map. Their records strongly illustrate the characteristics of each mode. His travel, using the underground, generates a set of disconnected urban islands, between which very little connection exists.

7.5.2 Mental Space

Following Lynch (1960) in using the mental maps or the cognitive map method, the Urban Diary participants were asked to draw maps of the urban environments as they know them from their everyday experience. The maps reflect, as outlined in the previous Chap. 4, the elements, movements and atmospheres the individual can access from memory and arrange as a cohesive drawing on a piece of paper. The resulting sketch contains spatial facts but is generally wrapped in narratives.

There is no direct access to the map: the story is as much a part of the map as are the objects and places others might recognise. It is the connection between the elements of the map that make up the full picture which is of interest (Wood 1973). In a series of examples, these *spaces from memory* illustrate the similarities and differences between the individuals' experiences as they recall their cities from a personal perspective, as discussed above in Sect. 7.2.1.

The participants were not given any map or plan to aid them nor did they prepare anything in advance. On one hand they were asked to draw the route they use most regularly to get to work, in as much detail as possible. On the other hand, they were asked to draw a representation of the entire city. This was usually at a larger scale and was therefore often sketched in a more abstract fashion. Examples of both types of sketch can be seen of London in Fig. 7.25 and of Basel in Fig. 7.26. Very little guidance was given: however, in some cases prompting questions helped develop more detail on individual maps.

Participants reacted differently to the task. Whilst some began drawing straight away, others thought about it first, planning ahead, as to how to represent their route and fit it on the sheet of paper. The given paper was prepared with a printed rectangle limiting the size of the map and providing additional orientation. Providing this additional box on the sheet of paper proved helpful. It was often referred to by the participants.

Whilst the participants were given a choice as to which of the two maps to draw first, they would often start with the everyday path between home and work. It seemed as if the familiarity of the route was easier to approach for most of the participants. Whilst the path sketch was almost always drawn as a step-by-step recall of the actual routine, the city sketch was conceptualised as a construction and built from key elements.

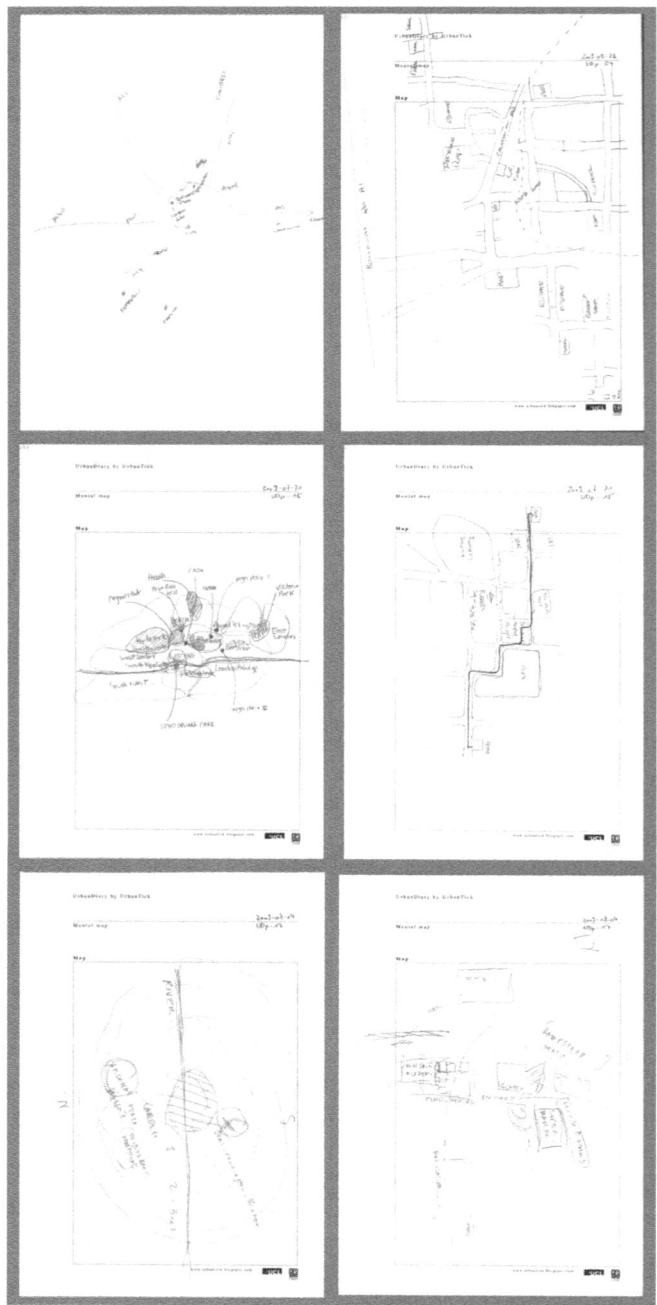

Fig. 7.25 A sample of three London participants sketching the urban context they know. The *top row* shows a London map, and the *bottom row* shows a detailed map of the commuting route from home to work. From left to right UDp-17, UDp-15 and UDp-04

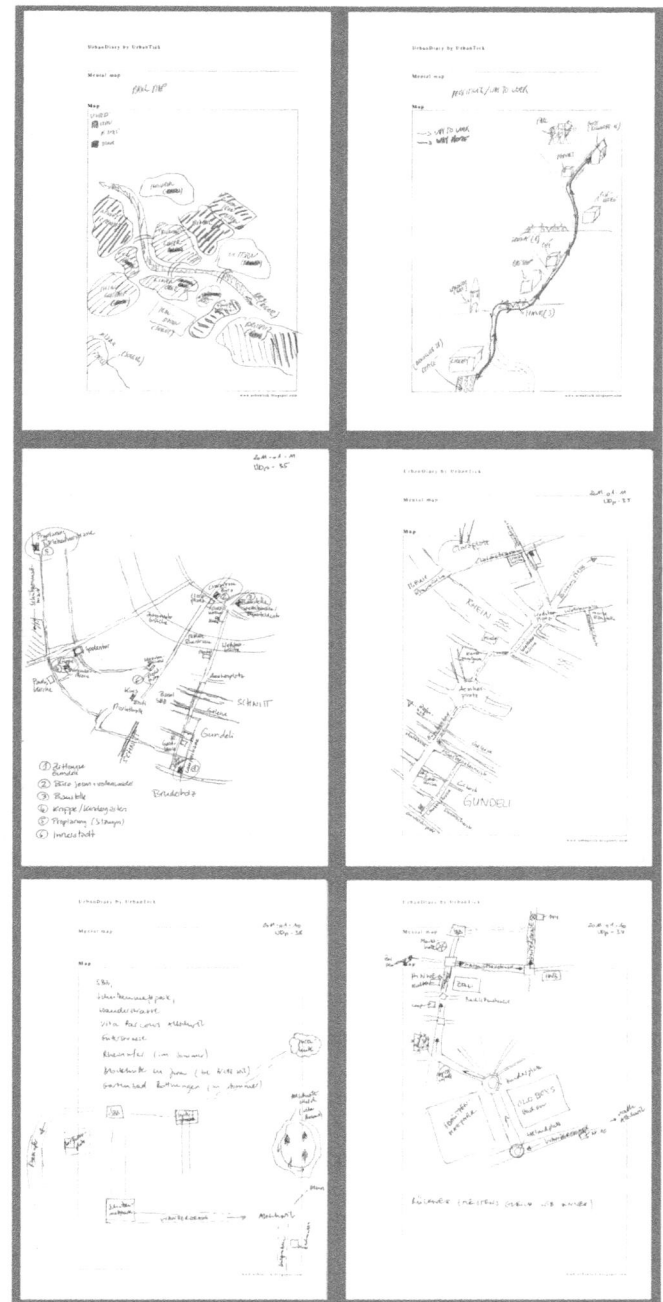

Fig. 7.26 A sample of three Basel participants sketching the urban context they know. The *top row* shows a Basel map, and the *bottom row* shows a detailed map of the commuting route from home to work. From left to right UDp-31, UDp-35 and UDp-38

This could be attributed either to the difference in scale or to participants' actual experience and first-hand knowledge of the city space. As we have seen earlier in Sect. 7.2 and the actual spaces visited in the city, only parts of the city are visited. The individual perspective in the sense of how one is related to the specific places is guided by a constrained handful of key locations. The wider context and the overall image of the city in these examples were pieced together from multiple sources and often second-hand knowledge.

In this sense, the maps diverge structurally; however, in most cases both maps still reflect a strong individual perspective. Whilst this seems obvious in the case of the path sketch, even the city sketch did focus on the aspects closer to the individual's interests. To illustrate this, three examples are discussed in detail in the following section.

Cognitive Map Sketches Examples by UDp-04—This participant works at a secondary school in central London. The patch sketch, in Fig. 7.27, shows him walking from his home to the school every morning and back again in the evening. Hence, he knows the neighbourhood well and important locations are fairly close in proximity.

The main element used for orientation in this example is the street network. The sketch shows a turn-by-turn representation of the street sections required to follow the path. As such this is the actual path and is drawn as if the participant was walking down the road. This is indicated in the way the map is oriented and develops across the paper, in the end not quite finding enough space to fit fully. The path develops against the geographic direction from the bottom left to the top right on the sheet (in preparation the portrait sheet was used in landscape orientation by this participant). On the other hand, it is visible in the way details are added, as if stumbled across. Examples are barriers closing off streets that have to be navigated or even indications on the quality of the surface or speed bumps.

These details can be attributed to the fact that the participant is a keen motorcyclist. Such elements in the urban street network are important or at least interesting knowledge if you are riding on a bike. They help one to choose routes, find connections which cars, for example, could not take, and know about street barriers, the quality of road surface and speed bumps in the local area.

Important locations are marked on the map as major buildings containing institutions, such as a post office, a prison, an underground station or a large recycling station. Whilst these are locations the participant has no qualitative connection to, he perceives them as important infrastructure, therefore includes them as objective points of orientation. However, there are also a number of places to which the participant has a specific relationship. This is the home, the school as the workplace, the local city farm where the participant helps out in his spare time, the pub he often goes to with colleagues from work and a number of housing estates he passes on this route.

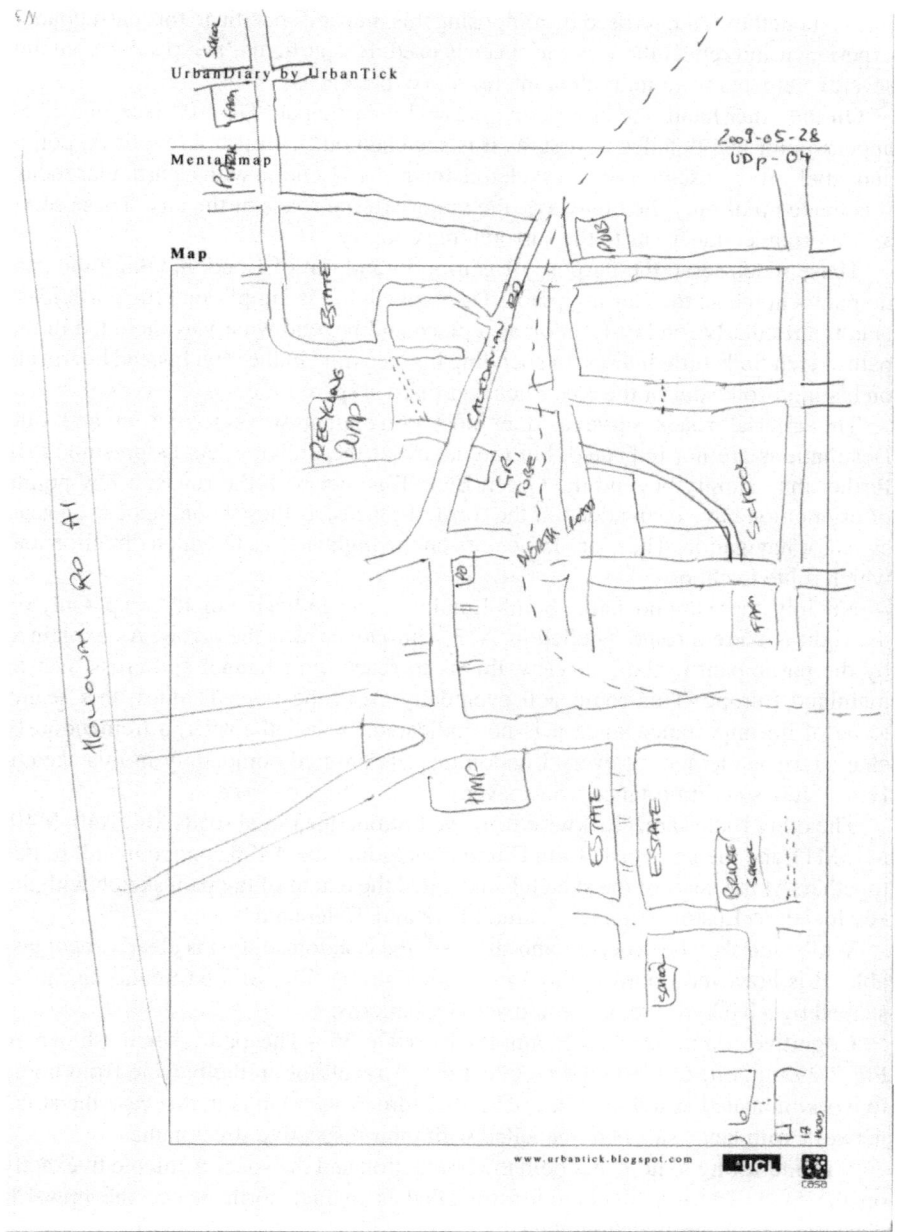

Fig. 7.27 The path sketch drawn by participant UDp-04

As such, the elements used in composing this map are specific to this participant's experience and reflect the way the space is used. His particular perspective contains several elements from individual interests to experiences.

On the other hand, the city sketch drawn by participant UDp-04 (see Fig. 7.28) appears a lot less detailed at first, as it is sketched out with just a few lines, points and labels. It represents a sort of skeleton diagram of London with a particular focus. It is made up of only the highways, the major arterial roads of the city. These serve as important connections to the surrounding country.

These roads serve the purpose of getting in and out of London, taking the participant's home as the starting point. The explanation is simply that the participant is not particularly fond of London as a place and beyond what was included in the path sketch finds little interest in spending his free time in the city. Instead he travels on his motorbike out of the city whenever there is spare time.

The arterial roads serve as the most convenient ways to get in and out. Destinations are not indicated, but the nature of the drawing lets us presume it is further than simply beyond the Green Belt. Together with the roads, a few points of orientation have been added in the form of places, as they would appear on road signs for navigation. These are indicators on the highways as to which direction and which route to choose.

Notably, there are no actual South London routes indicated on the map. Only on the right is there a route labelled as A13 with Dover near the arrow. As explained by the participant in the interview, this is to reach the Channel and cross over to mainland Europe. This route will eventually cross the river Thames; this seems to be of no importance since it is not indicated. In fact the river, a tremendously characteristic element of every London map, is omitted completely in this sketch. Hence, it has no importance whatsoever.

The other routes are, clockwise from the bottom, the radial roads M4, A40, M40, M1, M11 and the ring road North Circular including the A406 connecting all routes together. At the heart of the sketch is indicated the extent of the path sketch with the key locations Euston, Camden, Tufnell Park and Tottenham.

Whilst the sketches are bare and stripped, the London context is clearly recognisable. It is however a portrait through a rather strong filter of a particular narrative, shaped by a background of use and activity patterns.

Cognitive Map Sketches Examples by UDp-35—The path sketch, shown in Fig. 7.29 top left, is a detailed recollection of travelling on the bicycle from home to work indicated as a dotted line. The structure of the map is in this case the street network with key roads being labelled with their respective street names.

It is interesting to note that both the orientation and the space available fit exactly on the map. The overall direction travelled is in fact north, hence the upwards development of the sketch on the paper. Again, however, it is a linear map showing only the elements actually crossing the path. Two notable diversions from this concept are the indicating of the second bridge across the river and the *Münster*, the main cathedral indicated between the two bridges on the south bank of the river.

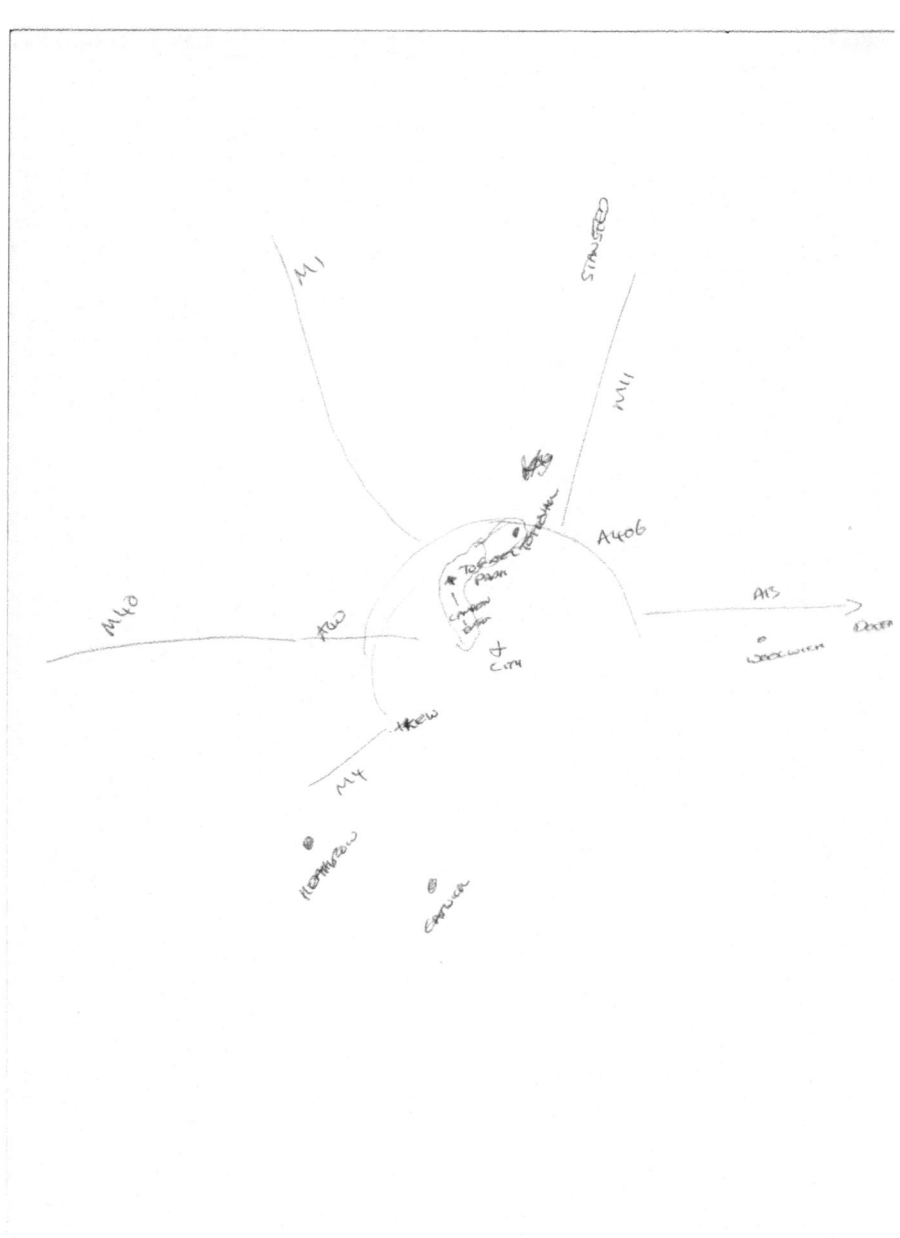

Fig. 7.28 The city sketch drawn by participant UDp-04

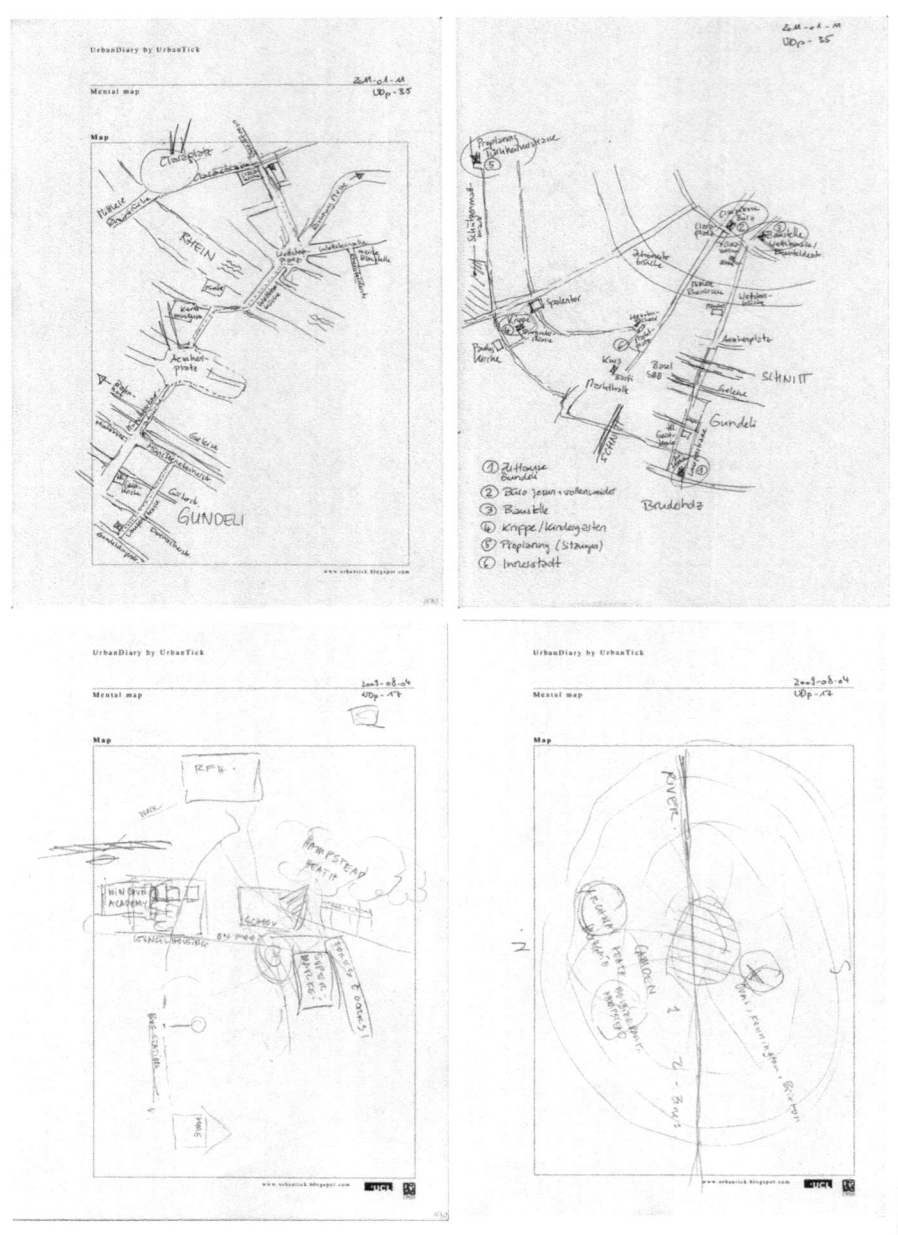

Fig. 7.29 The *top row* maps are by participant UDp-35 and the *bottom row* map by participant UDp-17. On the *left* we show the path sketch, the commute from home to work, and on the *right* we show the city sketch, a representation of the entire city the participants live in

7.5 Social Space

This is the origin of the city and the main point of orientation, especially in view from either of the bridges, hence an everyday impression.

The city sketch, shown in Fig. 7.29 top right, is essentially a slightly larger-scale reproduction of the path sketch. Its nature is similar using the same street detail representation. It only shows marginally more detail of the city. It extends however to the everyday context of important, key places. Many are located outside the actual commute but are tightly connected to it. The large-scale map offered the chance to include those and develop around these *anchor points*.

These additional points added to the map are places such as the nursery where this participant brings her children and picks them up, two locations where she often goes to for work-related meetings, as well as the location of evening courses that she attends. As such, the entire map is built solely around the *anchor points* specific to this participant.

Nevertheless, with a few additional key elements, some of the basic features of the city of Basel are represented rendering the location perfectly recognisable. As before, in the path sketch, the cathedral and the river Rhine are both represented at the centre of the map. The second element is the distinct road network, which is closely related to the shape of the river bend at the point where Basel is situated. Two roads are following three radial roads crossing the river and connecting through the city.[3]

Along these routes, some key elements are placed, mainly landmark buildings, such as the old city gate *Spalentor*, the market hall, the Paulus church or a large park.

It is interesting to note that even though the features are clear and understandable in the context of the city as a whole, again the subjective perspective dominates the map. The map is not developed as an overall image of all the elements but as procedural representation of experience. The objects are represented in relation to each other and not to the whole. The map is exact in the direction this participant connects the locations sequentially in an everyday location: but it is disjointed in the other directions.

Effectively, the map only *works* along the specific path sketched, the participant's path, and it is fragmented and disconnected otherwise: this is indicated in the sketch by the participant as cut-off lines. It is again this individual perspective of experience of the city that partially represents the city recognisably but at the same time highlights the particulars of the individual usage of spaces and understanding of the city from the perspective of the individual.

Cognitive Map Sketches Examples by UDp-17—The maps drawn by this participant are remarkable in the way they represent the spatial context. The path sketch, shown in Fig. 7.29 bottom left, is sketched out as a collection of places, stations visited during a normal working day. Single lines indicate the connections between the indicated *anchor points*. Two locations are positioned at either end

---

[3]This basic structure of the city of Basel was discussed and illustrated in Fig. 7.12 in the previous section on *path morphology*.

of the sketch (the paper is used in landscape), slightly separate from the other details. These two are the home location at the bottom and the workplace at the top. Everything else is happening in between.

This is broadly how the destinations are located in geographic space. However, the map itself is more a representation of how the day is unfolding for this participant in various places. Such an interpretation was confirmed in the interview: all the activities are taking place either on the way to work or from work. A great emphasis is thus put on the destination.

In contrast to this, there is little interest expressed in how the destination is reached. Even if prompted in the interview for more details regarding the actual journey between destinations, this participant was not able to recollect additional information such as landmarks. She explained that she travels on the bus, not paying attention how it travels the streets until she has to get off. Normally she is listening to music or reads a book on the public transport.

> ...because I listen to music and then I just look out of the window but without really looking... you know... so I just stay and wait till I arrive ......(laughs) (Interview UDp-17 2009-08-04).

This prompted the question as to how she does know where to get off, and with this she was not quite sure. There was no clear answer; it seemed more like a feeling, a sort of unconscious *sense of place*. She actually states:

> ...I stay on that can actually happen... but no, I watch out for the sign of the workplace (this participant's workplace has a large sign outside as it is a public institution) ... there is also a Marks & Spencer's just outside the bus stop ... (laughs). (Interview UDp-17 2009-08-04)

Later on she mentions that she reads the name of the stop on the sign outside at the bus stop or on the LED message board in the bus. This participant also walks a few routes. Those are reported verbally in the interview, but not included on the path sketch.

On the city sketch (see Fig. 7.29 bottom right), a similar observation of destinations disconnected from the route is presented. The entire map here is presented upside down, the way it was actually drawn.

The main element of the map is the zone plan of the underground network. Here again, this participant relies completely for orientation on the public transport network and represents the city map in this way. Zones 1–2 and 3 are shown, together with an indication that there are more zones further out. These inner zones are the ones mostly used by the participant. Along the tube lines, the connections are made to other areas of the city.

Remarkably the river is reduced to a straight line, running horizontally across the city. Especially on the tube map, the river has a distinct shape, a series of curves that seem of no relevance here. However it serves as a clear distinction between North and South London. These directions are indicated on the map, and it is clear that South London is an unknown territory.

In North London, two areas are indicated which are, together with the centre, the reference destinations of this participant. All three are connected with single lines. Due to the disengaged travelling on public transport and often on the underground, little spatial connection is actually experienced between the destinations. The results show vague plotting of areas. Everything is labelled with the names needed for orientation whilst travelling on the tube.

In this example the subjective focus on how the city is read is clearly in the foreground. In addition to the places featured, this example strongly shows the mode of transport which has developed into something like an orientation crutch and plays an important role.

All three examples illustrate how the individual perspective as discussed in the earlier Sect. 7.2 are transferred onto how the city is represented. However, from discussions and the interviews with the participants, it is clear that each of them clearly identifies with the city as an entity. They are all clear about where they live and identify themselves as citizens of that particular place.

Nevertheless, from the spatial investigations, it can be argued that they all experience and know a different, individual city. It is their own personal city, pieced together from places visited, situations experienced and connections discovered. As shown in Sect. 7.2.1, these can vary dramatically in shape or, as shown here, vary in detail and connectivities.

When the sketch, drawn as cognitive maps a subjective perspective, is met with an outsider's image of the city, both views are challenged and the successfulness of the cognitive map as a communication method tested. Whilst the outsider is interested to discover as many aspects as possible to recognise the places represented, the sketcher brings as many personal details into the picture as possible to make it his or her own city.

Whilst this concept of *personal city* sounds at first objective and holistic in reality, as pointed out in the examples, the representations are in fact mostly constructed along the sequences known and learned in everyday life. The *anchor points* play a fundamental role in the representation of both the path sketch and the city sketch and so does the sequencing. What is mainly relied on is continuity, as discussed in the previous Chap. 6. It can be argued now that continuity is an important concept, both in time and space.

Whilst the mode of transport and the way the individual is engaged in the process of moving influence the shaping of urban islands, it is arguably the routine of a sequence that shapes the image of the city. The places are known because they are relevant in connection to all other places known, and with this, we are moving a step beyond the mental maps as used by Lynch (1960). The image is built up through experience during daily routines. The way this image is put together is not an overall and objective viewpoint imposed on a city view but rather develops along the key reference points relevant to the individual. None of the maps provide a complete picture. The details develop in the making, one thing leads to another, as if stepping though the spaces mentally. The recollection was almost always structured along the sequence of the everyday routines as recorded from the GPS tracking, the interview and the schedules. The city is not just the city, it is everybody's personal city, and the image is not just some picture one has in mind. It could be described as a constructed recollection of remembered connections of activities, constantly updated with the latest experiences. Arguably there is no fixed and static image, just as there is no static shape, individually drawn, on the urban canvas. These social constructions result from everyday experiences shaping the *social space*.

7.6 Rhythmic Space

Rhythm and repetition are not an aspect of space to begin with. As discussed in the introduction to this chapter, *Cartesian* space is *objective*, all-embracing and absolute. It is a static construction to position everything according to one single point of reference. However, there is repetition in the way this construction is laid out and how it can be used in the sense of its mathematical and geometrical basics.

Similar to the construction of time as a repetitive sequence, space is based on units that can be divided and added up. However, space, as demonstrated in the various examples, can be a lot more than simple geometrical units. Space can be material, enabling individuals to express themselves and create identities and atmospheres or even become emotionally attached. Space can be territorial, collective, timed or social.

The investigation into the spatial aspects of the repetitive movement patterns observed in the urban context seems at first glance to be a counting of trips between destinations compared to the overall population. However, the discussion and the examples in this chapter have demonstrated that if we depart from a strict analysis of trips from an objective perspective and take subjective views into account, we open up new ways to see behind the pattern. This enables an in-depth discussion of *motivation* and *responsibility* as the mechanics leading to such a pattern.

Previously the term *constraint*, as introduced by Hägerstrand (1970), was used to describe the contextual limitations to individual actions. Given this, a shift towards a subjective perspective seems appropriate to use terms which put less emphasis on some overarching institution limiting the individual and stress the aspects of *agency*. The position taken here is that the pattern is influenced by a reciprocal relationship between the individual and the context. Both sides influence one another constantly shaping a unique condition in each case. As such the aspects are part of the here formulated *habitus*.

For example, Pile (1993, p. 125) quotes Ley (1978, p. 52), on the relationship of place and identity, to describe the interchange between place and identity as "dialectical", arguing that the social world is "the product of human creativity", but the social world has "a certain autonomy", though "their autonomy is always contingent".

This however does not contradict the discussion on constraints or negate the fact that constraints do affect most people's spatial as well as temporal radiuses. Many factors, such as socio-economic group, gender and age are influenced to various degrees, as, for example, discussed by Kwan (2000).

Since we are looking at a two-way system, arguably this changes the interpretation of the observations. Spatial conditions no longer guide individual action, but the individual motion contributes to the collective identity of the place. As such we

7.6 Rhythmic Space

have moved from an objective space construction to the social constitution of places. In this sense we can observe to some degree the *Social Logic of Space*, as presented by Hillier and Hanson (1984), by studying individual rhythms of movement.

As seen in Sect. 7.2, discussing the individual *path* as recorded using GPS technology, there is no random movement found in any of the tracks. Clear spatial parameters seem to guide the individual focus on how space in the urban context is explored and individually institutionalised as a *self-territory* and a *sense of place*, as used, for example, by Tuan (1990, 1977).

In this subjective shaping of context, the repetition plays an important role in shaping a territory. Even if in this case, territory is not an enforced entity, it is a contested unit of individually claimed room for action. Whilst the maps, as shown in Fig. 7.4, show the entire collection of trips claiming territory, only the animations, for example, shown in Fig. 7.21, provide a sense of repetitive claiming of space.

The *motivation* guiding the pattern consists of two main differences shaping territories on the level of repetition. There are clear benefits attached to this kind of spatial routine behaviour as observed in the Urban Diary GPS tracking study. For one thing, it is a source of security, with familiar environments being more predictable. At the same time, the familiarity in frequently visited spaces reduces the requirement for alertness. Less energy is spent trying to understand, orientate and navigate the spaces. As such, mobility is more energy efficient and allows one to start building up a relationship to places, resulting in different degrees of *place attachment*.

The second element in these *self-territories* is the aspect of time. As picked up with *What shape are you?*, the discussed *PPP* model of *place attachment* is crucially missing any aspect of time. *Place attachment* as well as *territory* in a classical sense is mainly used in connection to static places, such as homes and gardens (Altman 1975; Brown and Altman 1983; Brown and Werner 1985; Scannell and Gifford 2010). The term *process*, one of the three Ps in the concept *PPP*, as discussed before, could have the potential to cover this aspect. However, it is defined by Scannell and Gifford (2010) mainly in psychological terms, under the headings of affect, cognition and behaviour. Time as an aspect of dynamics is not incorporated. In addition there is no relationship between places mentioned as part of the PPP concept. *Place attachment* is discussed purely on the basis of a single place. Arguably defining a place against the background of *others* is a common way of distinguishing between places. This raises the question as to whether single and distinct places even exist and, more importantly, where one place ends and how it is defined as an entity.

It might well be that in psychology, the spatial context is not regarded as significant, but in the context of the city under the aspect of planning, the context and especially the connections between spaces are of foremost importance. Arguably the multitude of places is what makes the city (Borden 2001). This is specifically the case if discussing the city of the individual perspective as a result of *place attachment* to various places. The attachment builds up to various degrees for the *anchor points* and shapes a network of places. As such, the network has no direct spatial importance, but, as has been suggested earlier, the movement connecting the place has significant relevance. In fact it could be argued that the connections, as the collection of paths, are in fact part of the whole narrative.

It is the transformation of spaces whilst travelling that leads to a new place. Whilst Scannell and Gifford (2010) define the *PPP* framework for *place attachment* with reference to person, process and place, it is argued here that the aspects discussed in this work, such as time and travel, too are part of a concept of place. The usage of a new term, introduced earlier as *Habitarium*, does in fact cover this ground and can extend the *PPP* framework by adding the dimension of repetition, movement and time to a dynamic, large-scale, but still individual concept of everyday space.

Specifically on the level of the collective space, as shown in the New City Landscape maps, these connected spaces play an important role. Here too places of stronger *attachment* are being found and intensities vary. Similarly the discussion around *constraints*, as emphasised with regard to infrastructure or the landscape of natural context, for example, plays an important role.

Whilst the time analysis in the previous Chap. 6 has shown the similarities in activity pattern, spatially we have shown the differences. The specific morphology of each location leads to a different result in terms of density. However, similarities can be found in the way a few locations appear attractive and others are not. Each example has shown a great variety of densities: data from all of the cities could be articulated in terms of density differences. Clearly there are more and less active, and attractive, areas in all cities. This works as a spatial model.

Spatially, the question must be asked as to what extent the tracking sequence is a valid as a spatial model. There are parallels to the situation of networks. As discussed earlier in the context of the *Actor Network Theory* developed by (Latour 1996), and later retracted by him (Latour 1997). Retracting the theory was in part based on the observation that networks are very much non-spatial. Nodes and links graphically inscribe spatial configurations in a *Cartesian* sense. However, this should not result in a hasty interpretation of networks as being fundamentally spatial. It seems that networks are powerful exactly because they enjoy the luxury of not bowing to the Cartesian system in the first place.

Graphically, however, the network interpretation is a similar problem to the situation of the GPS tracks as a sequence of in Cartesian space describes locations. The difference however is the direct engagement at the moment of presence. As shown in the path maps as well as the cognitive maps, there is personal attachment resulting in an individual perspective or focus on space. This feeds into the personal city as a unique construction based on the activity pattern.

The important aspect here is the *sequence*. It was pointed out in the previous chapter that the sequence and in a time sense *continuity* are of relevance. This relates to the individual but especially to the fundamental discussion of time-space matters. Both are constraint born in the nature of their respective field, as *time*, one clock time unit follows the other and *space*, one Cartesian grid cell is adjacent to the next. With the introduced aspects, the meaning of a purely objective system is extended beyond the measuring of its entities.

It was shown in the various examples above, and in the previous chapters, cognitive maps and schedules carry the fundamental essence for the individual experience. Again in a dialectic way, this is between individual agency and contextual agency. It is here where *anchor points*, *place attachment*, *responsibilities* and *motivation* and *ambiance* create a complex condition shaping, based on repetition, the individual *habitarium* in a spatial sense.

The condition however is neither to be placed in the clock time nor the Cartesian framework. The dynamics and interplay do not fit any of the two static concepts. *Temporality*, in the sense of this dynamic shaping and reshaping of time-space, requires a set of additional dimensions that begin to bridge across the boundaries of either of the two concepts. Repetitive time-space patterns and their relevance for individual place making are in a way self-reflecting, creating a temporality of many dimensions.

We will now draw all these threads together in a theory of temporality which explains the rhythmic city, as a prelude to pointing the direction forward to further research on space-time pattern in large cities. As such this next chapter represents our conclusion.

Bibliography

Altman I (1975) The environment and social behavior: privacy, personal space, territory, and crowding. Brooks/Cole Publishing Company, Monterey

Altman I, Chemers MM (1984) Culture and environment. The Brooks/Cole basic concepts in environment and behavior series. Cambridge University Press, Cambridge

Ambrose PJ (1969) Analytical human geography: a collection and interpretation of some recent work. Number 2 in concepts in geography. Longmans/Harlow

Batty M, Hudson-Smith A (2005) Urban simulacra: London. Archit Des 75(6):42–47

Berry BJL, Garrison WL (1958) The functional bases of the central place hierarchy. Econ Geogr 34(2):145–154

Bisbee E (1937) Objectivity in the social sciences. Philos Sci 4(3):371–382

Borden I (ed) (2001) The unknown city: contesting architecture and social space: a strangely familiar project. MIT, Cambridge

Brown BB, Altman I (1983) Territoriality, defensible space and residential burglary: an environmental analysis. J Environ Psychol 3(3):203–220

Brown BB, Werner CM (1985) Social cohesiveness, territoriality, and holiday decorations the influence of Cul-De-Sacs. Environ Behav 17(5):539–565

Carlstein T, Parkes D, Thrift NJ (eds) (1978) Timing space and spacing time. Edward Arnold, London

Daston L, Galison P (1992) The image of objectivity. Representations 16(40):81–128

Debord G (1957) The naked city. (Reprinted in Sadler S (1998) The situationist city. MIT, Cambridge, MA, p 60)

Debord G (1981) Theory of the dérive. In: Knabb K (ed) Situationist international anthology. Bureau of Public Secrets, Berkeley, pp 50–54

Fredrickson LM, Anderson DH (1999) A qualitative exploration of the wilderness experience as a source of spiritual inspiration. J Environ Psychol 19:21–40

Golledge RG (1978) Learning about urban environments. In: Carlstein T, Parkes D, Thrift NJ (eds) Timing space and spacing time. Volume 1: making sense of time. Edward Arnold, London

Hägerstrand T (1970) What about people in regional science? Pap Reg Sci 24(1):7–24
Hägerstrand T (1982) Diorama, path and project. Tijdschr Econ Soc Geogr 73(6):323–339
Hillier B, Hanson J (1984) The social logic of space. Cambridge University Press, Cambridge
Kapler T, Wright W (2005) GeoTime information visualization. Inf Vis 4(2):136–146
Krausse J, Lichtenstein C (eds) Your private sky: R. Buckminster Fuller, art design science. L. Müller, Baden
Kwan M-P (2000) Gender differences in space-time constraints. Area 32(2):145–156
Kwan M-P (2004) GIS methods in Time-Geographic research: geocomputation and geovisualization of human activity patterns. Geogr Anna Ser A-Phys Geogr, 86B(4):267–280
Latour B (1996) On actor-network theory. Soz Welt 47(4):369–381
Latour B (1997) Keynote speech: on recalling ANT. In: Actor network and after, Lancaster, 1997. Department of Sociology, Lancaster University, pp 1–4
Ley D (1978) Social geography and social action. In: Ley D, Samuels MS (eds) Humanistic geography: prospects and problems. Croom Helm, London, pp 41–57
Lynch K (1960) The image of the city. MIT, Cambridge
Manzo LC, Perkins DD (2006) Finding common ground: the importance of place attachment to community participation and planning. J Plan Lit 20(4):335–350
Mosley B (2012) A plan for Plymouth, July 2012. http://www.plymouthdata.info/Plan%20for%20Plymouth.htm
Moudon AV (1997) Urban morphology as an emerging interdisciplinary field. Urban Morphol 1(1):3–10
Mulligan G, Partridge M, Carruthers J (2012) Central place theory and its reemergence in regional science. Ann Reg Sci 48(2):405–431
Pile S (1993) Human agency and human geography revisited: a critique of 'new models' of the self. Trans Inst Br Geogr 18:122–139
Pinder D (1996) Subverting cartography: the situationists and maps of the city. Environ Plan A 28(3):405–427
Proshansky HM (1978) The city and self-identity. Environ Behav 10(2):147
Rose-Redwood RS (2002) Rationalizing the landscape: superimposing the grid upon the island of Manhattan. PhD thesis, The Pennsylvania State University
Sadler S (1999) The situationist city. MIT, Cambridge
Scannell L, Gifford R (2010) Defining place attachment: a tripartite organizing framework. J Environ Psychol 30(1):1–10
Spiekermann K, Wegener M (1994) The shrinking continent: new time—space maps of europe. Environ Plan B Plan Des 21(6):653–673
Tuan Y-F (1977) Space and place: the perspective of experience. University of Minnesota Press, Minneapolis
Tuan Y-F (1990) Topophilia: a study of environmental perception, attitudes, and values, Morningside edn. Columbia University Press, New York
Ullman E (1941) A theory of location for cities. Am J Sociol 46(6):853–864
Wood D (1973) I don't want to but I will…'. PhD thesis, Clark University Cartographic Laboratory, Worcester
Wood D (2010) Lynch debord: about two psychogeographies. Cartogr Int J Geogr Inf Geovis 45(3):185–199
Zahavi Y (1979) The UMOT project. A report prepared for the US Department of Transportation, Research and Special Programs Administration, and the Federal Republic of Germany Ministry of Transport. Technical report, DOT-RSA-DPW-20-79-3, Wahington, DC

Chapter 8
Temporality: The Rhythmic City

Abstract In this chapter, the habitus is positioned in relation to the concepts of time and space, highlighting the production capacity of the habitus regarding these two concepts. It will also be positioned with regard to the existing constraint model. Besides the habitus, discussed earlier as the present past, this final chapter focuses on the term temporality as the description of in between time manufacturing room for agency and allows for the *production* of time and space. The previous findings are discussed in regard to a description of temporality based on time-space concepts, specifically as patterns of routine and rhythm. This will be based on the concept of the habitus, which has been filled with additional meaning from the findings of the fieldwork data. Examples drawn from the fieldwork will be discussed under the heading of both concepts time and space simultaneously, combining the aspects previously discussed separately.

IN THIS CHAPTER, THE *habitus* is positioned in relation to the concepts of *time* and *space*, highlighting the production capacity of the *habitus* regarding these two concepts. It will also be positioned with regard to the existing *constraints* model.

Besides the *habitus*, discussed earlier as *the present past*, this final chapter focuses on the term *temporality* as the description of *in between time* manufacturing room for *agency* and allows for the *production* of *time* and *space*.

The previous findings are discussed in regard to a description of *temporality* based on time-space concepts, specifically as patterns of routine and rhythm. This will be based on the concept of the *habitus*, which has been filled with additional meaning from the findings of the fieldwork data. Examples drawn from the fieldwork will be discussed under the heading of both concepts *time* and *space* simultaneously, combining the aspects previously discussed separately.

8.1 Introduction

We discussed the two aspects of *time* and *space* in the previous two Chaps. 6 and 7 where we analysed them separately. However, the examples detailed did draw from the same data pool. Under the two different aspects however, the examples did portray the content at times in different ways, allowing for the same aspects to be discussed from different perspectives.

The *time* analysis highlighted the importance of rhythms and repetitive structures in everyday life for individual organisation, arrangement and social interaction. However, the main finding was the consistency of the cyclical patterns observed both on the individual level (with the Urban Diary project) and on the collective level (in terms of the New City Landscape project).

With the chapter on *space*, the important aspects were aspects of *place attachment* and *anchor points* resulting from the pattern of activity and structuring the *self-territory* created through repetitive activities in space.

Both investigations highlighted a number of key terms in their respective contexts. *Continuity* is a term describing the connectedness of sequential activities within either the individual or collective narrative. It has become apparent in all the examples that an important aspect of everyday life is the way it continues. It is surprisingly an aspect that was not covered by Hägerstrand in his space model, as quoted in Giddens (1986). The aspect of continuity has to be added.

The second term that emerged from the examples was the *sequence* in the sense of having an order to different activities. However, it is important to point out that this does not necessarily follow *time* or *space* constraints but can be defined by the activity in the social or cultural context. The key to *sequencing* is the memory as the present past and the narrative.

The third key term is *agency*, which describes the individual capacity to take decisions and influence the world around them. This also requires the ability to process information and respects both the past, present and future context of the decision. *Agency* is consequently embedded in its context and cannot be detached from it.

The following discussion will present examples that combine the two aspects of *time* and *space* and analyse the implications. The focus will be on how *time* and *space* are experienced in everyday life.

8.2 The Experience of Time and Space

As seen in the section Schedules and Planning, the dominating aspect is the activity and the time it takes place. The spatial aspect is not present as an organising factor and neither is place. Where something is happening is secondary. If we were to consider only the schedules, it is almost as if space does not exist. Nevertheless, space is always implicit, and all participants are well aware of this fact. It is thus not communicated and can be interpreted as something that is taken for granted. Going to work is a journey from A to B, and it does not matter that much where B is actually located.

This can be interpreted as a neglect of one factor or – and this is more relevant here – as a hint at the importance of repetition. What we are talking about in schedules are repetitive patterns, and inscribed in the pattern is a certain spatial consistency. The observation that space is not an active but an implicit part of the schedule could also mean that repetition has to have a certain degree of spatial consistency.

A location in geographic terms might seem too static to fit with the directional idea of a schedule and the passage of time. On the other hand, there are activities, or rather events, that feature in the schedule that are not so much time based but more influenced by their spatial features. This is of course in relation to the rest of the spatial locations, such as short activities that require a detour or a specific route to get to. The school run, for example, is often an important feature of the commute. The activity is not the school but getting there. It has a strong spatial definition, much more so than other activities. Through this, it becomes rather more important and stands out.

Picking up on the discussion of context as introduced in Chap. 6, in the Sect. 6.3 *Schedules and Planning* in connection to time budgets, the structure behind ordering times is discussed. The examples have demonstrated the importance of the context. Similar to *space*, *time* is influenced by a set of parameters that can change, and as such it is flexible and fluid. It looks like a pattern, but there is something that links the two into a background order. This is what *habitus* describes: it is in fact the mechanical reproduction pattern we will discuss in detail later on in Sect. 8.5.

The visualisation of activity in linear time to a large extent neglects the repetitive pattern of the data. Since it is only one dimensional, the superimposition is always layered. This is mainly down to the absence of a secondary indicator to put the linear progression into context. Its visualisation is presented in Chap. 6 in Sect. 6.2.2.

As such, a measure of distance is introduced. Distance is measured from the base location for each individual participant, relying on the fact that the home location is the truly stable location acting as a hub, where at least once in 24 h the individual checks in. With such a change of parameters, and rather than amount of activity the spatial dimension is taken into account, the expression of temporality becomes the main focus. Temporality is thus becoming spatial in a repetitive sense.

For the working week, the distance starts to increase just after seven o'clock as participants leave the house to travel to work. Generally the distance then stays more or less the same throughout the day, sometimes with a little bit of movement around the lunch time break. In the evening, the distance changes again until it is back to zero as the participants get back home. See Fig. 8.1 for illustration.

However, compared to the morning, the evening is a lot less precise. The morning fits across the sample into a timeframe of around 1 h. The evenings are more diverse, and different activities take place opening a timeframe of up to 4 h.

8.2.1 Experience Time and Temporality

With the discussion of *time* and *space* separately in the previous chapters, it is now interesting to discuss examples that bring these aspects together and allow us to discuss them as part of the same observation. To do this, the data collected in the

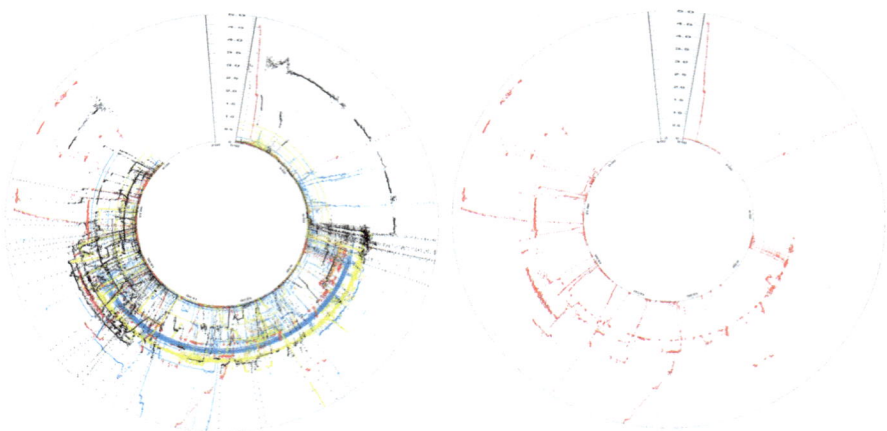

Fig. 8.1 Showing two time-distance rose graphs over 24 h circular plotting the distance from home. The base distance is on the central ring with increased distance being plotted outwards. On the *left* is a summary with multiple participants in different colours. On the *right* is shown a 24 h rose graph for an individual participant. The pattern of distance increase, followed by a stable plateau with a decrease in transit, is characteristic for a working day. See http://www.flickr.com/photos/40984848@N04/8702487788/ for an online version

interviews and cognitive maps are combined with the GPS tracking over the same time and space frame. What we are specifically focusing on is how the experience of *time* and *space* varies in the mental maps in comparison to the GPS records as the technical representation of both clock time and Cartesian space.

The mental maps were discussed in regard to *space* in Chap. 7. One of the observations was the linearity that characterised the description given by the participants. The maps represented straight narratives sequenced along activities and place. Such linear or *strip maps* are also discussed in the literature where they are often described as opposed to *comprehensive maps*.

The linear way of movement and the simultaneous cognitive reality create a mental replica of the context. As Spring (2006) outlines in her book contribution *The Linear City: Touring Vienna in the Nineteenth Century*, the way individuals move as an activity is linear in space.

The terms *comprehensive maps* and *strip maps* as introduced by Tolman (1948) in his paper *Cognitive Maps in Rats and Humans* discuss three aspects influencing the type of cognitive map produced:

> I can merely summarize it by saying that narrow strip maps rather than broad comprehensive maps seem to be induced: (1) by a damaged brain, (2) by an inadequate array of environmentally presented cues, (3) by an overdose of repetitions on the original trained-on path and (4) by the presence of too strongly motivational or of too strongly frustrating conditions. (Tolman 1948, p. 9)

8.2 The Experience of Time and Space

And:

> By way of illustration, let me suggest that at least the three dynamisms called, respectively, 'regression' , 'fixation' , and 'displacement of aggression onto outgroups' are expressions of cognitive maps which are too narrow and which get built up in us as a result of too violent motivation or of too intense frustration. (Tolman 1948, p. 10)

Time has been described preferably in nontechnical or subjective terms. Only in connection with the technical dimension of clock time has this subject become rationalised and objectively ordered. Besides the dimension of linear or clock time, other dimensions of time exist and have been proposed, analysed and discussed in most areas of science (Giedion 1982; Glennie and Thrift 2009; Andersen and Grush 2009; Bryson 2007; Elias 1992; Miller 1992; Zerubavel 1989). As presented in Chap. 3 Sect. 2.2, Time Space, a range of concepts have been presented. The topics of social time and individual time both feed into the topic of time being experienced as a context relevant impression of time for both the individual and the collective.

The comparison presented below draws on data collection from the Urban Diary project. On one side is the data recorded technically via the GPS device over time and on the other are the collection of cognitive maps drawn by the participants of the Urban Diary study. Both data sets cover the same spatial area and activity but from different angles using different methods. The comparison is set up to highlight key aspects but in no way to judge one method against the other. It is not the aim to point out a specific weakness on one side and a success on the other nor is one right or the other wrong. Instead, the comparison is expected to result in a new view revealing aspects of *time* and *space* that neither of the two could do separately.

The technical aspect of the raw GPS data is focused by the information gathered through the cognitive map. Conversely, the cognitive map is detailed by the objectification of the GPS data. To some extent, the comparison also allows for additional aspects to be considered in a cross-comparison between the participants. On one hand, the conceptual and individual cognitive map is in specific places being put into a Cartesian context, and the seemingly globally anchored GPS data is augmented with personal narratives.

On this basis, the experience of time, as a specific aspect of how activity shapes the spatial context, is presented in the following. The resulting visualisations are presented with a preceding representative example, two drawn from each case study in London and Basel. The following examples will be discussed as a cross-analysis, placing the cognitive map sketches beside the technical GPS records as track logs of places visited. It is a comparison focusing on the commute between home and work.

As part of the interview, the participants were specifically asked to draw a mental map of the usual commute between home and work. The recorded GPS data has been extracted from the entire data set on the basis of the work pattern, both in time, as weekdays, and space, for the route indicated by the participants.

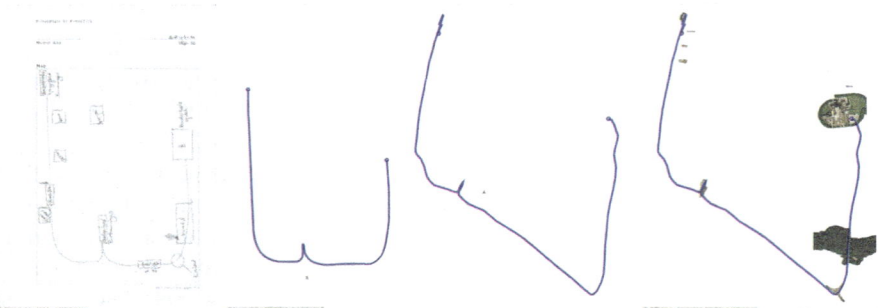

Fig. 8.2 Three sketches showing the commute of UDp-36. On the *left* is the cognitive map, on the *right* the GPS record within the cognitive map indicated points of reference. The middle section shows the direct comparison of the two representations simplified with important areas highlighted on both sketches. See http://www.flickr.com/photos/40984848@N04/8701369703/ for an online version

8.2.2 Basel Experienced in Time

The first example visualises commutes as recorded by UDp-36, illustrated in Fig. 8.2. This participant's route forms a distinctive W shape. This is graphically dominant element in both the cognitive map and the GPS path.

The shape starts on the left where the home location is. The middle section is where this participant changes on to another bus at the train station. The first bus drives into the main square in front of the train station where also the second bus leaves, creating this small loop. The second bus journey continues out towards the workplace, shown on the right.

Whilst the entire commute is approximately 3 km in distance, the distance as the crow flies between home and work is only about 1 km. However, topographically—something that is not shown on either of the maps but was an important element in the interview—the workplace is on top of the hill whilst home is in the village at the bottom of the valley.

From the elements added to the cognitive map it is clear that the home location has greater significance in its context than the workplace has. The participant has added details such as the buildings around the home that play an important role in everyday life. These are, for example, the supermarket as well as the presence of noisy neighbours.

The work location on the other hand is a destination with no context. However, it is interesting to note how the participant places the key elements on the route, in terms of actually describing the route. Clearly the interchange is of importance, but also both times the route takes a major change of direction, an element of reference is provided. At the first turn is a building that has been replaced and which was a construction site for a long time in the past. At the second turn, the information is given that there is a roundabout.

8.2 The Experience of Time and Space

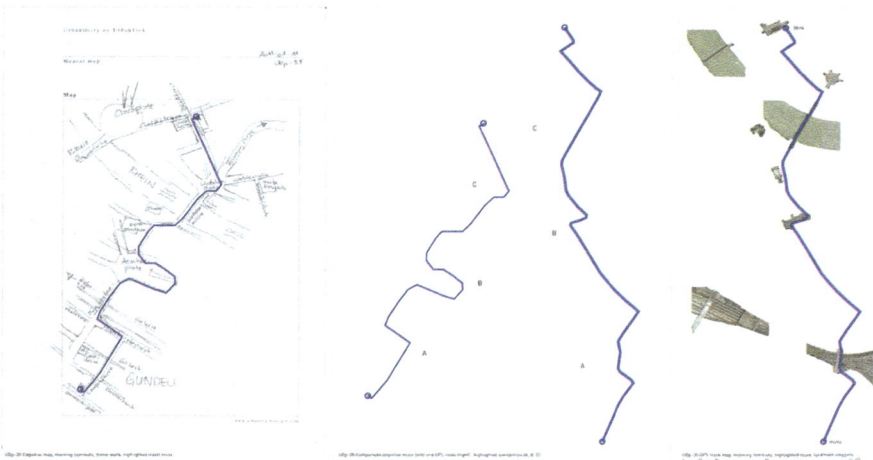

Fig. 8.3 Three sketches visualising the commute of UDp-35. On the *left* is the cognitive map, on the *right* the GPS record with in the cognitive map indicated points of reference. The middle section shows the direct comparison of the two representations simplified with important areas highlighted on both sketches. See http://www.flickr.com/photos/40984848@N04/8701369805/ for an online version

Whilst the rest of the route is only sparsely commented upon these points of reference are given. They are not only of importance to describe the route, but it becomes clear that they are reference points for the participant. On the bus, they are important, not for navigation, but to orient and check on progress as to when to get off.

As for the aspect of time, it seems that there is a rhythm to the arrangement of elements. At a rough interval, there are objects as points of orientation to navigate between destinations.

This example demonstrates how the complexity of a multilayered task, such as commuting, is reduced and folded into a handful of references that still make sense of the narrative as a whole.

The examples of UDp-35, as shown in Fig. 8.3, was discussed earlier in the previous chapter in Sect. 7.5.2. In comparison to the GPS recordings, however, it is interesting to come back to this example and discuss the aspects of space-time compression in the path sketch as the representation of the experienced space. As a reminder, the route starts at home, shown at the bottom of the map, and ends at work, represented at the top.

Two areas are of specific interest. The first is where the main section is folded in between points of orientation on a long linear section between the areas *A* and *B*. The cognitive map and the track log show rather different scales. On the cognitive map this section seems to have no relevance, whereas on the track log, there are specific details such as bends and turns.

This part of the route is travelled on one-way streets and on cycle paths. Hence, there is little navigation and negotiation required. The areas before and after, in the map marked as A and B, are both difficult to navigate on a bike because of the complexity of the crossroads with numerous lanes and tramways. These details are obviously more important and require higher levels of attention and subsequently show up as dominant location of the cognitive map.

The second example is a section right at the end of the route, towards the top of the map. Whilst the path sketch shows a straight line towards the destination, the track log shows a distinct double turn before heading for the destination. Whilst the path sketch shows the direct route to the office, the actual route chosen follows the cycle path, a quieter and safer route.

Whilst there are occasions the direct route was chosen, usually if travelled on public transport, it seems that the direct connection is cognitively more important. It seems that this section is more relevant in the way it connects the points of reference, in this case the workplace and the square indicated *Wettsteinplatz*. These two elements belong together as an entity in the sequence, but not in the way they are navigated.

Under the aspect of time, this example shows how the time dimension is represented in terms of intensity and decision making. It could be argued that where a high attention span and a density of decision making is required, time passes more slowly, allowing one to fold in the more *quiet* sections, reducing their representation on the sketch both in size and detail.

8.2.3 London Experienced in Time

In comparison two examples from the London sample have been chosen, shown in Figs. 8.4 and 8.5, using the same analysis of side-by-side data visualisation as we detailed for Basel.

For participant UDp-02, the modes of transport are bus journeys, shown in the visualisation, Fig. 8.4, as bold red lines, and walking, shown as thin red lines, to and from the bus stop.

Comparing the length of both modes as represented in the path sketch to the track log, it is clearly visible that the length of the bus journey is different. Similarly, the walked part is represented in a lot more detail in the mental map than the bus journey, indicating specific buildings or points of reference.

This suggests that the perception of space changes with time, mode of transport and especially with speed. Less detail is registered by some commuters on the bus as they *know* the route and focus on the destination. Some of the participants, as discussed in Chap. 7, have explicitly stated in the interview that, for example, on the bus they ignore the route and concentrate on a book, with music playing through their headphones or they simply just look out the window without registering anything. This leaves them with little knowledge that could be retrieved later and used to describe the journey.

8.2 The Experience of Time and Space

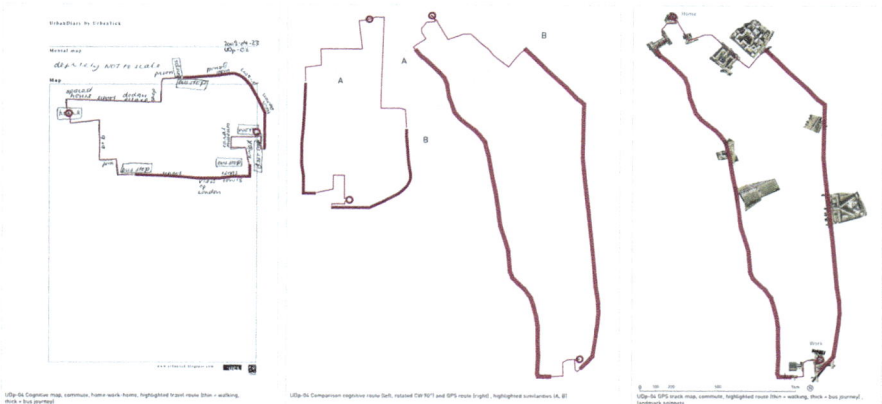

Fig. 8.4 Three sketches showing the commute of UDp-02. On the *left* is the cognitive map, on the *right* the GPS record within the cognitive map indicated points of reference. The middle section shows the direct comparison of the two representations simplified with important areas highlighted on both sketches. See http://www.flickr.com/photos/40984848@N04/8702492408/ for an online version

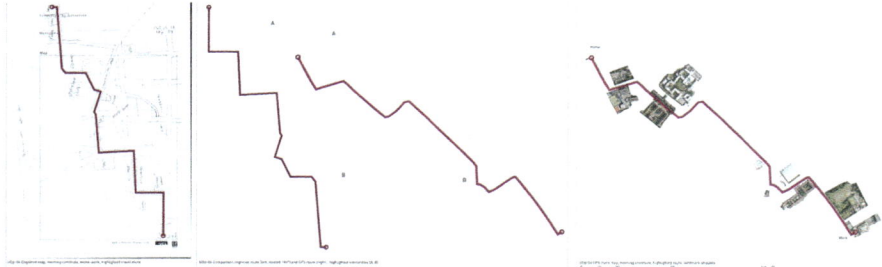

Fig. 8.5 Three sketches showing the commute of UDp-04. On the *left* is the cognitive map, on the *right* the GPS record with in the cognitive map indicated points of reference. The middle section shows the direct comparison of the two representations simplified with important areas highlighted on both sketches. See http://www.flickr.com/photos/40984848@N04/8701369741/ for an online version

What can be observed is that time changes in connection to the experience and circumstances of the activity. It is compressed when engagement is low at the same time folding-in of space, and in extended areas, experience is more intense due to the physical and mental engagement of the participant.

Participant UDp-04 has, as discussed previously in Sect. 7.5.2, a particular focus on the spatial context. His commute is on foot between home and work. Details are subjective and often related to specific interests. As a keen motorcyclist, he is interested in the quality of the road network, which enables him to navigate easily.

In comparison, between the path sketch and the track log, there is a specific section between the indicated areas *A* and *B* of significant difference regarding length and detail. The path sketch shows a short connection between two detailed areas, whilst the track log on the Cartesian reference system shows a long road between two sections with bends and corners.

This part is, as picked up in an earlier example, a one-way street. The way the path sketch is constructed is as a continuous sequence, a step-by-step reconstruction of the morning commute. In this context, since this section of the road is passed against the direction of the actual traffic flow, it is of little relevance for navigation. Because it can only be accessed as a pedestrian, it is, in the larger framework, also shown earlier as discussed in Fig. 7.28, thus not useful.

Regarding the detailing of the sketch, it can be noted that there is tremendous detail around the points of destination. These are also the areas where this participant spends his time. The context around the home is familiar as the local neighbourhood with shops and other reference elements. The workplace is significant, as the participant spends some time after school around the area as he has an engagement at the local farm, just across from the workplace, and he also spends time after work with colleagues at the local pub. All these details are shown on the map.

Time in this example is connected to detail and intensity of experience, whilst it is folded in with less exciting areas represented on the sketch.

The examples demonstrate the relationship between the topics of space, time and experience. As discussed above, an aspect of folding occurs, where important locations are brought closer together and space and time are folded into one another to fit the sequencing of the experience. There is a clear dominance of the sequence as a representation on relevance, previously discussed as an individual focus. This experiential specialisation provides a clear guide to the subjective space.

There is a significant amount of detail concentrated around destination points. As predicted in the discussion around *anchor points*, the immediate context of a key location is of importance proportional to the destination. It is boosted if in the vicinity of different focal directions lead to it. The elements have relevance in more than just one way. As shown in example UDp-36, these are local shops and the local pub. These are located in the vicinity of different activities and play a role in different focal activities.

The concept of *anchor points* then has an additional relevance in experience spaces. It is connected to the rhythm of everyday activity where under the aspect of time, the location is experienced again and again in sequence according to activity. In this sense, the *anchor points* develop over time and interact according to the rhythm.

The connection, previously highlighted, between destinations is of great relevance to the individual and plays a major role both in the construction of and identification with key elements. As seen in the example UDp-04, the direction of travel plays a role. This however leads to conflict with the static reference frames of *time* and *space*. As a consequence, they both are usually neglected or adapted in favour of the accuracy of the subjective relevance and the narrative. Mental maps are sketches visualising the inner image for a third person. However in their expression, they are subjective and developed in a sort of soliloquy where self-recognition is valued highly.

Time as such is an aspect of this method of sequencing. The experience represented as a step-by-step guide captures the way time is part of the experience directly. It encapsulates the value according to intensity of involvement. As such, the elements cannot be subdivided and play their individual part in the narrative of time-space stories.

Whilst the abstract concepts of both *time*, a continuous count of units, and *space*, a geometrical description of units, are helpful constructions to objectively quantify, subjectively these concepts break down. They are used in a muddled up form, where parts are borrowed from both concepts, whilst other elements are left unused.

Subjective space, it can be noted, is built around the agency and the activity of the individual's *habitus*.

8.3 The Concept of Clock Time and Cartesian Space

The examples in the previous section have shown how both time and space vary in subjective expressions of experiences and memories. It could be argued that, in fact, the two aspects are not the essential elements of these expressions. If they play a role in their development and representation at all, it is mainly as a secondary framework to interpret the created sketch and to communicate it.

It is only through this interpretation and the explanations of consequences that *time* and *space* build importance. On no occasion has a spatial or indeed time experience led the narrative.

Both terms, *time* and *space*, in their technical sense as *clock time* and *Cartesian space*, are part of the foundations of science (Hawking and Ellis 1975; Disalle 2006; Davies 1995). They are also deeply integrated in everyday use, as argued before and discussed for example by Glennie and Thrift (2009) and accepted as a natural state for which we have some sort of feeling or sense.

From the observations and findings discussed before it can be argued that neither *time* nor *space* is a natural phenomenon that plays a major role in everyday experience. However, both are becoming essential in descriptions and are employed to measure and quantify in order to communicate subjective observations.

Both terms represent theoretical *concepts* used to characterise, qualify and explain phenomena both in abstract examples and in the real world. Questions regarding these *concepts* have been raised from early on, and examples are *Zeno's Paradoxes*, *The Dichotomy*, *Achilles and the Tortoise* and *The Flying Arrow* (Ray 2002). Aristotle has added some more paradoxes, for examples *The Paradox of Places*, where he argues that each place has a place, has a place and so on, hence there are infinite places plus objects can be in different places at the same time (Huggett 2010).

Whilst many answers have been given to these paradoxes to date, including some explanation attempts by Aristotle himself, there remain doubts, especially amongst philosophers, as to whether purely mathematical answers can solve the raised questions in their entirety. The discussions around the *concepts* will continue and extent with new discoveries and inventions. From *experience* however, we know that a race can be won, and the arrow does eventually reach its target. It thus becomes clear that *concept* and *experience* might not be entirely the same and one cannot always explain the other.

Interestingly Hägerstrand (1970), also discussed by Pred (1977), in presenting his three points about time-space, does in fact rely on a mix of the two, *experience* and *concept*. He mentions the *limited stacking capacity of space*: in other words, *a body can only be in one space at the time*. Already in his time, doubts may have been raised; for example, how these requirements cover real-world activity, with innovative information and communication tools being in place, and as discussed by Buckminster Fuller (Krause and Lichtenstein 1999) in 1963, previously discussed in Chap. 7, Sect. 7.5, on how increased travel speed made the world shrink. Similarly, the case of simultaneity is argued by Foucault (1986, p. 22)[1]:

> We are in the epoch of simultaneity: we are in the epoch of juxtaposition, the epoch of the near and far, of the side-by-side, of the dispersed.

In his conditions, Hägerstrand relies on *experience* rather than *concept*. Subsequently however, he uses *concept* alone to describe, measure and quantify the observed phenomenon.

The distinction between the two is important in the way the problem is formulated, and the findings should be interpreted. The argument, to accept the terms *time* and *space* as concepts, changes the perspective in ways to allow for *experience* to become part of the argument. Arguably some of the problems around objective perspectives on individual experience, agency or power tackled by time geography stem from the acceptance of the concepts as *natural*, in the sense of both clock time and Cartesian space as naturally existing. However, positioning *time* and *space* as concepts does not at all question them in their consistency or relevance as for example, in Zeno's paradoxes or indeed Hägerstrand's base description relying on experience. The *habitus* in fact describes the structuring practice thereof.

The two *concepts* are accepted as widely as they are across the sciences, as a logical description and quantification of many observed phenomena. These are agreements or social constructions as argued by Glennie and Thrift (2009). However, this view should be interpreted as a result of the understanding of contextual relationships of our times, rather than a dogma as basis for yet another *concept*.

As Foucault (1986) argues in *Of Other Spaces*, each time has its understanding of the *concepts*. The seventeenth century had a fixed and hierarchical understanding of space, with each type of space having a clear and distinct position. This changed with Galileo's discovery that the earth rotates and all objects are in fact in movement and transition, the static interpretation of the space *concept* burst and made way for a more dynamic understanding of *space* as a series of relations of proximity which can be explained as sequences, trees or grids.

---

[1] This paper was translated in 1986 by Jay Miskowiec for Diacritics. The original paper was published in as *Des Espaces Autres* by the French journal *Architecture-Mouvement-Continuite* in October, 1984, based on notes for a lecture given by Michel Foucault in March 1967.

Our current times might already have moved beyond the simultaneous, and we can observe that individualisation is dominating our culture. Whilst it is often expressed in terms of the masses—mass events, large networks and circles of friends—it is in fact practised as lonely and isolated individuals acting for themselves. This is promoted by technologies supporting this trend, since communication methods make it possible to be part of the whole world via the web and digital social networks but physically remain isolated. Also, mobile technology is strongly emphasising the individual, serving as a portable connection delivering anonymous real-time information to any location as interactive communication and entertainment. This includes services such as instant location information with mapping services, SatNav GPS navigation and direction services, as well as news and real-time updates from a network of related services and contributors.

The fact that the cultural practice no longer has a location, or has relocated to the chair at home in front of the computer, TV or indeed to any location using mobile technologies, does transform spatial practice, in terms of the physical importance and identity of space in a urban context.

This new *subjective spatial practice*, with a constantly increasing focus on the individual, develops a changing understanding and meaning of time and space. The discussed conceptual problems in earlier understandings of space are seen in the context of such shifts. The background of rapid individualisation demands new descriptions as to how the shifting values are playing their part in the whole—in the case of this book, the urban context.

The urban context, or indeed the city, has earlier been identified as one of the environments where both concepts, *time* and *space*, coincide. This has been discussed in examples such as the market or a religious place, where time and place are the fundamental entities, for example, enabling trading and social interaction. Both parties have to be at the same place at the same time to exchange goods. In fact it could be argued that the city is the birth place of the two concepts *time* and *space*. The essential aspects for living closely together in densely populated areas like cities, such as food, goods, waste and materials, are social phenomena, which have become organisationally possible due to the discovery or invention of the two concepts *time* and *space*.

Organising individuals is perfectly possible utilising the two concepts governing conventions. Organisationally it allows for the identification of the subjects in relation to other subjects. Between them, clear claims as well as arrangements can be made. In this conception however, the definition is as a consequence against the context and the collective. It is the in between that is of importance not the subject itself: as such, both *time* and *space* are relational. For example, a single plot of land does not make much sense, as in a group of plots it needs defining, and consequently its *spatial* parameters can be valued. Similarly for social interaction, the time of day is only important if there are other individuals, living different lives, engaged in different activities, playing a part in the same group with whom one needs to socially interact.

It seems plausible, as argued by, for example, by Reichholf (2008), that repetitive activities such as religious practices are in fact early implementations of these concepts and have, in the long term, led to permanent settlements. In the activity of rhythm, the space gains importance as it is repetitively used at a given time, for example, a tribe that travels to a certain spot on the hill once a year for the celebration of midsummer night. This creates in its production the spot as a *place*. The ritual establishes the context for the concepts, for other tribes can join and meet at the same place at the same time. From such early forms, the concepts of calendars were developed to the detailed, accurate and sophisticated calendar system we practise today (Zerubavel 1985; Richards 2000).

In turn, this arguably requires the definition of temporality as a duration rather than a simple attribute in the sense of *fluctuating* or *not of persistence*. The movements and the activities are a manifestation of such a temporality. It is not just temporal, in terms of being in the present, it also has duration. There is a beginning and an end, even if they are fading away. This is the long moment where the creation of space is part of the activity. There is not only the action as a temporal activity in the city, but there is the time for this action that enables the city to be part of this action.

In many ways, this is where the Hägerstrand model ignores the dimensions of a temporal activity, and as in the traditional *time-space aquarium*, it is crushed into other dimensions, hence ending up as a line. Ignoring the nature of the activity renders it meaningless for the context in two ways, both in terms of the experience of the individual and the influence it can have towards a collective space. What makes the city is really the manifestation in space rather than the trajectory. It is the interaction and the presence that can give meaning and shape the identity of place and individual.

Kevin Lynch has captured some of these moments in his *Image of the City* (Lynch 1960) as a one-off frozen picture. However, he ignores any dynamics or time aspects of space in his resulting image of the city. He does talk about experience and how individuals have a memory of the city, but he does not address how this memory comes about or how the activity of being present in the urban space shapes this image.

These aspects are detailed in *Rhythmanalysis* by Lefebvre (2004). However, this does not feature the spatial aspect to the routine and the repetition. Even though Lefebvre describes activities in space in great detail, there seems to be no link between the activity and the spatial configuration as such. Lefebvre does not, like Hägerstrand, simplify the temporality of the activity, although he does acknowledge it to some extent.

The spatial conditions, as discussed in the earlier chapters and as shown in the examples, are strongly influenced by activities. The spatial and urban morphology conditions are influenced by the individual and the group. The city as an artefact is the product of the society building and living it. As Dalton et al. (2012, p. 3) put it:

> However, the relationship between space and society is a two-way relationship: not only does a society create the spaces that it uses, but a group of people (be it the inhabitants of a settlement, an urban neighbourhood, or the users of a complex building) are directly affected and influenced by the spaces they inhabit.

8.4 Reflection on the Time and Space Context

With these final examples and in order to start drawing the pieces and paths of investigation of this research together, it makes sense to look back on the different key observations and findings. Most of these have been discussed to some extent in the previous chapters in a more embedded context, but here they shall be recited to densify the research findings.

The main research interest from the beginning was the question regarding the existence and impact of cyclical patterns in individual everyday activities especially in an urban context. A certain rhythm was expected based mainly on the experience of institutional patterns such as shop opening hours, working hours and public transport schedules but also natural phenomena such as day and night, sleep and feeding patterns or seasonal changes.

The literature in various fields especially from the late 1960s, such as writings by Halberg et al. (1994) in the field of chronobiology, suggested a much more fundamental embeddedness of repetitive patterns both in the environment but especially within the workings and functions of the human biology which is in the form of activities translated into individual activities (Capra 1997).

What is striking in the findings based on the exchange with the study participants is the clear absence of rhythms as an aspect of individual decision making. The individuals were not or not fully aware of the patterns in their everyday activities. As it is often referred to in the literature, humans do not have a sense of routine as we do have a sense of vision or hearing or smell. Once they were introduced to the concept, it was easy for them to relate to it, but it did not appear as an organising principle in the first place. On the contrary the presence of routines was something that was not desired, if not to be avoided. As discussed in the Sects. 4.6 and 6.3, flexibility and spontaneous behaviour was desired.

Whilst the perception of repetition and routine was only vague, the presence of repetitive patterns in the data was obvious. The routines represent in the data have been discussed in the various chapters before; however, the discrepancy between the perception of routine and the recorded routine was not expected to such an extent.

It appears that the individual activities are masking the existence of the routine with a layer of decision-making processes. As argued before, the routine carries energy benefits but mainly social benefits in the sense of synchronising groups of people and enabling social interaction.

Whilst the routine are barely recognised by the individual, they are exceptionally rigid and persistent. Beyond the detection of cyclical patterns in the data, the rhythms are very hard to beat on a daily basis by the individual.

One of the questions regarding the data collection and the use of the GPS technology is the truthfulness of it all or to what extent the individual participant can manipulate the data collection by changing the behaviour. This is a valid question since the individuals are not supervised. However, there are two factors that strictly limit the scope for made up behaviour on the participant side.

One is the duration of the recording period. It is challenging to keep up a change of behaviour over a long period consistently. With the recording period of 2 months, it is safe to say that variations in the activity are visible and consistent pattern stands out clear.

Secondly and even more importantly, there is very limited scope for free improvisation along the daily routine. The schedules are tightly packed with activities, carefully balanced in regard to effort and gain, for there to be room to negotiate variety. This is perfectly visible for one in the schedules in Sect. 6.3 and also was evident in the interviews.

Whilst this is true for aspects of schedule and activities, this is only partly true for the aspect of space and route choice. It is however of a different significance, and it can be expected that there are various reasons for a change of route beside the impact of the study.

As such, it is safe to say that the nature of the activity observed in this study is very stable for various reasons, and hence the data is true in the sense that it represents a participants routine well (Fig. 8.6).

The difference between measured space and perceived space is significant. In fact it could be argued they are hardly the same. As argued in the previous Sect. 8.2, both follow very different logics. Whilst Cartesian space is geometrical and static, the perceived space is structures around sequence continuity and *anchor points*. This was demonstrated in the Sects. 8.2.2 and 8.2.3.

Whilst we do use Cartesian-based concepts and clock time for both recording and representation, it proves important to pair them with the individual mental image and the experience to create a more complete picture.

From these findings, it becomes clear that the terms time and space as separate and possibly opposing terms are not able to explain the observed phenomenon of rhythms to any satisfaction. We are unable to describe the dynamics of the patterns discussed above and hence have to look for alternatives to be able to focus on activity

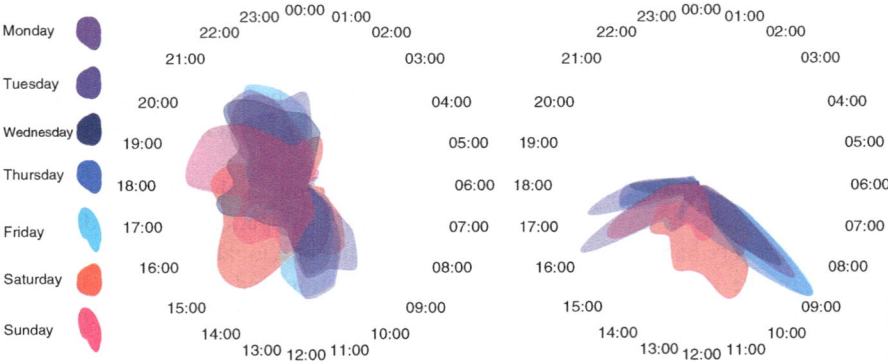

Fig. 8.6 Two of the study participants compared regarding their routine in schedules based on weekdays. Both show a significant structure in themselves, but in comparison the times and frequency are rather different. See http://www.flickr.com/photos/40984848@N04/4562770273/ for an online version

and behaviour. This is why the *habitus* is proposed here as a third aspect. With this the patterns as well as the aspects of agency, continuity and sequence of the observed and previously discussed behaviour.

As hinted in the introduction, this change in approach poses problems regarding the tools of investigation themselves. Both GPS technology and GIS technology are the Cartesian space representatives per se, being born out of this time-space concept these methods themselves objectify the results. It is argued here that this objectification is where the crucial information about the activity is being lost.

However, the GPS and GIS do provide a boost for the argument in the way that they make the results seem reliable and *true*. This helps individuals to accept the data, especially because the subject of investigation is something not in reach of our human senses as discussed above.

8.5 Routine in Time and Space as Habitus

Aspects of time have previously been introduced related to both the individual and the social as individual aspects, for example, in Chap. 6. The separation of the two time perceptions was foremost promoted by Durkheim, Hubert and Mauss (Tabboni 2001). Their research characteristic is:

> A distinct preponderance of social aspects over individual aspects, with the two set up in opposition to each other (Tabboni 2001, p. 18).

Whilst there are tensions between the social and the individual perspective on time, the starting point for the discussion in the following will not be the two aspects in opposition. Rather they are seen as complementary. Or as Tabboni (2001, p. 18) puts it:

> ...the aspect of time which is sociologically most interesting, its capacity to constitute a link between the social, individual and natural worlds in the individual choices that are made at the different levels of a single human experience.

This is very much the definition given by Elias (1992) of the different aspects of time, as he argues against the opposition between the individual and the social and describes the individual, social and natural time as simply three different levels of the same human experience, that of change and choice (Tabboni 2001). This is, if taken further, a crucial point as it adds the additional dimension to the continuous clock time. Time is no longer merely a collective rhythm of different activities but becomes a social construction. This links back to Glennie and Thrift (2009), describing it as an abstract symbol of a cognitive instrument of the *habitus* (Tabboni 2001).

The concept of the *habitus* was already introduced in Chap. 1 in regard to the cyclical pattern and routines in an urban context. Its key aspect is the presence of the past or the constant referral to the past as the reference for the present activity. As such, the *habitus* is created by the past conditions but defined together with its application in the present condition. It governs the practice by referring to past experiences in the past and present context and allowing for their *reproduction*. The limitations of this reproduction process are set by the historical and socially situated conditions, and therefore does not produce any random or unpredictable novelty (Bourdieu 1990). In this sense the *habitus* is responsible for recreating successful practices or, as Bourdieu calls them, *positively sanctioned* activities for each task.

The observation of rhythms and repetitive patterns is an observation of the result rather than the cause. However, in the *habitus*, we describe the structure and motivation leading to the results. As Bourdieu (1990, p. 54) argues:

> The *habitus*, a product of history, produces individual and collective practices – more history – in accordance with the schemes generated by history.

The concept of *habitus* thus comprehensively summarises the phenomena observed and described in the examples in the previous chapters. It does, however, seem to stand against the occasionally used time geography concept of *constraints* and will be discussed separately in the following section.

The *habitus* is the driver of activities. It structures the routines and cycles of everyday life which in turn lead to the production of space. In this sense it is the *habitus* that guides the production of what Hillier (1996) describes as *space*. Hillier argues that:

> Human behaviour does not simply happen in space. It has its own spatial forms (Hillier 1996, p. 29).

Here we propose to go a step further and argue that the *human behaviour*, or what we call *habitus*, is in fact responsible for the production of space. It is in this way that human *agency*, or what Hillier (1996) calls *social constitution*, leads to the spatial configurations, the built environment and architecture of what we generally describe as *space*. Whilst Hillier (1996) argues, that *space is the machine*, the argument here is that the *habitus* is the structural logic and *space* conceptualises the production result.

This study and our examples can be positioned in the context of such discussion. The emphasis is upon establishing the concept of the *habitus* as the core term to describe and conceptualise the observed pattern of repetition and routines in everyday practice and to discuss the consequences for the understating of the resulting two concepts *time* and *space*. Out of this we develop the term temporality. To do this, *temporality* as a concept is re-examined as a state that effectively transcends both terms practically and theoretically as a *concept* and as *experience*.

8.5.1 Constraints vs. Habitus

It could be argued that, for example, the lunch break is a *capability constraint* in the sense of Hägerstrand's constraints concept, limiting the range and duration of activities. On the other hand, the lunch break can also be interpreted as a *coupling constraint* since lunch is a social event where groups or families meet. And finally, the lunch break could be an *authority constraint* with, for example, factories' schedules of working hours and fixed lunch breaks.

Can an event as a result fit all three categories? Without going too deeply into the discussion around the constraint problem, it can be noted here, and it has been demonstrated in Chaps. 6 and 7 that there is much more to something like the lunch break than being a simple *constraint*. It is much more of a social event, meeting people and participating in an activity with a group or family. There are many sides to it, directed by a complex mix of influences guided both by time (moment) and space (location) in the sense of the German word *Zeitpunkt*, translated as *point in time*, featuring a distinct point of spatial reference.

Time geography has built its concept around the idea of constraints. For example, Miller (2004, p. 648) describes it, referring to Hägerstrand, as:

> Time geography focuses on the interrelationships between activities in time and space and the constraints imposed by these interrelationships. Rather than attempting to explain or predict an individual's allocation of time amongst potential activities in space, time geography highlights the factors that restrict an individual's choice. Although often applied to daily and weekly time frames at the urban scale, time geography can also accommodate scales as extensive as a person's lifetime (Hägerstrand 1970).

The questions this position raise, which were raised before in Chaps. 6 and 7, concern the relationship between the individual and the *constraint*. Arguably with *constraint*, time geography is referring also to society even if this is not explicit. However, Hägerstrand also includes natural aspects, groups and modes (Hägerstrand 1970). Nevertheless, what is being ignored is how these constraints come about. What makes constraints?

This in turn will help to address the question as to how the *constraint* influences the individual. To address these identified points of critique, it is here proposed to position the constraints model as part of the *habitus*. In many ways, the constraint model is already integrated in the *habitus*, in a more neutral sort of way. Whilst with *constraint* there is always a negative cognition as limiting, excluding, hindering or even strangling, *habitus* is a much more practical matter of course. Furthermore *habitus* has a clearly defined theoretical background, as discussed in the introduction chapter of this book. *Constraint* can be criticised for not having this rich contextual connection. In this context nevertheless, the two terms can be merged, and constraint can become part of the habitus.

This will arguably make it clearer as constraints are explained in a wider context, including the origin, and the habitus now contains a model to describe the actual instant. Whilst we will continue to use *habitus*, it becomes clearer that this term is not only singular and does in fact cover both groups and institutions (Hägerstrand 1970). On the other hand, the context of culture and society inherent to the *habitus* is being imposed on all *constraints*.

8.5.2 Towards the Adaptation of **Habitus** *in an Urban Context*

In a positive sense, the term has the capacity to transcend the limitation of *constraints* and turn them into potentials. Whilst *constraints* are objectively conditioned by clock time and Cartesian space, the *habitus* is conditioned by the context of the previous circumstances. It is defined out of the moment, its conditioning history and not by a static frame of reference. The repetition allows it to play to its strength in reducing the energy required to complete tasks whilst increasing the safety and success aspect by repeating a successful solution.

One of the examples to illustrate this point is the house or the place individuals *create* for living in. As seen in the examples in the previous chapters, *home* is a remarkable point of reference for all study participants. Whilst it could be defined in time and space, it has other qualities too. The perspective of *constraints* would argue, for example, that it limits the ability of the individual to roam the urban landscape with the requirement to return *home* for certain activities such as sleeping, eating and hygiene. This is, of course, on the basis of *time* and *space*.

This is true if compared to the random possibilities offered by the box space concept, where all spaces in the city are constant. However, we have seen, and *habitus* describes this, that the creation of individual spaces and their repetitive use reduces energy requirements for navigation and orientation and improves safety, health and success rates for the individual. Quite simply the building of a home, a familiar space, simplifies the tasks of everyday living by allowing knowledge to build up and organisation to unfold. At the same time, it acts as framework for the projection of other habits that can be reproduced within it.

Such a definition of the home, as a place according to one's self, in the sense of arranged in the way it is suitable for being oneself and in the sense of living as stimulating a life as possible, might shed new light on the desire to settle. Whilst previously, we have argued that the settlement has coincided with the intellectual development of abstract concepts of *time* and *space* allowing for the organisation of densely populated places. The origin could however, and this comes before any form of organisation is actually necessary, be the basic desire to create homes, in order to minimise energy spent arranging and organising, orientation and navigating and free up these resources for production, and support these with relationships to other homes. On the one hand, the context of other homes impacts on the qualities of one's own home, by using them as reference points both for physical goods and intellectual reference and for the exchange of the products being created. In this context stands the requirement for the forms of organisation, for example, regarding a market, a graveyard, a prison or a funfair, elements of coexistence of *time* and *space*.

8.6 Towards an Integrated Temporality

Both Durkheim et al. (2001, original 1915) and Bourdieu (1990) argue in regard to *habitus* and history that there are numerous pasts. This however seems complicated and unlikely. As quoted in the introduction of this book, Durkheim argues that the presence is a product of the many personas one has been in the past. The question this provokes is: what is a person if one has been numerous personas? How would

8.6 Towards an Integrated Temporality

such a persona be defined, when does a persona start and when does it end? Is it based on a day? And if so is this a daylight day (experience) or a 24 h day (clock time)? The question is: " how does the *habitus* represent *time* and *space*?"

As argued in Chap. 6, based on time structured examples, and in Chap. 7 on space structured examples, continuity is an important feature in the way activities are experienced. It does not just end and start anew. There is a certain flow; activity might change its pace or its density, but there is no interruption. It is also something that Hägerstrand (1970) tried to capture in his constraint model. However, it can be argued that this was limited to the physical body, and did not include the activity or the intention or indeed its historical context.

The argument was made earlier for the nonexistence of *agency* in Hägerstrand's time-space model, and this is where this continuity applies to, the human *agency* per se. As Bourdieu (1990) has argued, *habitus* is the result of history. Whilst there are arguably not many but only one continuous trail, or indeed *path*, of a single history, the immediate past is, as Bourdieu (Durkheim, 1977. p. 11, as cited in 1990, p. 56) further argues, unconsciously present.[2] Neither Durkheim nor Bourdieu discusses the question as to why the *habitus* is close to repetition and still able to adapt to new circumstances at any moment. Why consistency and routine reproduction of contextual historical events does not stand in the way of evolution and change is still an open question.

Arguably one of the strongest points questioning the relevance of repetition is a missing concept to cover for spontaneous changes and abrupt jumps of evolution, in fact, any progress. All we were able to present in the introduction was a vague concept of slight adaptation in each iteration of the cycle whilst referring to concepts of feedback. This in itself requires a process at each moment of the cycle, which is not a completely satisfactory setup. Whilst there are a majority of cases, the cycle can continue with minor alteration. This is, for example, a new day, similar to the previous one, but the 29th instead of the 28th. However, the question is legitimate in as far as there are plenty of examples that express in their nature a sudden shift, abrupt change or jump in activity. Evolution as well as general adaptation to a changing context is key to existence and continued existence.

The key is arguably to be found with continuity in the presence of the history. The rhythm is dictated by the unconscious being of the past, just as Durkheim above argued. In many ways the *habitus* could be described as the self, lagging behind the very moment, re-enacting the previous successful solution. This allows for a relaxed and energy-saving approach to agency. The self is in control, executing what are perceived as individually made decisions, even though most of the time actions are strongly guided by previous experiences as well as learned and observed models. Nevertheless this, as a consequence, creates individuals' feeling of one's own world which one controls. Arguably such a model is the reason why routine and repetition

---

[2]For reference see the footnote with the quote by Durkheim on page 5.

or the *habitus* itself are not perceived as defined in a wider context, as a result of everything else. It truly feels individual, unique to each individual and specific to each and every situation.

There is at any time room for adjustment in case the context demands adaptation and immediate changes. This can be described as the conscious kicking in and taking over to adjust the course. In this case, the energy requirements rise sharply, the alertness rises accordingly, but the individual can adapt and settle into a new context. Amongst other things, the past is rewritten according to the new conditions. The *habitus* thus is extremely adaptable and flexible to adjust to changing circumstances. This has been noted by Bourdieu (1990, p. 291) as:

> ... the *habitus*, which, in new situations, can invent new ways of fulfilling the old functions.

The adaptability and flexibility of the *habitus* stem from its independence from static points of reference. It is only referring to its previous iterations, which in turn were set in the same conditions. This setting is in numerous ways the most sustainable setup for the demanding task of individual human agency. However, it shall not be understood as only being economical arguments. This is merely a singular aspect. Much more important is the social aspect of *habitus*. The setting allows for a relaxed and simplified form of living together. It is arguably as such the basis of dense urban settlements governing cultural and social behaviours. Whilst a lot of rules and rule enforcement is still required, the basis is hardwired into everyday life as a masked collective decision system, making use of both, the individual and the collective memory whilst in fact writing both in the process. After all, this stresses the importance of experience and how this influences the future. Being children and growing up, individuals learn from previous instances and re-enact them when the moment presents itself.

Most importantly, however, this conclusion demands a partial rethinking of urban conceptualisation in this context. The individual is no longer the focus but has to be regarded as the exception. The focus has to be on the *habitus*, the general applicable model of collective guidance. The starting point is no longer the single case, individual decision or action. The whole has to be conceptualised as an entity functioning in its entirety as a collective guidance system. This is demanding, since there is no such experience that could be drawn upon to design such an approach. Here the *habitus*, visible in routines and rhythms, strongly agrees with earlier proposed urban context description models such as the space syntax concepts as described by Hillier and Hanson (1984). They developed the understanding of spatial configuration as a result of social organisation. *Habitus*, however, goes a step further in that it describes the formulation of social organisation in rhythms and routines. As such *habitus* is physical through the activities where *time* and *space* always remain intellectual and conceptual.

This sequencing of a new approach especially regarding space is of the utmost interest. Concepts of space diverge from a strictly Cartesian conceptualisation of space as a box. As introduced by Lefebvre and Nicholson-Smith (1991) in *The Production of Space*, the emphasis is a productive aspect of space and the involvement of activity as the defining aspect.

To be understood in this sense is the reading of the GPS track records as *paths* and visualisations of continuous spatial production. Furthermore with regard to the conceptualisation of temporality built around the clear role definition of *habitus*, the conditioning of space out of *practice*, in close connection to *agency*, gains new meaning.

We can propose to read space as being similar to a delayed reaction process with a present history as the guideline. If such a *temporality* with an actual dimension is in place, space has a dimension in the moment of its production, and it is no longer impossible to think of it as produced.

Previously the main problem with an understanding of space as being *produced* was the reduction of space to a temporal point. The Cartesian space is always in the very moment and has no dimension: it is a static piece of representation. The geometry allows only for point clouds as a semi 3D construct. Effectively this reduces spatiality to a single dimension entity. This excludes any movement, any change, any process that could play the role of agents in a *productive* concept.

With an extended concept of temporality however, both *time* and *space* have a dimension, being embedded in what was previously called continuity. Having no frozen state, the concept of *production* can become effective in that it unfolds and fills the dimension with an activity.

This places *time* and *space* clearly in the realm of cultural artefacts. Earlier we have speculated over their nature and claimed they were concepts accepted as conventions. However, it can be argued that in fact *time* and *space* really are concepts born out of the decision-making process in the *moment* or simply part of our proposed *temporality*, this *temporality* being a result of the *habitus*, which describes the process of the present past.

8.7 The *Habitus*

> Town plans are thus no mere diagrams; they are a system of hieroglyphics in with man has written the history of civilisation, and the more tangled their apparent confusion, the more we may be rewarded in deciphering it (Geddes 1949).

Such a model is able to explain the quantity of time based on cycles and routines in the urban context. The pattern we have set out to study in this book led to a description of urban and social phenomenon entirely based on the reproduction of moments and a new interpretation of generally accepted concepts of *time* and *space*. What was previously explained as constraints in fact is the *habitus*, as in its context-related presence of the immediate past. What was simply an individual experience is in fact *temporality* as a duration of the moment in its open connection to any previous and following instances.

It was hinted in the introduction that this book to some extend also represents a journey during which a number of aspects have developed and the standpoint has transformed.

The point of departure was an understanding of *time* and *space* as absolute and separated entities. This understanding was influenced by the education in architecture and math under the strong influence of the Euclidean vision of space and the introduction to technologies founded on the same principles such as GPS and GIS. The logical and hierarchical structure of these tools strongly influenced the shaping of research questions and methods based on their capabilities and limitations.

The motivation however was a curiosity regarding the workings of *time* and *space* in everyday life initially described as cycles and routines. The interest in activity pattern is not new. Under different key words, this topic has been fascinating researchers over the past 50 or so years. The presented book has relied on the literature around the key readings introduced in Chap. 1, for example, Hägerstrand. Other fields however, namely, crime research has a long-standing interest in similar patterns. For examples Cohen and Felson (1979) in *Social Change and Crime Rate Trends: A Routine Activity Approach* focuses on the term *routine activity*.

Recently, as introduced in Chap. 5, the available masses of individual data, also called *big data*, have renewed interest in routines, and models are being developed to mine these available data sets using computer algorithms to detect pattern. Furthermore these projects often go a step further and work towards predictions.

As earlier quote from Barabási et al. (2010), human mobility predictions can be as high as 80 % if based on their recent history of activities. Similarly researchers at UCLA and Santa Clara University have developed a tool to predict crime hotspots as *boxes* as small as 500 meter square. The tool is being used by the Los Angeles police and has delivered improving results so far by assigning officers to these boxes and, as the officials claim, through their presence deterring crime in the process. One of the researchers behind the project Professor Jeffrey Brantingham from UCLA explains:

> ... the notion that criminals tend not to stray too far from areas they know best. (Risling 2012)

Or as Professor Brantingham explains, a place of crime is more likely to attract other crimes. This is a similar framing of the understanding of patterns as the here presented *habitus*. It is a similar presence of the recent past that Professor Brantingham describes.

Whilst the tools still rely on the use of Cartesian analysis and representation using GIS, the notion of routine and what here is called *habitus* as a certain degree of repetition is inherent in the projects not only as a result but essentially as a hypothesis. It can be said that a certain repetitiveness and routine in human activities in the latest research are often an assumed or at least investigated parameter. However, most of these projects do not challenge the static universal frameworks of *time* and *space*. This is to say that the static understanding of *time* and *space* not necessarily have to be changed or even put aside to make space for a new interpretation. Instead as proposed here, the two terms can be extended by a third aspect, the *habitus* to ensure a more dynamic interpretation of real-world phenomena.

8.7 The *Habitus*

In this sense, the chosen methods to undertake the research proposed was retrospectively a suitable decision even if it at first glance might surprise as somehow old and inappropriate. It proved to be a strong foundation and something the research work could support itself upon. It was however essential to open the methods and the closed field of Euclidean space and introduce additional aspects such as the individual perspective, experience and memory. This setting allowed for the foundations of *time* and *space* to questioned and examined without destroying it and propose an extension in the form of the *habitus*.

It is important to note here that the dimension of *temporality* varies. This makes it challenging to grasp and visualise. The variation however is defined by the relationships between the task at hand and the records of previous tasks. For example, the morning commute is, in this model, related to the commute on any of the days before but specifically to the most recent or most important commuting reference. The *temporality* in this case is more or less a day. However if the commute is to the swimming pool, we only go once a week, and the timeframe of the *temporality* is a week. If it is a trip to the airport, the *temporality* might be in the order of a few months and so on. The frame of reference varies according to the present past produced by the *habitus*.

The *temporality* is always related to the most recent similar event for reference. This is however an abstract, reduced description. Of course there are many details of the whole summary of previous events if it is a recurring activity, where a number of different experiences play an important role. In this sense it is a momentary connection to the entire history of similar activities.

This model of *temporality* and decision making aims to find, by and for the individual, an acceptable solution rather than an optimal one. Constrained by imperfect information and allowing for the largest degree of flexibility, the *habitus* guides this open structure with unconscious production ties to the past.

The *habitarium* then is the inhabited individual space resulting from all everyday activities in their connection to the historic, present, future, social, cultural and natural context. It is opposed to the time-space cube proposed by traditional time geography which includes the conditions, the context and the constraints as aspects of continuous negotiation. Whilst it consists of an infinite generative capacity conditioned by the historical and social conditions, it is far from a creation of some unpredictable novelty since it is near mechanically reproduced from its previous original setting.

These essential aspects are two-way connections, and the *habitarium* is constantly changing. Where the *environment* however is objectively defined from an external perspective, the *habitarium* is strongly driven by the internal individual *agency*. Its defining ties to the conditioning original or *seeding* situation not only lets it transcend both *time* and *space*, but it defines a new structural element: a reflective presence with the capacity to simply reproduce complex patterns.

The acknowledgement of the *habitarium* puts a name to the actual role of individuals that can be conceptualised as a part of the city certainly in regard to the *production* of *time* and *space*. Before now, the individual path was a mere object, loosely connected by the individual. Using the term results in a different perspective on urban space, promoting *agency* which, together with activity, is at the centre of what these spaces actually are made of, as opposed to the old concept of urban space as a service or as discussed in Chap. 2 the *Urban Machine*.

The city is as a consequence a produced process as a result of individual and collective activities. The technology we are nowadays using as part of everyday lives opens up the opportunity to transform the static urban concept into a dynamic understanding of processes. Real-time information and updates on the go are already part of such a practical production of city. Functions are being transformed and reinterpreted able to adapt both in terms of identity and usage. The urban spaces are capable to merge social activities where leisure, entertainment, business and service go hand in hand. The times we live in are at our fingertips.

Increasingly it is a nomad's life we are living, and all that matters is the being and the rhythm of being places as the establishment relationship to reality. In this context, the *habitus* is the timely interpretation of *time* and *space*, the description of the subject to its context and the collective as the making of the city.

8.7.1 Concluding Remarks

My hope is that this contribution to the broad field of time-space discussion across many disciplines will offer a fresh approach to discuss the phenomena we observe, especially in the built environment of our cities. By deciding to shift the focus of practice to something other than *time* or *space*, we create the possibility of describing the production of the two aspects using a third logic, rather than inventing separate quick fixes regarding their structure or indeed trying to explain them on their own terms.

The structuring capacity of individual *agency* has long been acknowledged and was put forward in parts several times, for example, Hillier and Hanson (1984), Lefebvre (2004), Lefebvre and Nicholson-Smith (1991), Bourdieu (1990), and de Certeau (1984). These were, however, not taken out of the *time* and *space* context, preventing the *agency* from developing relevance. For example Hillier and Hanson (1984) specifically discusses the relevance of social aspects in regard to *space* in *The Social Logic of Space*. The discussed logic however remains unexplored. On the other hand, Bourdieu (1990) does not put his *habitus* in a relevant *time* or *space* context. Bringing these aspects together as *habitus* is no big move but an essential stepping stone to understand the mechanics of our everyday lives which we organise and govern using the concepts of *time* and *space*.

A wish in regard to the discussed *habitus* is for it to be regarded in connection to its value for the discussion around sustainability. It was Buckminster Fuller who, early on, highlighted the importance of the topic and promoted a worldview emphasising the limited extent of our existence in regard to resources but mainly also regarding the physical extent of our planet earth (Krausse and Lichtenstein 1999). Illustrating it with an early photograph of today known as the *blue marble*, he stressed the implication of the acknowledgement of the finite planet we inhabit.

The *habitus* similarly stresses the importance of our routines as the production of our identities and presence but also the related consumption of energy and production of waste. It stresses the responsibility that has to be executed by claiming individual and collective agency with regard to the production of *time* and *space*. The sustainable benefit lies in the acknowledgment of finite resources, not just physically but similarly regarding *habitus*. The clear conditioning guidance inherited from the past and present context can help to value the production.

Regarding sustainability as a holistic concept, the *habitus* extends the understanding of it in relation to our environment, in the way it highlights how our activities as routines are shaping the physical results as conceptualised by *time* and *space*. Whilst Fuller showed we are part of the whole (Krausse and Lichtenstein 1999), the *habitus* shows we also are the collective producers. In the consequence, this means we are able to take the step from the machine-determined functionalism towards the practical- and action-led production.

This is especially important in the urban context, where most responsibility is still offloaded onto smart technological systems and anonymous institutions as discussed in *Urban Machine* Sect. 2.1. The urban areas are forecasted to grow and host three-quarters of the population, and utilising their *agency*, both individually and collectively, will help to create a fundamentally responsible environment whilst at the same time improving temporal-spatial conditions.

Bibliography

Andersen HK, Grush R (2009) A brief history of Time-Consciousness: historical precursors to James and Husserl. J Hist Philos 47(2):277–307

Barabási AL, Song C, Qu Z, Blumm N (2010) Limits of predictability in human mobility. Science 327(5968):1018–1021

Bourdieu P (1990) The logic of practice. Polity Press, Cambridge

Bryson V (2007) Gender and the politics of time: feminist theory and contemporary debates. Policy Press, Bristol

Capra F (1997) Web of life, Flamingo, London

Cohen LE, Felson M (1979) Social change and crime rate trends: a routine activity approach. Am Sociol Rev 44(4):588–608

Dalton RC, Hölscher C, Turner A (2012) Understanding space: the nascent synthesis of cognition and the syntax of spatial morphologies. Environ Plan B 39(1):7

Davies PCW (1995) About time: Einstein's unfinished revolution. Viking, London

de Certeau M (1984) The practice of everyday life. University of California Press, Berkeley

Disalle R (2006) Understanding space-time: the philosophical development of physics from Newton to Einstein. Cambridge University Press, Cambridge

Durkheim É, Cosman C, Cladis MS (2001) The elementary forms of religious life. Oxford world's classics. Oxford University Press, Oxford

Elias N (1992) Time: an essay. Blackwell, Oxford

Foucault M (1986) Of other spaces (1967), heterotopias. Architecture /Mouvement/ Continuité, 21(2):22–27

Geddes P (1949) Cities in evolution. Williams and Norgate, London

Giddens A (1986) The constitution of society: outline of the theory of structuration. Polity Press, Cambridge

Giedion S (1982) Space, time and architecture: the growth of a new tradition, 5th edn., rev. and enl edition. Harvard University Press, Cambridge

Glennie P, Thrift NJ (2009) Shaping the day: a history of timekeeping in England and Wales, 1300–1800. Oxford University Press, Oxford

Hägerstrand T (1970) What about people in regional science? Pap Reg Sci 24(1):7–24

Halberg F, Cornélissen G, Carandente A (1994) Introduction to chronobiology – variability: from foe to friend, of mice and men. Volume 7 of chronobiology seminar. Medtronic, Minesota

Hawking SW, Ellis GFR (1975) The large scale structure of space-time. Cambridge University Press, Cambridge, MA

Hillier B (1996) Space is the machine: a configurational theory of architecture. Cambridge University Press, New York

Hillier B, Hanson J (1984) The social logic of space. Cambridge University Press, Cambridge

Huggett N (2010) Zenoś paradoxes. In: Zalta EN (ed) The Stanford encyclopedia of philosophy. Winter 2010 edn.
http://plato.stanford.edu/archives/win2010/entries/paradox-zeno/

Krausse J, Lichtenstein C (eds) (1999) Your private sky: R. Buckminster Fuller, art design science. L. Müller, Baden

Lefebvre H (2004) Rhythmanalysis: space, time and everyday life. Continuum, London

Lefebvre H, Nicholson-Smith D (1991) The production of space. Basil Blackwell, Oxford

Lynch K (1960) The image of the city. MIT, Cambridge

Miller CR (1992) Kairos in the rhetoric of science. In: Kinneavy JL, Witte SP, Nakadate N, Cherry RD (eds) A rhetoric of doing: essays on written discourse in honor of James L. Kinneavy. Southern Illinoise University Press, Carbondale

Miller HJ (2004) Activities in space and time. In: Hensher DA, Button KJ, Haynes KE, Stopher PR (eds) Handbook of transport geography and spatial systems. Volume 5 of handbooks in transport. Elsevier, Oxford

Pred A (1977) The choreography of existence: comments on hägerstrand's time-geography and its usefulness. Econ Geogr 53(2):207–221

Ray C (2002) Time, space and philosophy. Routledge, London

Reichholf J (2008) Warum die Menschen sesshaft wurden: Das größte Rätsel unserer Geschichte, 2nd edn. Fischer, Frankfurt

Richards EG (2000) Mapping time: the calendar and its history. Oxford University Press, New York

Risling G (2012) 'Predictive policing' technology lowers crime in los angeles. Huffington Post.
http://www.huffingtonpost.com/2012/07/01/predictive-policing-technology-los-angeles_n_1641276.html

Spring U (2006) The linear city: touring Vienna in the nineteenth century. In: Sheller M, Urry J (eds) Mobile technologies of the city. The networked cities series. Routledge, Abingdon

Tabboni S (2001) The idea of social time in norbert elias. Time Soc 10(1):5–27

Tolman EC (1948) Cognitive maps in rats and men. Psychol Rev 55(4):189–208

Zerubavel E (1985) Hidden rhythms: schedules and calendars in social life. University of California Press, Berkeley

Zerubavel E (1989) The seven day circle. University of Chicago Press, Chicago

Chapter 9
Appendix

9.1 Glossary

Habitus is the presence of the past or the constant referral to the past as the reference for the present activity. As such, the *habitus* is created by the past conditions, but defined together with its application in the present condition. It governs the practice by referring to past experiences in the past and present context and allowing for their *reproduction*. The limitations of this reproduction process are set by the historical and socially situated conditions and therefore does not produce random or unpredictable novelty (Bourdieu 1990). In this sense the *habitus* is responsible for recreating successful practices or, as Bourdieu calls them, *positively sanctioned* activities for each task. The concept of the *habitus* is further detailed in Chap. 1 in regard to the cyclical pattern and routines the thesis is focusing on in an urban context.

Time-space cube is a visualisation developed by time geography as a way to combine temporal and spatial data in one illustration. The trajectory, of an individual moving through space for examples, is plotted in a 3D cube, with X an Y representing horizontally the spatial location and the z-axis plotting the time vertically. See Fig. 2.10 for illustration of the concept. As time passes and the object moves in space, the location rises upwards on the z-axis and moved horizontally in the corresponding x/y direction. This cube or aquarium maps the location of objects or individuals in space time. In the 1970s, it was a completely new visualisation unknown to social scientists.

Habitarium is the inhabited individual space resulting from all everyday activities in their connection to the historic, present, future, social, cultural and natural context. It is much more than the time-space cube proposed by traditional time geography which includes the conditions, the context and the constraints as aspects of continuous negotiation. Whilst it consists of an infinite generative capacity conditioned by the historical and social conditions, it is far from a creation of some unpredictable novelty since it is near mechanically reproduced from its previous original setting.

Continuity is a term describing the connectedness of sequential activities within either the individual or collective narrative. It has become apparent in all the examples that an important aspect of everyday life is the way it continues. It is surprisingly an aspect that was not covered by Hägerstrand in his space model, as quoted in Giddens (1986). The aspect of continuity has to be added.

Sequence in the sense of having an order to different activities. However, it is important to point out that this does not necessarily follow *time* or *space* constraints, but can be defined by the activity in the social or cultural context. The key to *sequencing* is the memory as the present past and the narrative.

Agency describes the individual capacity to take decisions and influence the world around them. This also requires the ability to process information and respects both the past, present and future context of the decision. *Agency* is consequently embedded in its context and cannot be detached from it.

Time geography is going back to the early works of Thorsten Hägerstrand, for example, his *Innovation Diffusion as a Spatial Process.* 1967 published in English (originally published in 1953 as *Innovationsförloppet ur korologisk synpunkt.* Medd. från Lunds Universitets Geografiska Institutioner 25.). The work studies the live line study of 10,000 individuals in rural Sweden. The developed geography discipline as time geography concerns the unfolding of daily life in time and space. However it has been extended over time in the last 60 or so years and has grown to be an important part of human geography. A definition cited by summarising a definition given by Hagerstrand:

> Time-geography constitutes a foundation for a general geographical perspective. It represents a new structure of thought under development, which attempts to consolidate the spatial and temporal perspectives of different disciplines on a more solid basis than has thus far taken place. Time-geography is not a subject area per se, or a theory in its narrow sense, but rather an attempt to construct a broad structure of thought which may form a framework capable of fulfilling two tasks. The first is to receive and bring into contact knowledge from highly distinct scientific areas and from everyday praxis. The second is to reveal relations, the nature of which escape researchers as soon as the object of research is separated from its given milieu in order to study it in isolation, experimentally or in some other way distilled.

Urban Diary is the title of the GPS tracking project to investigate personal rhythms and cycles in an urban context. Over a period of 2 months, the spatial extension of a group of participants' everyday life was recorded. This data was used as the basis for this thesis. The project started in 2009, and until 2011 a sample of 56 participants' data was gathered in different cities. Method and details are described in Chap. 4, Sect. 4.1.

Urban Tick is the working title for the research undertaken that leads to this thesis. It concisely references the two main aspects of city or urban context and the aspect of rhythms or ticking pattern. The urbanTick (intended spelling) research work started in 2008 and is ongoing. The progress is continuously published on the dedicated blog at http://www.urbantick.blogspot.com and has led to several publications including *Studies in Temporal Urbanism: The urbanTick Experiment* published by Springer.

New City Landscape   is the title of the Twitter data mapping project investigating temporal and spatial aspects of urban life. Geolocated tweets are mapped resulting in a collective surface of activity density acting as a virtual landscape of the city. The project was initiated together with Andrew Hudson-Smith, Director of CASA at UCL, and the code for the data collection was written by Steven Gray at CASA, UCL as part of a JISC grant. The work started in 2010, and the resulting maps have been widely published online and in print. The data forms part of this thesis, and methods and context are described in detail in Chap. 5, Sect. 5.4.

Digital Social Network   (DSN) describes a range of Web-based networking platforms that allow the personalisation of online platforms and the building of personal connections between different users, including the sharing of digital content. In essence, it is a digitised circle of friends and contacts. Central to the social network is the aspect of sharing and the creation of *shared experience*. This practice of sharing and collaboration in the widest sense is what give the platforms the name social networking. The individual data put online at the portal is being shared with a selected group of *friends* and they are able to comment, *like*, resend or respond to and extend the content in some other form (Carroll and Romano 2011). This practice creates an entangled and interaction-like environment, where content is very fluid. It is in constant change every instant, as it is being newly linked to other content or individuals.

Cognitive Map   goes, according to Portugali (2005), back to Tolman (1948) who, in his experiments in the 1930s and 1940s, demonstrated how animals and humans are capable of constructing representations in their minds about the external environment they have experienced.

Time Rose Diagram   Over time it shows the data for each time period. By plotting the time around the circle, a section of the circle is assigned. Within this section the data values are plotted from the centre outwards. The circular arrangement, with its similarity to the familiar clock face, makes the time dimension comprehensible and, with its implied repetition pattern, indicates that a rhythm is inscribed in the data. The invention of rose diagrams is widely associated with Florence Nightingale (1820–1910) in her famous *Diagram of the causes of mortality in the army in the East* printed in the publication *Notes on matters affecting the health, efficiency, and hospital administration of the British Army: founded chiefly on the experience of the late war* (Nightingale 1858). See the Fig. 6.23 for illustration.

Clock time   or sometimes described as *physical time* is the time measured by clocks or the time represented by discrete quantities. It refers to time as being a measurable thing with clear objective qualities. This is a term used also, for example, by Glennie and Thrift (2009) in *Shaping the Day: A History of*

Timekeeping in England and Wales, 1300–1800, Zerubavel (1989) in *The Seven Day Circle* and Richards (2000) in *Mapping Time: The Calendar and its History* describing specifically the time as counted by the clock.

Social time   the aspects of time based on individual or collective experience. Often this is related to duration. It refers to sequences of events, the process of continuous change in relationship between people. As Elias (1992) describe it with the capacity to connect to one another.

Body Space   is a concept developed under two main headings. The first aspect is the natural, biological and functional aspects of rhythms and patterns directly related to the human body, both internal and external. This includes all body functions such as heart beat, breathing, the blinking of an eye, and also cell activity and nutrition transport. Furthermore it includes eating and drinking and resting and many more. This draws direct lines to natural *constraints* resulting from the way we as humans experience the environment. The second aspect explores the experience, meaning and creation of space through these functions and the body as a whole. The aspect of the body as a physical object in space is investigated in relation to how this is the main basis of space creation. This draws direct lines to the Urban Diary tracking project where the movements of individual *bodies* are recorded, resulting in a spatial inscription on the city.

Spatial Narrative   is the result of the combination of body function, rhythm and routine behaviour. In the sequence of tasks and activities resulting in a daily story of movement and activities, both individually and also in the city collectively Urban Narrative are constantly created. The city is at the same time the stage and a result of the opera. The individual as well as the environment both have their identity created through and from this. This draws from both action (in the present) and memory (from the past). The body, being the vehicle to receive, process and deliver information, must be regarded as the central element for this choreography to take place, even if it is extended by machines or absent from memory or in virtual worlds (Hudson-Smith 2003). It is the processing of the fundamental experience that lets us visualise the essential information. Aspects of mobility are important in the preliminary conception of Urban Narrative as a succession or sequence.

Urban Machine   describes the notion of cities as multidimensional constructs of social activities, processes and configurations taking shape or shaping through relationships. Cities are also places where flows and power networks intensify and manifest in physical form. The multitude of activities is far beyond a single person's comprehension, whether as an actor or as a receptor. This chaos of interwoven functions resembles a black box, a machine rattling and creaking beyond an individual's impact reach. It just works, and it works along our routines, whilst it has its own cycles of flow and production. Metaphors such as *it works* or *it breaks down* are widely used in everyday language referring to services, functions and events.

9.2 External Resources

As part of the data management and analysis process a number of animated visualisations have been produced. These are available online under the following links.

9.2.1 Urban Diary GPS Track Animations

Is a series of animations visualising the Urban Diary GPS data collected by the participants in London and Plymouth. Here Google Earth is used as the rendering engine and in some cases the CASA 3d virtual London model is used as context for the data.

- UD TwoMonth Detail Bloomsbury – https://vimeo.com/4469668
- UDtwoMonth London – https://vimeo.com/4277201
- plymouth365 24 h sun – https://vimeo.com/2436280
- plymouth365 plyCentre – https://vimeo.com/2436298

9.2.2 aNCL

Some of the Twitter data collected for the New City Landscape project has been animated. The visualisations show individual urban areas as they live and breathe on the social networking platform Twitter over 24 h. The data is in fact collated over seven days. Each message shows as a yellow dot. Interactions between users are shown as yellow lines with a small dot travelling in the direction of the receiver. Interaction is based on RT (retweets) and (messages directed at specific users).

- aNCL – London Twitter Traffic – https://vimeo.com/28018319
- aNCL – Den Haag Twitter Traffic – https://vimeo.com/25777243
- aNCL – Calgary Twitter Traffic – https://vimeo.com/29607900
- aNCL – San Francisco Twitter Traffic – https://vimeo.com/22029533
- aNCL – Zurich Twitter Traffic – https://vimeo.com/22029533
- aNCL Geneva Twitter Traffic – https://vimeo.com/22447109
- aNCL Barcelona Twitter Traffic – https://vimeo.com/26752902
- aNCL – Ljubljana Twitter Traffic – https://vimeo.com/26361339
- aNCL – Brussels Twitter Traffic – https://vimeo.com/25806772
- London Twitter Cloud 2 – http://www.youtube.com/watch?v=V42JiVEABOY

Bibliography

Bourdieu P (1990) The logic of practice. Polity Press, Cambridge
Carroll E, Romano J (2011) Your digital afterlife: when Facebook, Flickr and Twitter are your estate, what's your legacy? New Riders, Berkeley
Elias N (1992) Time: an essay. Blackwell, Oxford
Giddens A (1986) The constitution of society: outline of the theory of structuration. Polity Press, Cambridge
Glennie P, Thrift NJ (2009) Shaping the day: a history of timekeeping in England and Wales, 1300–1800. Oxford University Press, Oxford
Hudson-Smith A (2003) Digitally distributed urban environments: the prospects for online planning. PhD thesis, the Bartlett School of Architecture, University College London
Nightingale F (1858) Notes on matters affecting the health, efficiency, and hospital administration of the British army: founded chiefly on the experience of the late war. Harrison and Sons, High Wycombe
Portugali J (2005) Cognitive maps are over 60. In: Cohn A, Mark D (eds) Spatial information theory. Volume 3693 of lecture notes in computer science. Springer, Berlin/Heidelberg, pp 251–264
Richards EG (2000) Mapping time: the calendar and its history. Oxford University Press, New York
Tolman EC (1948) Cognitive maps in rats and men. Psychol Rev 55(4):189–208
Zerubavel E (1989) The seven day circle. University of Chicago Press, Chicago

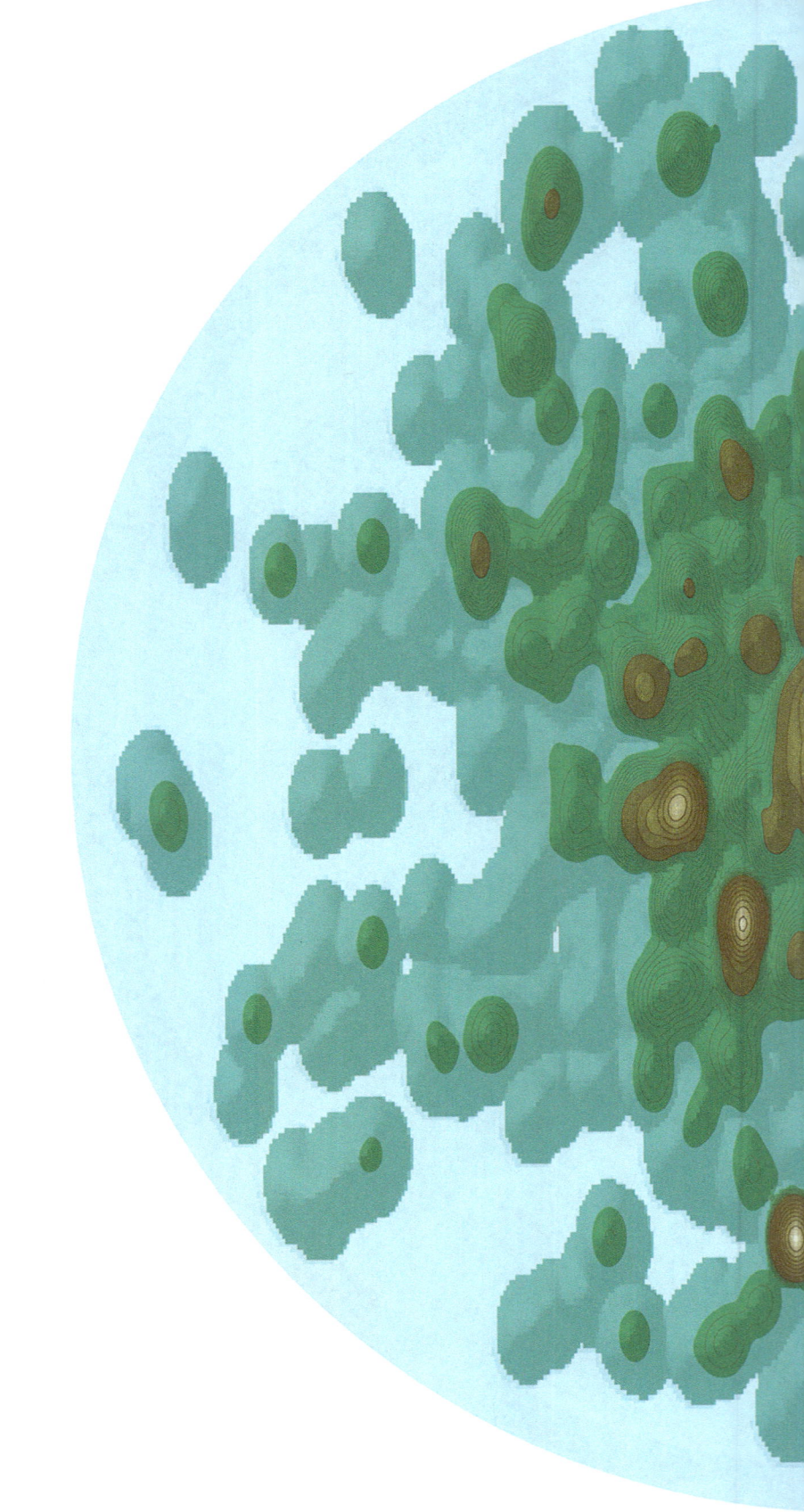

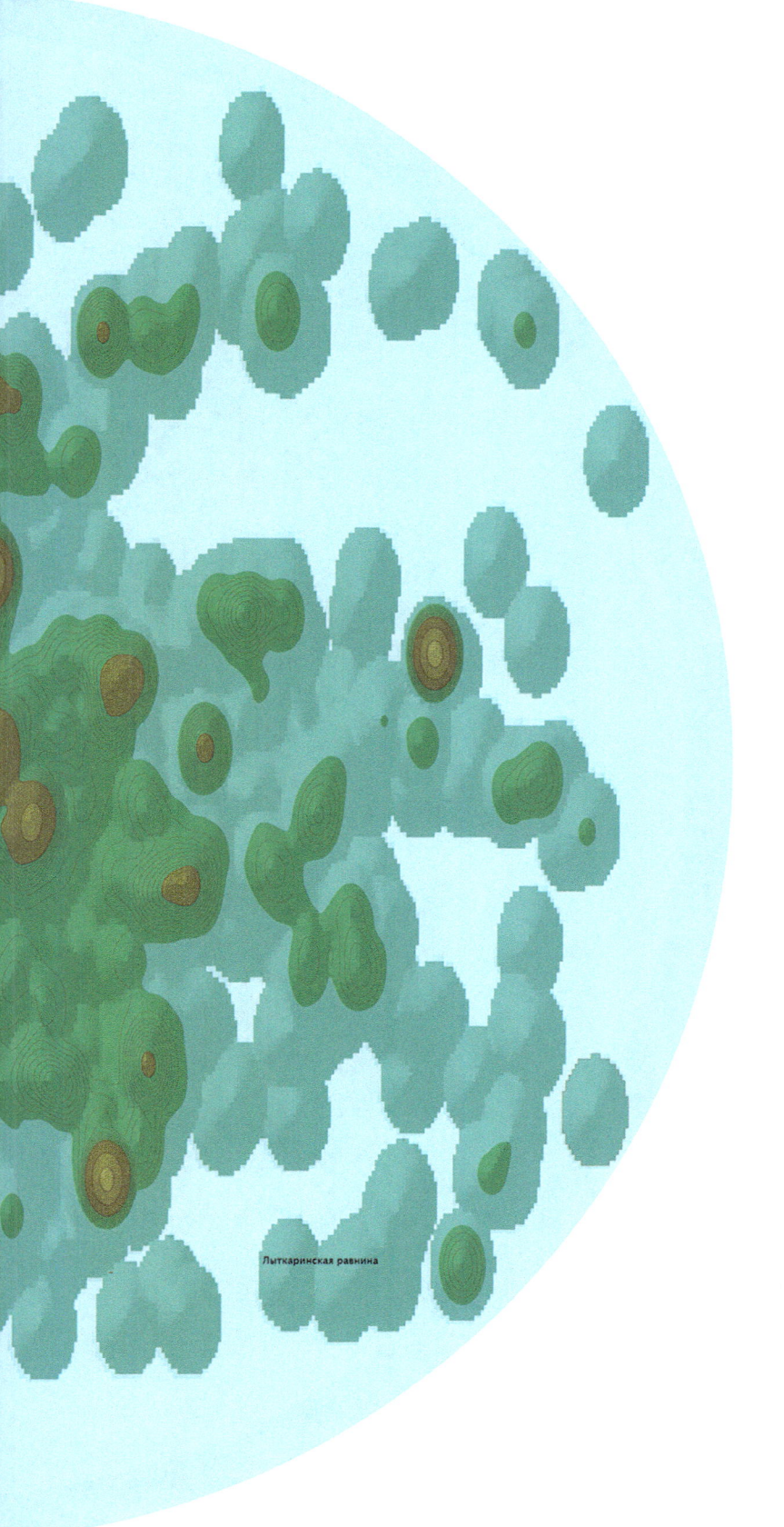
Лыткаринская равнина

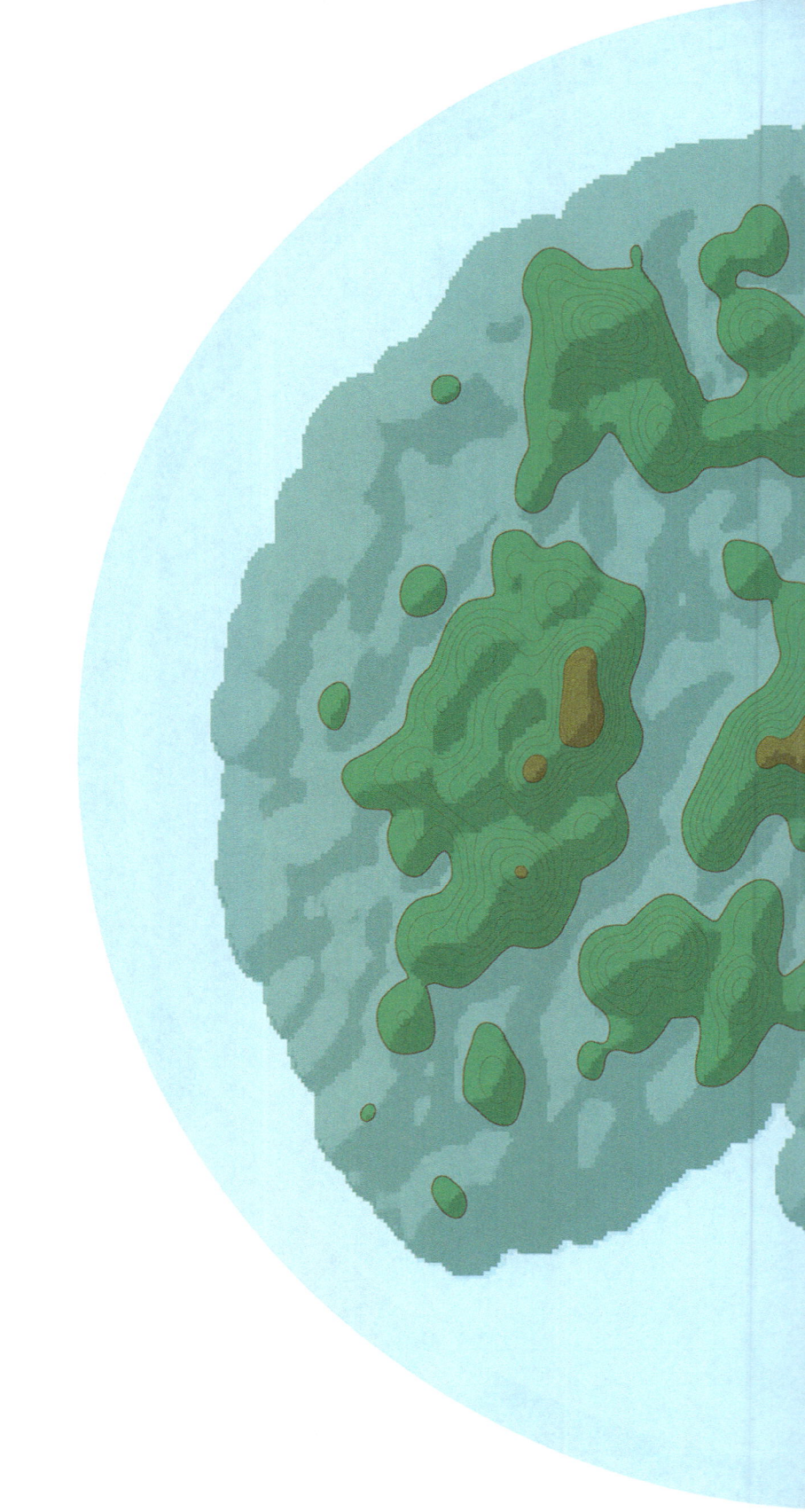

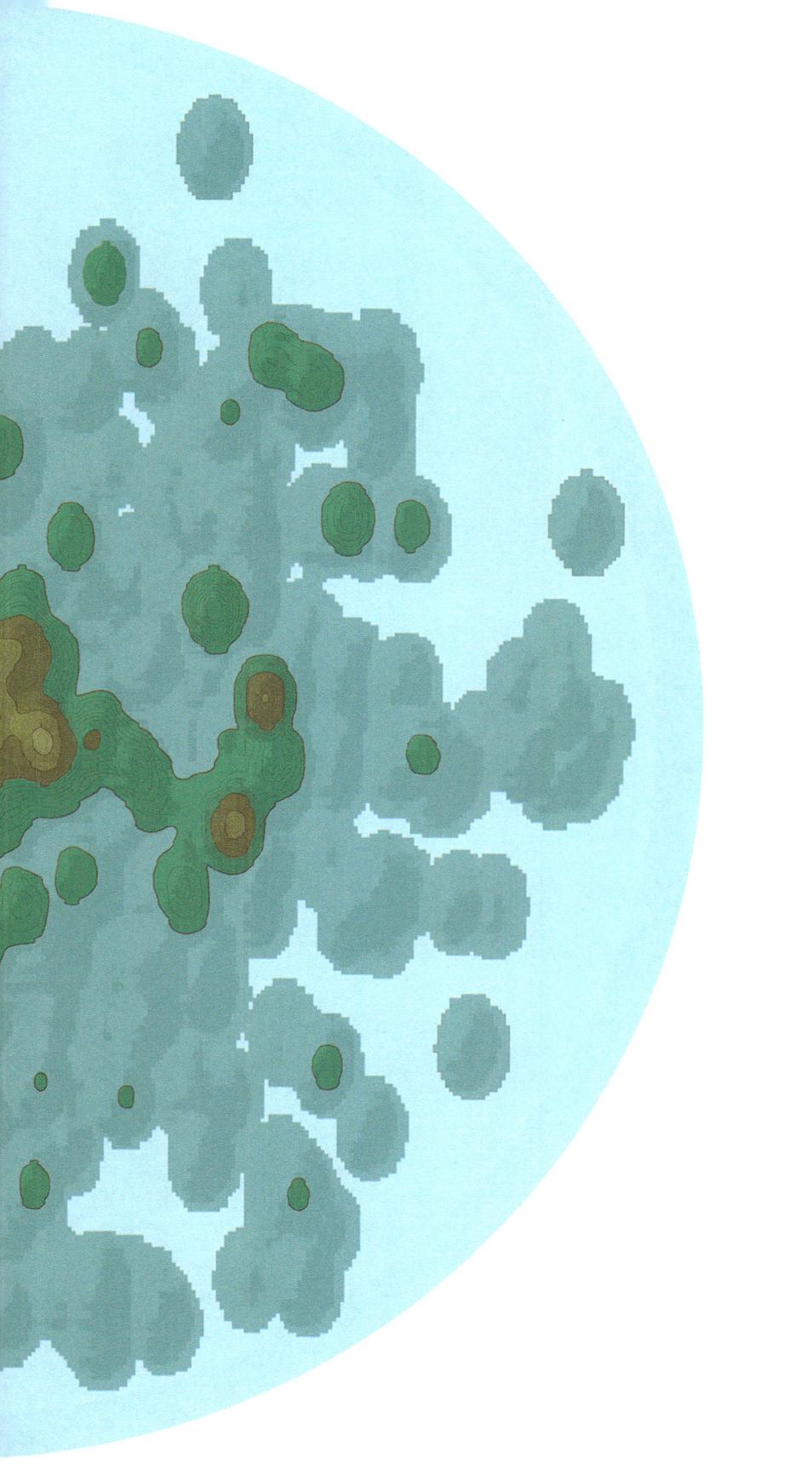

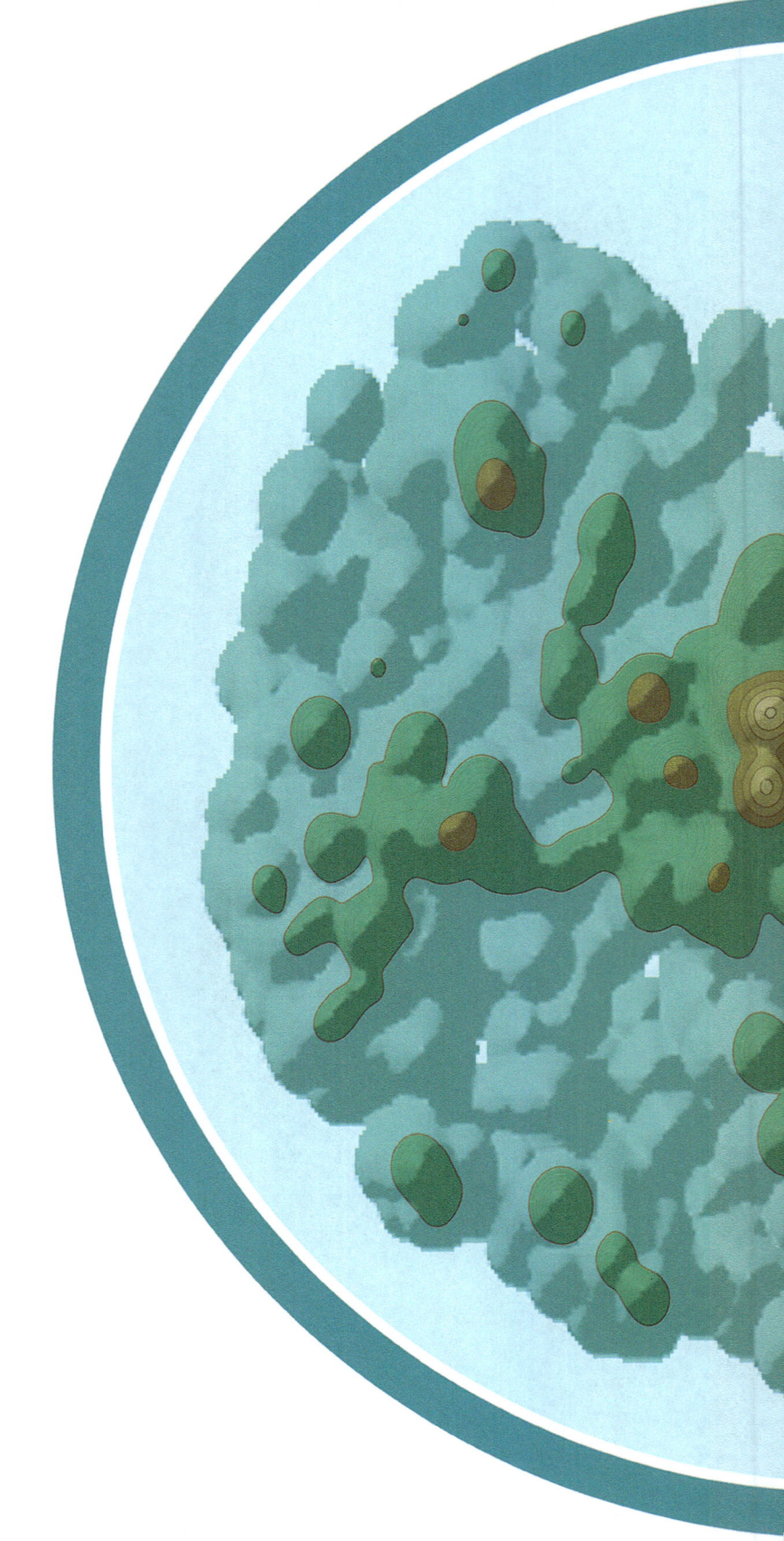

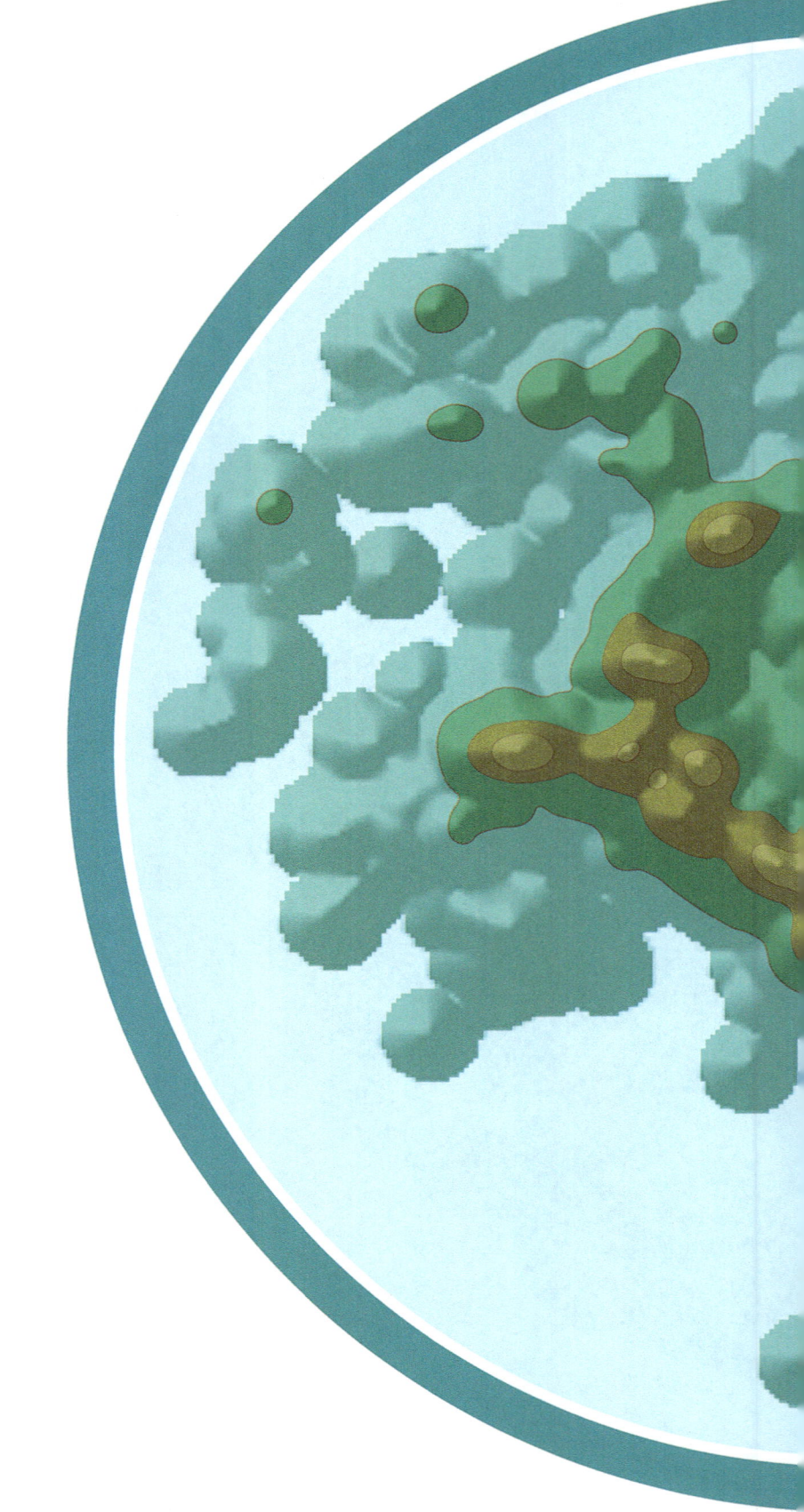

Index

A
Abercrombie, P., 166, 204, 205
Activity, 1, 14, 23, 37, 56, 84, 131, 194, 238, 265
Agency, 2, 3, 10, 49–51, 56, 140, 199, 202, 203, 232, 235, 238, 247, 248, 253, 254, 257–259, 261–263, 266
Ahmed, N., 32
Alexander, C., 14
Altman, I., 203, 233
Ambrose, P.J., 192
Anderson, D.H., 203
Ante, S.E., 109
Appleyard, D., 46, 47, 86
Aristotle, 247

B
Barabási, A.L., 6, 85, 119, 260
Barnett, E., 169
Basel, 8, 58, 63, 72, 73, 136, 140, 141, 143, 148, 149, 152–163, 198, 199, 204–206, 218, 221, 223, 229, 241–244
Batty, M., 146, 165, 214
Baxmann, I., 24
Besten, O.N., 69
Bisbee, E., 191, 192
Body space, 7, 8, 15, 37–52, 268
Boissevain, J., 85
Bonnett, A., 19, 74
Borden, I., 42
Bourdieu, P., 4, 5, 25, 254, 256–258, 262, 265
Boyd, D.M., 89
Brake, D.R., 89

Brantingham, J., 260
Brown, B.B., 203
Bryant, M., 89
Butler, P., 103

C
Carlstein, T., 3, 14, 29, 30
Cartesian space, 10, 189–192, 203, 211, 214, 215, 217, 218, 232, 234, 240, 247–250, 252, 253, 256, 259
Chemers, M.M., 203
City planning, 7, 13, 21
Clock time, 9, 25, 60, 62, 125, 132–135, 149, 150, 158, 165, 181, 183, 193, 203, 235, 240, 241, 247–250, 252, 253, 256, 267
Cognitive maps, 3, 8, 11, 52, 55, 58, 67, 198, 221, 224, 226, 229, 231, 234, 240–245, 267
Cohen, L.E., 260
Collective, 2, 4, 9, 10, 24, 45, 51, 56, 57, 60, 61, 64, 69, 77, 80, 84–86, 88, 124–126, 135, 141, 151, 163, 164, 167–169, 184, 190, 196, 211, 217, 232, 234, 238, 241, 249, 250, 253, 254, 258, 262, 263, 266–268
Constraints, 3, 9, 10, 14, 30, 31, 33, 34, 38, 50, 58, 68, 111, 131, 140, 150, 151, 154, 158, 168, 199, 202, 207, 215, 216, 232, 234, 238, 254–257, 259, 261, 265, 266, 268
Continuity, 9, 132, 149, 150, 168, 183, 185, 194, 202, 231, 234, 238, 252, 253, 257, 259, 266
Conzen, M.R.G., 206

Crandal, D.J., 99
Crang, M., 150
Crowd sourcing, 9, 83
Cycle, 1–2, 4, 5, 7, 8, 14–16, 26–28, 37–40, 43, 52, 58, 59, 80, 87, 134, 135, 151, 173, 185, 226, 244, 254, 257, 259, 260, 266, 268

D
Dalton, R.C., 250
Daston, L., 190
Data mining, 5, 88, 90, 93–104
Debord, G., 10, 190, 218–220
de Certeau, M., 25, 57, 150, 262
Digital social networks (DSN), 7, 9, 83–92, 94, 95, 97–106, 112, 249, 267
Durkheim, E., 133, 253, 256

E
Einstein, A., 44
Elias, N., 25, 132, 133, 154, 183, 253, 268
Ellison, N.B., 89
Ethics, 6–7, 9, 84, 90, 92, 112, 121–124
Everyday, 2, 4–6, 8, 14, 37, 47–51, 56, 57, 59–61, 63, 74, 75, 79, 85, 88, 104, 106, 119, 126, 132–134, 140, 151, 164, 168, 173, 183–185, 191, 194, 196, 199, 202, 206, 207, 211, 218, 220, 221, 229, 231, 232, 234, 238, 242, 246, 247, 251, 254, 256, 258, 260–262, 265, 266, 268
Experience, 2–5, 7–10, 13, 17, 18, 25, 33, 34, 37–39, 44–52, 56–59, 61, 65, 67, 70, 73–77, 88, 91–92, 95–97, 107, 125, 131, 132, 167, 170, 181, 183, 189, 193, 198, 199, 203, 213, 216, 218, 220, 221, 224, 226, 229, 231, 232, 235, 238–247, 250–254, 257–259, 261, 265, 267, 268
Experienced time, 9, 131

F
Felson, M., 260
Fieldwork, 2, 3, 5–8, 10, 55, 56, 77, 131, 132, 192, 237
Fischer, N.I., 172
Fischli, P., 18, 19
Fisher, N.I., 171
Foucault, M., 248
Fredrickson, L.M., 203
Fuller, B., 218, 248, 262
Fuller, W., 263

G
Gabaglio, A., 171
Galileo, 25, 134
Galison, P., 190
Gauntlett, D., 67, 69
Giddens, A., 30–33, 56, 238
Gifford, R., 202, 203, 216, 233, 234
Gilbertson, S., 90
Glaser, B., 72
Glennie, P., 25, 132, 134, 247, 248, 253, 267
Goffman, E., 97
Golledge, R.G., 78, 194
Gould, P., 66, 68
Gray, S., 108
Grosz, E., 38, 41, 43
Guare, J., 119

H
Habitarium, 47, 48, 52, 199, 234, 235, 261, 265
Habitus, 3–5, 9–11, 48, 49, 59, 126, 131, 132, 140, 164–166, 168, 185, 197, 198, 202, 232, 237, 239, 247, 248, 253–263, 265
Hadfield, P., 178
Hägerstrand, T., 3, 14, 26, 29–33, 47, 49, 50, 56, 58, 60, 150, 151, 154, 192, 202, 213, 215, 218, 232, 238, 248, 250, 254, 255, 257, 266
Halberg, F., 251
Halfpenny, P., 92
Hall, P., 42
Hamm, B., 56
Han, J., 94
Hanson, J., 5, 86, 233, 258, 262
Hanson, S., 55
Hardt, M., 52
Harvey, A.S., 167
Hein, A., 34
Held, R., 34
Hillier, B., 5, 55, 86, 233, 254, 258, 262
Hockney, D., 133
Holloway, I., 72
Howard, E., 165
Huberman, D.M., 107
Hubert, H., 133, 253
Hudson-Smith, A., 214, 267

I
In between time, 10, 136, 185, 237
Individual, 1, 14, 55, 84, 131, 189, 238, 265
Interviews, 8, 9, 11, 14, 55, 57, 58, 67, 70, 72–77, 92, 94–96, 131, 150, 151, 153, 154, 158, 160–163, 183, 226, 231–232, 240–242, 244, 252

Index

J
Jackson, J.B., 46
Jacobs, J., 164, 178
Johnson, D.K., 166

K
Kelsey, T., 91
Kempf, P., 41
Kilroy, B., 62
Kincaid, J., 90
Klanten, R., 99
Knoke, D., 119, 120
Kraak, M.J., 117
Kwan, M.-P., 32, 33, 48, 213, 232

L
Larsen, J., 60
Lee, J., 32
Lefebvre, H., 2, 3, 15, 25, 38, 132, 154, 250, 258, 262
Ley, D., 68, 232
Ling, A., 166
Livingstone, S., 88
Location-based services, 9, 83, 112
Loew, M., 56
London, 6, 14, 58, 99, 136, 191, 241, 269
Longley, P., 146
Longstaff, G.B., 139
Lynch, K., 2, 3, 22, 26, 50, 51, 56, 66, 217, 221, 231, 250

M
Manzo, L.C., 202
Massey, D., 14, 51
Matei, S., 68
Mauss, M., 133, 253
Mental maps, 2, 8, 9, 34, 56, 65–71, 73–77, 126, 131, 189, 217, 221, 231, 240, 244, 246
Merrill, D., 106
Miller, C.C., 106
Miller, C.R., 255
Miller, H.J., 32, 60
Miskowiec, J., 248
Moudon, A.V., 206
Muratori, S., 206
Muybridge, E., 43, 44

N
Negri, A., 52
Neuhaus, F., 95, 114

New City Landscape, 7, 9, 10, 80, 83–126, 132, 168, 169, 180, 182, 189, 207–211, 214, 216, 238, 267, 269
Nicholson-Smith, D., 2, 25, 258, 262
Nightingale, F., 170, 171, 267

O
Ormeling, F., 117

P
Papacharissi, Z., 89
Participants, 2, 48, 57, 85, 133, 189, 238, 266
Path, 2, 3, 9, 12, 31, 33, 46–48, 50, 60, 62, 67, 106, 124, 131, 149, 181, 192–196, 199, 201, 204–208, 220, 221, 224–226, 228–231, 233, 234, 240, 243–246, 251, 257, 259, 261
Pattern, 2–10, 14, 15, 20, 23, 26, 27, 33, 37–40, 52, 55–61, 73–75, 77–80, 83–86, 110, 112, 115, 125, 126, 131, 132, 135–142, 144, 146, 149, 151, 155, 156, 159, 160, 162–164, 166–168, 170, 173–175, 178, 179, 181, 184, 190, 194–197, 199, 202, 204, 207, 211, 213, 214, 216, 217, 219, 220, 226, 232–235, 237–241, 251–254, 259–261, 265–268
Patton, M.Q., 72
Pentland, W.E., 167
Perkins, D.D., 202
Picasso, P., 133
Pile, S., 50, 232
Pinder, D., 217
Place, 2, 4, 6, 8, 13–16, 18, 19, 25–28, 31, 42, 44, 45, 47, 48, 51–52, 56, 57, 59, 61, 62, 73, 86, 88, 89, 97, 99, 103, 106, 108, 110, 119, 124, 125, 138, 143, 144, 146, 148, 149, 154, 156, 159, 162, 164, 168, 173, 178, 180, 181, 183–185, 191, 193, 196, 199, 202–204, 207, 208, 211, 215–221, 224, 226, 229–235, 238–242, 247–251, 256, 259, 260, 262, 266, 268
Portugali, J., 67, 133, 267
Pred, A., 248
Present past, 10, 135, 237, 238, 259, 261, 266
Privacy, 6, 9, 84, 88–90, 92, 112, 121–124
Procter, R., 92
Psarra, S., 47

R
Rainie, L., 111
Ravenstein, E.G., 139
Reichholf, J., 164, 249

Rhythm, 1–11, 13–15, 21, 24–26, 37–40, 45, 52, 56, 57, 61, 77, 85, 87, 88, 132, 135, 138, 151, 154, 164, 167, 170, 173, 181–185, 198, 199, 232–235, 237, 238, 243, 246, 250–254, 257, 258, 262, 266–268
Richards, E.G., 26, 34, 132, 268
Rose-Redwood, R.S., 190
Routine, 1–6, 8, 10, 14–16, 28, 45, 47, 48, 52, 57–61, 65–71, 73–80, 87, 97, 106, 126, 132, 134, 139, 146, 150, 158, 167, 168, 183–185, 194, 198, 199, 201–203, 218, 221, 231, 233, 237, 250–260, 263, 265, 268

S
Scannell, L., 96, 202, 203, 216, 233, 234
Scott Brown, D., 48
Self-territory, 10, 189, 202–204, 218, 233, 238
Sequence, 4, 8, 15, 20, 26, 28, 31, 37, 45–47, 59, 60, 62, 76, 77, 93, 99, 132, 140, 146, 148, 149, 158, 159, 162, 169–171, 193, 194, 197, 214, 231, 232, 234, 238, 244, 246, 248, 252, 253, 266, 268
Shane, D.G., 18, 46
Sheller, M., 60
Smith, A., 111
Smithson, A., 46
Social network, 85, 86, 88–91, 119, 125, 181, 207, 267, 269
Social time, 9, 25, 132, 241, 268
Space, 1, 13, 37–52, 55, 83, 132, 189–235, 237, 265
Spencer, J., 62
Spiekermann, K., 218
Spuybroek, L., 34, 39
Stanley, M., 119
Strauss, A., 72
Strogatz, S.H., 14, 39, 119
Sullivan, L.H., 16

T
Tabboni, S., 133
Temporality, 10, 25, 87, 88, 94, 125, 126, 181, 185, 235, 237–263
Thrift, N.J., 25, 132, 134, 247, 248, 253, 267
Time, 1, 13, 38, 56, 83, 131–185, 189, 237, 265
Time-geography, 3, 7–9, 13, 14, 25, 29–33, 49, 50, 60, 131, 140, 191, 213, 254, 255, 265, 266
Time Rose diagram, 9, 125, 170–172, 267

Time-space aquarium, 3, 29, 33, 49, 52, 250
Time-space cube, 3, 26, 30, 31, 47, 261, 265
Tolman, E.C., 67, 240, 267
Tracking, 6–9, 21, 38, 40, 55–61, 63, 73, 74, 79, 97, 126, 131, 136, 139, 192, 195, 231, 233, 234, 240, 266, 268
Tschumi, B., 47
Tuan, Y.-F., 44, 52, 203, 233
Tufte, E.R., 171
Twitter, 7–9, 32, 55, 83, 88, 90, 94, 97, 98, 103–112, 114, 115, 117, 120–125, 168–170, 173, 175, 179–181, 207, 208, 210, 211, 267, 269

U
Urban Diary, 7, 38, 40, 47, 48, 51, 52, 55–80, 86, 88, 97, 110, 126, 132–150, 169, 189, 192–207, 213, 218, 221, 233, 238, 241, 266, 268, 269
Urban Islands, 10, 190, 219–221, 231
Urban Machine, 7, 13–35, 84, 262, 263, 268
Urban Narrative, 15, 32, 44, 45, 47, 51, 268
Urry, J., 60

V
Venturi, R., 46

W
Watson, J.P., 204
Watts, D.J., 120
Webmoor, T., 95
Wegener, M., 218
Weiss, D., 18
Wellman, B., 88
Werner, C.M., 203
White, R., 66, 68
Whorf, B.L., 28
Wilde, E., 106
Wood, D., 217
Wu, F., 107

Y
Yang, S., 119

Z
Zahavi, Y., 218
Zerubavel, E., 27, 132, 154, 268

Printed by Printforce, the Netherlands